A Practical Guide to Atmospheric Simulation Chambers

Jean-François Doussin · Hendrik Fuchs ·
Astrid Kiendler-Scharr · Paul Seakins ·
John Wenger

Editors

A Practical Guide to Atmospheric Simulation Chambers

 Springer

Editors
Jean-François Doussin
Department of Sciences and Technology
Paris-Est Créteil University—CNRS
Créteil, France

Astrid Kiendler-Scharr
Department of Energy and Climate
Research
Forschungszentrum Jülich
Jülich, Germany

John Wenger
School of Chemistry
University College Cork
Cork, Ireland

Hendrik Fuchs
Department of Energy and Climate
Research
Forschungszentrum Jülich
Jülich, Germany

Paul Seakins
Department of Chemistry
University of Leeds
Leeds, UK

This work has received funding from the European Union's Horizon 2020 research and innovation programme through the EUROCHAMP-2020 Infrastructure Activity under grant agreement No 730997.

ISBN 978-3-031-22279-5   ISBN 978-3-031-22277-1   (eBook)
https://doi.org/10.1007/978-3-031-22277-1

This Springer imprint is published by the registered company Springer Nature Switzerland AG
The registered company address is: Gewerbestrasse 11, 6330 Cham, Switzerland

This book is dedicated to Dr. Ian Barnes and to Prof. Astrid Kiendler-Scharr

Preface

The ability to predict the evolution of the atmosphere over a wide range of time scales (hours to decades) brings great benefits to society. Examples include short-term public warnings of hazardous air quality and the long-term evaluation of climate warming and policy effectiveness. Atmospheric predictions use complex models that are underpinned by observations and a sound understanding of the underlying processes, which include the interactions between atmospheric components and their environment. Atmospheric simulation chambers are among the most advanced tools for studying and quantifying atmospheric processes and are used to provide many of the parameters incorporated in air quality and climate models. Without chamber-derived parameters to constrain predictive models, any physico-chemical forecasts of the atmosphere are highly unreliable, both in the short- and long-term.

The largest uncertainties in our current knowledge of atmospheric processes and their impact on air quality and climate change are associated with complex feedback mechanisms in the Earth System. Understanding and quantifying those mechanisms that are just becoming measurable are only possible through a synergistic approach that combines atmospheric observations, detailed simulation experiments and modelling. This methodology is the most efficient means of obtaining a quantitative understanding of physico-chemical transformations in the atmosphere and chamber studies are a key component of that approach.

The level of scientific understanding of climate drivers, the health impacts of complex mixtures of air pollutants, and the interaction between the two is still evolving. Simulation chambers were originally created to study the impact of atmospheric processes on regional photochemistry. This approach has since been extended to understand particles formation, cloud microphysics and global warming. More recently, atmospheric simulation chambers have been applied to a wider range of research areas such as human health and cultural heritage. In all cases, a key objective is to work under conditions that are as realistic as possible.

In the real atmosphere, it is difficult to separate chemistry from meteorology, emissions, transport and other variables. Since the late 1960s, closed systems have been developed to provide a controlled atmosphere to study the formation and the evolution of air pollutants by isolating specific compounds of interest and controlling the oxidizing environment.

Initially, chamber experiments were mainly focused on understanding the chemical processes governing the formation of photochemical smog. These "smog chamber" experiments enabled investigations into the formation of secondary pollutants such as ozone and oxygenated reactive nitrate species, as well as the associated atmospheric oxidation mechanisms. This approach has been extremely useful in producing gas-phase kinetic data, branching ratios and product distributions of chemical reactions. Together with data arising from flow tubes and flash photolysis experiments, this knowledge allowed the scientific community to develop numerical chemical models that are used to predict the chemical evolution of the atmosphere. Nowadays, chambers are also essential tools for evaluating these chemistry modules and for predicting the formation of secondary pollutants in the absence of uncertainties associated with emissions, meteorology and mixing effects.

Over the past few decades, simulation chamber studies have been extended to include processes related to the formation, chemical aging and physico-chemical properties of secondary organic aerosol (SOA). This research has compelled the atmospheric science community to pay even more attention to the design of simulation chambers, e.g. their size or wall material, and the experiments carried out in them. Nevertheless, the general aim of chamber experiments remains focused on the simulation of processes occurring in ambient air and under controlled conditions.

Our understanding of atmospheric chemical and physical processes has evolved considerably over the last two decades, but new challenges are expected to arise from responses of the Earth System to changes in the climate due to anthropogenic activities.

To respond to such challenges, the atmospheric simulation chamber community in Europe has organized itself within the EUROCHAMP consortium. Three consecutive EU funded projects—EUROCHAMP, EUROCHAMP-2 and EUROCHAMP-2020—spanning the period 2004–2021, have enabled most of the European research groups involved in experimental atmospheric simulations to adapt their research platforms to the emerging research needs. A wide range of topics have been addressed including unexpected gas-phase chemistry, understanding oxidative capacity, secondary organic aerosol formation and growth, along with studies of the cryosphere, air-sea exchange and real-world emissions. The integrated suite of state-of-the-art simulation chambers within the EUROCHAMP consortium provides unprecedented opportunities for atmospheric scientists to perform experiments that address the most important questions in air quality and climate research.

Innovative methodological research carried out in the EUROCHAMP projects, as well as best practices and standard protocols are reported in the present volume. With the production of the first-ever "Practical Guide to Atmospheric Simulation Chambers", we are not only aiming at producing a key tool for knowledge transfer within the EUROCHAMP community, but also provide the global atmospheric science community with a unique resource that outlines best practice in the operation of simulation chambers and in related data exploitation.

Moreover, at a time when the provision of open data has become the standard approach in scientific publishing and reporting, EUROCHAMP provides a sustainable Data Center that includes results from thousands of simulation chamber experiments, as well as a range of advanced data products. As Europe looks to efficiently re-organize its atmospheric research infrastructures, many EUROCHAMP facilities and consortium members will become part of the Aerosols, Clouds and Trace Gases Research Infrastructure (ACTRIS). The Data Center will continue to grow and be available for access within the framework of the ACTRIS Research Infrastructure.

This guide is based on the work carried out in the EUROCHAMP community. Its scope, though, is much larger, as it aims to provide a broader scientific audience with the knowledge needed to analyze, reuse, review or combine simulation chamber data.

Créteil, France — Jean-François Doussin
Jülich, Germany — Hendrik Fuchs
Jülich, Germany — Astrid Kiendler-Scharr
Leeds, UK — Paul Seakins
Cork, Ireland — John Wenger

Contents

Chapter 1
Introduction to Atmospheric Simulation Chambers and Their Applications

Astrid Kiendler-Scharr, Karl-Heinz Becker, Jean-François Doussin, Hendrik Fuchs, Paul Seakins, John Wenger, and Peter Wiesen

Abstract Atmospheric simulation chambers have been deployed with various research goals for more than 80 years. In this chapter, an overview of the various applications, including emerging new applications, is given. The chapter starts with a brief historical overview of atmospheric simulation chambers. It also provides an overview of how simulation chambers complement field observations and more classical laboratory experiments. The chapter is concluded with an introduction to the different aspects requiring consideration when designing an atmospheric simulation chamber.

Atmospheric simulation chambers, such as those in the EUROCHAMP network, are highly valuable research tools for investigating chemical and physical processes that occur in air. They are used in a large number of applications, ranging from air quality and climate change to cloud microphysics, cultural heritage and human health. Chambers were originally developed as laboratory-based systems to investigate the formation of clouds or photochemical smog and hence, were called cloud chambers or smog chambers, respectively. Their ability to provide a controlled environment to study the formation and evolution of atmospheric pollutants, by isolating

Astrid Kiendler-Scharr passed away before publication of this chapter.

A. Kiendler-Scharr · H. Fuchs
Forschungszentrum Jülich, Jülich, Germany
e-mail: a.kiendler-scharr@fz-juelich.de

K.-H. Becker · P. Wiesen
Bergische Universität Wuppertal, Wuppertal, Germany

J.-F. Doussin (✉)
Centre National de la Recherche Scientifique, Paris, France
e-mail: jean-francois.doussin@lisa.ipsl.fr

P. Seakins
University of Leeds, Leeds, UK

J. Wenger
University College Cork, Cork, Ireland

© The Author(s) 2023
J.-F. Doussin et al. (eds.), *A Practical Guide to Atmospheric Simulation Chambers*,
https://doi.org/10.1007/978-3-031-22277-1_1

specific compounds of interest and controlling the oxidizing environment, made them especially useful in elucidating the key factors governing photochemical smog formation on a local to regional scale. Within EUROCHAMP-2020 and across the world, chambers dedicated to the exploration of atmospheric chemistry outnumber the atmospheric physics and cloud chambers. For this reason, this guide has an emphasis on atmospheric chemistry related aspects of simulation chambers.

Initially, smog chamber experiments were focused on elucidating the processes responsible for the observed increase in atmospheric secondary pollutants such as ozone and peroxyacyl nitrates (PAN-type compounds). This approach was later broadened to include studies of the kinetics and mechanisms of gas phase atmospheric oxidation and chambers have been extremely useful in producing kinetic data, branching ratio and product distributions (Becker 2006). Together with data arising from flow tubes and flash photolysis experiments, this knowledge allowed the scientific community to build complex numerical chemical codes that have led to the development of the models used to predict ozone formation. Nowadays, chambers are also essential tools for evaluating these chemistry models and for predicting the formation of secondary pollutants in the absence of uncertainties associated with emissions, meteorology and mixing effects (Carter and Lurmann 1991; Dodge 2000; Hynes et al. 2005). Experimental chamber data have been key to the development and optimisation (e.g. Gery et al. 1989; Carter 2010; Bloss et al. 2005a), as well as the evaluation (e.g. Saunders et al. 2003; Goliff et al. 2013; Jenkin et al. 2012; Bloss et al. 2005b; Metzger et al. 2008; McVay et al. 2016) of chemical mechanisms used in a wide range of science and air quality policy models. Today, chamber-derived data remains a key component in the development and evaluation of future atmospheric chemical mechanisms (Kaduwela et al. 2015; Stockwell et al. 2020).

In the past few decades, chamber facilities have been increasingly used to investigate processes leading to secondary organic aerosol (SOA), an important component of atmospheric aerosol (Finlayson-Pitts and Pitts 1986; Dodge 2000; Finlayson-Pitts and Pitts 2000; Kanakidou et al. 2005; Barnes and Rudzinski 2006; Hallquist et al. 2009). The general methodology which has been (and still is) useful for gaseous pollutants is now providing valuable data related to SOA formation (e.g. Hatakeyama et al. 2002; Pankow 1994; Odum et al. 1996; Cocker et al. 2001; Pun et al. 2003; Takekawa et al. 2003; Martin-Reviejo and Wirtz 2005; Baltensperger et al. 2005; Donahue et al. 2005; Pathak et al. 2007; McFiggans et al. 2019; Zhao et al. 2018, Ciarelli et al. 2017) as well as the physico-chemical properties of aerosols and their changes during atmospheric transport and processing (De Haan et al. 1999; Kalberer et al. 2006; Field et al. 2006; Linke et al. 2006; Meyer et al. 2009; D'Ambro et al. 2017; Huang et al. 2018; Zhao et al. 2017).

Furthermore, due to the wide range of experimental requirements, simulation chamber designs vary considerably. As pointed out by Finlayson-Pitts and Pitts (2000), although the general aims of all chamber studies are similar–i.e. to simulate processes in ambient air under controlled conditions–the chamber designs and capabilities to meet these goals vary widely. This in turn means that chambers and their associated measurement technologies are being adapted to a growing number of applications.

This chapter provides a short history of atmospheric simulation chambers Sect. 1.1, investigations of atmospheric processes Sect. 1.2, approaches for bridging the gap between laboratory and field studies Sect. 1.3, emerging new applications Sect. 1.4, and considerations on the design and instrumentation of atmospheric simulation chambers Sect. 1.5. Respective references to the more detailed discussion in Chaps. 2–8 are provided in each of the subsections.

1.1 A Short History of Atmospheric Simulation Chambers

Atmospheric simulation chambers have been used for more than 80 years. As early as the 1930s, Findeisen performed studies on cloud droplet size distributions and conducted cloud chamber experiments, which was a highly novel approach at the time. Findeisen's cloud chamber was approximately 2 m³ in volume and connected to a vacuum pump, which allowed the process of adiabatic expansion and atmospheric cloud formation to be mimicked in the chamber (Storelvmo and Tan 2015).

Photochemical smog formation, first observed in the Los Angeles area in the 1940s and 1950s stimulated study in large chambers to simulate plant damage and health effects such as eye and lung irritation (Haagen-Smit 1952). Europe followed suit in chamber construction and application to atmospheric processes and through a range of national and European Union funding streams, Europe now leads the world in the use of large, highly instrumented chambers for atmospheric model development and evaluation. These large facilities are complemented by a range of smaller chambers that have been designed for specific purposes.

The first large European chamber was the "Große Bonner Kugel" (Groth et al. 1972), constructed at the University of Bonn and completed in 1968. The programme led by Groth and Harteck initially focused on air glow reactions at the low pressures pertaining to the *upper* atmosphere. However, studies of tropospheric interest were also undertaken, but at a very basic level and without the use of photolytic sources. Radicals were generated by discharge flow techniques, and this limited the range of conditions that could be used.

The facility, which was operated by Becker, Fink, Kley and Schurath for several years (Groth et al. 1972), had the following properties as indicated in Table 1.1.

At that time dark OH radical sources and the importance of OH reactions were not known. Figures 1.1 and 1.2 show the facility installed at the Institute of Physical Chemistry, Bonn University. The chamber has not been used since the mid-1980s because of its enormous operational cost and has since been completely dismantled.

In the mid-1970s, as our understanding of the basics of tropospheric chemistry increased and particularly the role of photolysis, the Pitts group at Riverside (Finlayson-Pitts and Pitts 1986, 2000) started to construct an indoor chamber with the objective of exploring photochemical smog formation. Advances in the understanding of photochemical processes had been slow because appropriate analytical techniques still had to be developed at that time. However, activity soon increased with the construction of a similar chamber in Japan (Akimoto et al. 1979a, b), while

Table. 1.1 Key properties of the "Bonner Kugel"

Volume	$221 \ m^3$
Surface	$177 \ m^2$
Inner diameter	7.5 m
Heatable	$T_{max} = 350 \ °C$ with 233 kW power, cooling of baffles between pumps and chamber with liquid hydrogen
Material	Stainless steel, 10 mm wall thickness
Pumping speed	240 000 l/s with 8 diffusion pumps
Lowest pressure	10^{-12} bar

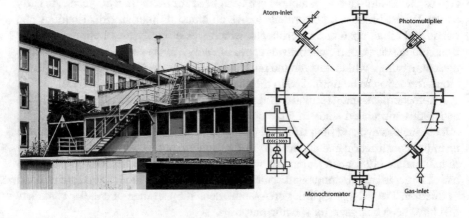

Fig. 1.1 The housing of the "Große Bonner Kugel" (left) and cross-section through the spherical reaction chamber "Große Bonner Kugel" (right). Courtesy of K.H. Becker, Bonn, Germany

Atkinson in the Pitts group started to successfully investigate the kinetics of the initiation reactions of OH, O_3 and NO_3 with volatile organic compounds (VOC). Concurrently, other groups used Teflon bags to study smog-forming reactions under irradiation by natural sunlight, but their results were limited to the Los Angeles conditions.

The importance of the OH radical in atmospheric chemistry had been promoted by Weinstock (1969), working at the Ford Motor Company research laboratories at Dearborn. In this laboratory, Niki used a relatively small photoreactor to develop the application of FTIR spectroscopy for quantitative investigation of atmospheric reactions (Niki et al. 1972, 1981; Wu et al. 1976). IR absorption spectroscopy had been used for a number of years to study atmospherically relevant chemical reactions (Stephens 1958; Hanst 1971), based mainly on mirror systems which allowed long path light absorption (White 1942, 1976; Herriott et al. 1964; Herriott and Schulte 1965). However, it was the use of FTIR methods by Niki et al. (1981) and additional work in the Pitts' group to quantitatively measure rate coefficients and products in photoreactors by long path FTIR absorption spectroscopy that really accelerated and

Fig. 1.2   The pipe system by which the chamber could be heated to 300 °C (left), enormous pumping capacity needed to reach the vacuum of 10–9 Torr (middle), the platform at which the experiments were prepared and carried out (right). Courtesy of K.H. Becker, Bonn, Germany

promoted the use of the technique and FTIR has been one of the work-horses of chemical simulation chambers ever since.

In the 1960s and '70s, the understanding of atmospheric reactions developed as first the key role of the OH radical was recognised as the dominant oxidizing agent in the troposphere, based on the analysis of the CO budget (Heicklen et al. 1969; Weinstock 1969; Stedman et al. 1970; Levy 1971), and the measurement of the OH + CO rate coefficient two years earlier (Greiner 1967). The propagation of an OH radical chain was understood 10 years later when the rate coefficient of the fast reaction $HO_2 + NO \rightarrow OH + NO_2$ was measured by several groups (Howard and Evenson 1977; Leu 1979; Howard 1979, 1980; Glaschick-Schimpf et al. 1979; Hack et al. 1980; Thrush and Wilkinson 1981), initiated by studies of Crutzen and Howard (1978) that showed the importance of this reaction in stratospheric ozone chemistry.

In Europe in the 1970s, several groups e.g., Becker and co-workers in Bonn and Cox and co-workers in Harwell, started studies on tropospheric chemistry based on either the technique of long path FTIR absorption spectroscopy in simulation chambers by Becker and co-workers in Wuppertal or molecular modulation studies focusing more on elementary reactions by Cox. Becker and co-workers constructed a multiple reflection mirror system in a 420 L photoreactor, which could be operated between 223 and 323 K to determine the OH reaction rate coefficients in combination with product analyses in the ppm range. Subsequent developments involved the construction of a 6 m long quartz glass reactor of 1000 L volume, the QUAREC chamber, which enabled measurements to be extended down to the ppbV level. Over the years, other European laboratories started to use indoor chambers of larger volume irradiated by a range of photolysis sources (Baltensperger et al., in Villigen/Zürich, Carlier and Doussin in Paris, Hjorth et al., in Ispra, Herrmann et al., in Leipzig, Le Bras et al., in Orléans, Treacy et al., in Dublin, Wenger et al., in Cork). Tables 1.2 and 1.3 lists the larger indoor and outdoor reactors, respectively, that have been built up to 2000.

Large outdoor simulation chambers have many advantages in terms of photochemical smog simulation and several large outdoor chambers have been built in the US, with support from the EPA. A major objective of these studies was to determine ozone formation isopleths under chemical conditions representative of conditions observed in major US cities. These chambers were made from FEP Teflon foil, with volumes up to 25 m^3. Whilst they lead to improvements in the empirical understanding of smog formation, the results could not be generalised because of the limited range of conditions requested by the US EPA. In Riverside, Carter and co-workers developed a method to define the ozone formation potential of VOCs by determining maximum incremental reactivity (MIR) factors using chamber data and chemical modelling (Carter 1994). A similar method was introduced by Jeffries in Chapel Hill, who also used an outdoor chamber (Fox et al. 1975).

Other approaches involved the injection of real engine exhaust directly into a smog chamber and studying the formation of ozone. However, the data were still very US specific in terms of the VOC/NO_X ratios and so could not be generalised and applied in other countries. In parallel, with the simulation studies mentioned above, Atkinson and co-workers refined their method to determine the OH reactivity from relative rate measurements in chambers and developed structure reactivity relationships to calculate rate coefficients for OH radical reactions with VOCs (Atkinson 1986, 1987; Kwok and Atkinson 1995). Further developments in simulation work included work by Seinfeld and co-workers in the mid-1980s, to study secondary organic aerosol formation from the oxidation of aromatic and biogenic hydrocarbons via the use of a 65 m^3 outdoor chamber made of FEP Teflon (Pandis et al. 1991).

In Europe, the first development of a large, highly instrumented chamber was led in the mid-1990s, by Becker, Millán and co-workers who built the EUPHORE (European Photoreactor) outdoor chamber in Valencia, Spain. In fact, EUPHORE consists of two chambers made of FEP Teflon foil, each of which has a volume of 200 m^3 (Becker 1996). This facility became a centre for European laboratories to work co-operatively on mechanistic, kinetic and ozone formation studies using either controlled starting materials or real exhaust gases from gasoline and Diesel engines. The EUPHORE chambers were equipped with a comprehensive suite of analytical instrumentation, including in situ detection of the key radicals HO_2 and OH using laser-induced fluorescence measurements.

In 2000, the group of Wahner at Forschungszentrum Jülich, Germany, built a new double walled outdoor chamber called SAPHIR (Brauers et al. 2003), which has a volume of 280 m^3, see Fig. 1.3. The double wall made of FEP Teflon foil allows studies of oxidation processes at low NO_x concentrations (below 1 ppbV). The Jülich

Table 1.2 Indoor chambers without light sources or irradiated by black lamps or solar simulators up to the year 2000

Year	References and location	Description	Application
1968	Groth et al. (1972) Bonn	Dark chamber 220 m³, stainless steel (high vacuum)	Without light sources, for low pressure studies
1972	Niki et al. (1972), Dearborn	150 l Pyrex	Kinetic and mechanistic studies
1975	e.g. Doyle et al. (1975), Riverside	6 m³, evacuable, thermostated, FEP[a] coated aluminium	Photooxidant and kinetic studies
1979	Akimoto et al. (1979a), Tsukuba	6 m³ evacuable, thermostated, FEP coated aluminium	Photooxidant studies
1980	Winer et al. (1980), Los Angeles	6 m³ evacuable, thermostated, FEP coated aluminium	Photooxidant studies
1981	Barnes et al. (1979), Wuppertal	420 l Duran glass, evacuable, thermostated − 50 to + 50 °C	Gas phase studies
1982	Joshi et al. (EPA), Research Triangle Park	440 glass reactor	Photooxidant studies
1986	Barnes et al., Wuppertal	Quartz glass 1100 l, evacuable, thermostated 0 to + 25 °C	Gas phase and aerosol kinetic and mechanistic studies
1986	Evans et al. (1986), Australia	4 × 200 l FEP bags	Photooxidants studies
1988	Behnke et al. (1988), Germany, Hannover, now Bayreuth	ca. 3000 l, Duran glass, thermostated −25 °C to ambient temperature	Aerosol studies
1997	Möhler et al. (2001), Karlsruhe	84 m³, thermostated –90 to + 60 °C, *AIDA*	For trace gas, aerosol and cloud studies
1996	Wahner et al. (1998), Jülich	256 m³, FEP wall cover of a lab room	Without light source, for NO$_Y$ chemistry
1997	Doussin et al. (1997), Paris	977 l, glass	Gas phase mechanistic studies
1998	Cocker et al. (2001), Pasadena	2 × 28 m³, 10–40 °C	Aerosol studies
2000	Carter et al. (2005), Riverside	FEP, double wall	Low NO$_x$ studies

[a] FEP fluorinated ethylene propylene

Table 1.3 Outdoor chambers irradiated by sunlight up to the year 2000

Year	References and location	Volume, wall material
1976	Jeffries et al. (2013), Chapel Hill	25 m^3, FEP
1981	Fitz et al. (1981)	40 m^3, FEP
1983	Spicer (1983)	17.3 m^3, FEP
1985	Kelly (1982)	450–2000 l, FEP bags
1985	Jeffries et al. (1976), Chapel Hill	25 m^3, FEP
1985	Leone et al. (1985), Pasadena	65 m^3, FEP
1995	Becker (1996), Valencia	2 × 200 m^3, FEP, *EUPHORE*
2000	Wahner (2002), Jülich	270 m^3, FEP, double wall, *SAPHIR*

group did pioneering work in field measurements of OH and HO$_2$ concentrations (Hofzumahaus et al. 2009), so SAPHIR is fully equipped with the most advanced in situ radical measurement techniques (Fuchs et al. 2012a, b). A smaller double wall indoor chamber was recently built by Carter in Riverside, to study tropospheric oxidation processes at low NO$_x$ concentrations.

Two other chambers were built in Germany, at the same time, for the study of aerosol processes. In 1986, Zetzsch and co-workers built a 3000 l Duran glass indoor chamber in Hannover, covered inside with FEP, and irradiated by solar simulators. This facility has been moved to Bayreuth. In 1987, Schurath and co-workers started to

Fig. 1.3 The double wall outdoor chamber SAPHIR in Jülich, Germany (© "Forschungszentrum Jülich/Sascha Kreklau")

operate the 84 m^3 aluminium chamber AIDA (Aerosol Interaction and Dynamics in the Atmosphere) in Karlsruhe, which has homogeneous temperature control between + 60 °C and −90 °C for trace gas, aerosol and cloud process studies. Other groups also now operate medium sized chambers.

A milestone for the European landscape of atmospheric simulation chambers was the implementation of the EUROCHAMP initiative, which started in May 2004 with the goal of joining together the existing European facilities into one integrated infrastructure of atmospheric simulation chambers.

The integration of all these chamber facilities within the framework of EUROCHAMP, followed by the EUROCHAMP-2 and EUROCHAMP-2020 projects, promoted the retention of Europe's international position of excellence in this area and it is unique in its kind worldwide. The mobilization of a large number of stakeholders dealing with environmental chamber techniques provided an infrastructure to the research community at a European level, which offers maximum support for a broad community of researchers from different disciplines. Overall, the EUROCHAMP projects fostered the structuring effect of atmospheric chemistry activities performed in European chambers and initiated wider international collaborations by supporting transnational access activities. Nowadays these facilities are fully available for the whole European scientific community and are exploratory platforms within the new Aerosol, Clouds and Trace Gases Research Infrastructure (ACTRIS). The following tables summarize current chambers across the world (Table 1.4) starting with the chambers of the EUROCHAMP consortium.

1.2   Investigations of Atmospheric Processes

1.2.1   Reaction Kinetics and Product Studies

Being the building blocks of the general atmospheric chemical mechanism, the study of the kinetics of elementary steps and the related product distribution has been the main application of simulation chambers. Involving pure gas phase conditions this has been–and is still–often carried out in small photoreactors of a few hundred litres or in small indoor simulation chambers. In the case of kinetics studies, Teflon bags of several litres to a few cubic-meters working under atmospheric pressure and ambient temperature under artificial irradiation (generally UV fluorescent tube) were often used to apply relative rate methods (Brauers and Finlayson-Pitts 1997). Nevertheless, the atmospheric fate of hundreds of various volatile organic compounds (VOC) was also studied–and is still–in rigid chambers such as the one displayed in Fig. 1.4 (Barnes et al. 1987; Doussin et al. 1997; Etzkorn et al. 1999; Picquet-Varrault et al. 2001; Atkinson 2000). This systematic kinetic and mechanistic work has produced over time a comprehensive database that has established the foundations of most chemical schemes used in numerical models.

Table 1.4 List of current chambers across the world; The chambers of the EUROCHAMP consortium are listed in alphabetic order of chamber acronym, and chambers across the world outside the EUROCHAMP consortium are listed in alphabetic order of country

EUROCHAMP chambers

Chamber name and location	Indoor/Outdoor	Wall material	Volume (m³)	Light source	Temperature	Comment	References
AIDA KIT, Karlsruhe, Germany	Indoor	Aluminium	85	LED	+ 60 °C to − 90 °C	Evacuable and developed for aerosol and cloud simulation	Möhler et al. (2003), Wagner et al. (2006)
CESAM CNRS-UPEC, Créteil, France	Indoor	Stainless steel	4,2	Xenon arc lamps	−10 to + 50 °C	Evacuable and developed multiphase atmospheric processes	Wang et al. (2011)
CHAMBRe INFN, Genoa, Italy	Indoor	Stainless steel	3	Arc	Ambient	Evacuable and suitable for bioaerosols	Massabò et al. (2018)
CSA CNRS-UPEC, Créteil, France	Indoor	Pyrex	0,97	Fluorescent Tubes	Ambient	Evacuable Gas phase chemistry	Doussin et al. (1997)
ESC-Q-UAIC CERNESIM, Iaisi, Romania	Indoor	Quartz	0,76	Fluorescent Tubes	Ambient	Gas phase chemistry	
EUPHORE CEAM, Paterna, Spain	Outdoor	FEP	200	Sun	Ambient	Dual chambers gas phase and aerosol	Zádor et al. (2006)
FORTH-ASC FORTH, Patras, Greece	Indoor	FEP	10	Fluorescent tubes	Ambient	Indoor simulation chamber	Kostenidou et al. (2013), Kaltsonoudis et al. (2017)
FORTH, Patras, Greece	Outdoor	FEP	2	Sun/Fluorescent tubes	Ambient	Mobile	Kaltsonoudis et al. (2019)

(continued)

Table 1.4 (continued)

Chamber name and location	Indoor/Outdoor	Wall material	Volume (m³)	Light source	Temperature	Comment	References
HELIOS CNRS_Orléans, Orléans, France	Outdoor	FEP	90	Sun	Ambient		Ren et al. (2017)
CNRS_Orléans, Orléans, France	Indoor	FEP	16	Fluorescent tubes	Ambient		Le Person et al. (2008)
HIRAC University of Leeds, Leeds, UK	Indoor	Stainless steel	2	Fluorescent tubes	−25 to + 70 °C	Evacuable gas phase chemistry and radicals	Glowacki et al. (2007)
IASC UCC, Cork, Ireland	Indoor	FEP	27	Fluorescent tubes	15–25 °C	Atmospheric chemistry	
ILMARI University of Eastern Finland, Kuopio, Finland	Indoor	FEP	29	Fluorescent tubes	16–25 °C	Combustion emission and health	Leskinen et al. (2015)
ISAC CNRS-IrceLyon, Villeurbanne, France	Indoor	FEP	2	Fluorescent tubes	Ambient		Bernard et al. (2016)
ACD-C TROPOS, Leipzig, Germany	Indoor	FEP	19	Fluorescent tubes	Ambient	Dual chamber	

(continued)

Table 1.4 (continued)

Chamber name and location	Indoor/Outdoor	Wall material	Volume (m³)	Light source	Temperature	Comment	References
MAC University of Manchester, Manchester, UK	Indoor	FEP	18	Xenon arc lamp	Ambient	Can be coupled with the MICC cloud chamber	Alfarra et al. (2012)
QUAREC Wuppertal University, Wuppertal, Germany	Indoor	Quartz	1.1	Fluorescent tubes	298 ± 5 K	Evacuable	Barnes et al. (1994)
RvG- ASIC University of East Anglia, Norwich, UK	Indoor	FEP and borosilicate glass	3.3	Fluorescent tubes	-55 to $+30$ °C	Atmosphere–ocean–sea-ice–snow simulation chamber	Thomas et al. (2021)
PACS-C3 PSI, Villigen, Switzerland	Indoor	FEP	27	Arc	Range 15–30 °C		Paulsen et al. (2005)
PACS-C3 PSI, Villigen, Switzerland	Indoor	FEP	9	Fluorescent tubes	Ambient	Mobile	Platt et al. (2013)
PACS-C3 PSI, Villigen, Switzerland	Indoor	FEP	9	Fluorescent tubes	Range 10–30 °C		Platt et al. (2013)
SAPHIR FZJ, Jülich, Germany	Outdoor	FEP	270	Sun	Ambient	Atmospheric chemistry at low concentrations, radical budgets	Rohrer et al. (2005a)

(continued)

Table 1.4 (continued)

Chamber name and location	Indoor/Outdoor	Wall material	Volume (m^3)	Light source	Temperature	Comment	References
SAPHIR-PLUS FZJ, Jülich, Germany	Indoor	FEP	9	LED	0–50 °C	Plant chamber	Hohaus et al. (2016)
SAPHIR-STAR FZJ, Jülich, Germany	Indoor	Quartz	2	Fluorescent tubes	Ambient	Continuous stirred reactor	
Large Aerosol Chamber FZJ, Jülich, Germany	Indoor	Teflon	250	None	Ambient		Zhao et al. (2010)
Other chambers worldwide							
CSIRO, Sidney, Australia	Indoor	FEP	25	Fluorescent tubes	Ambient		Hynes et al. (2005)
Fudan University, Shanghai, China	Indoor	Teflon coated Stainless steel	4.5	No	Ambient		Li et al. (2015, 2017), Zhang et al. (2011a, b)
IAP, Beijing, China	Indoor	FEP	1	Fluorescent tubes	Ambient		Zhang et al. (2019)
CRAES, Beijing, China	Outdoor	FEP	56	Sun	Ambient		Li et al. (2021)

(continued)

Table 1.4 (continued)

Chamber name and location	Indoor/Outdoor	Wall material	Volume (m³)	Light source	Temperature	Comment	References
Zhejiang University Smog chamber (CAPS-ZJU), Hangzhou, China	Indoor	FEP	3	Fluorescent tubes	Ambient	Dual	Li et al. (2020)
Guangdong University of Technology–dual-reactor chamber (GDUT-DRC), Guangzhou, China	Indoor	FEP	2	Fluorescent tubes	Ambient	Dual	Luo et al. (2020)
CAS, Guangzhou, China	Indoor	FEP	30	Fluorescent tubes	−10 to 40 °C		Wang et al. (2014)
RCEES, Beijing, China	Indoor	FEP	30	Fluorescent tubes	Controlled		Chen et al. (2019)
SHUSC, Shanghai University Smog Chamber, Shanghai, China	Indoor	FEP	1.2	Fluorescent tubes	Ambient		Qi et al. (2020)

(continued)

Table 1.4 (continued)

Chamber name and location	Indoor/Outdoor	Wall material	Volume (m³)	Light source	Temperature	Comment	References
Tsinghua University, Beijing, China	Indoor	FEP	2	Fluorescent tubes	10–60 °C,		Wu et al. (2007)
Aarhus University, Aarhus, Denmark	Indoor	FEP	2	Fluorescent tubes	−15 to 26 °C		Kristensen et al. (2017)
Toyota, Aichi, Japan	Indoor	FEP	5	Fluorescent tubes	Ambient		Takekawa et al. (2003)
NIES, Tsukuba, Japan	Indoor	Teflon coated Stainless steel	6	Xenon arc lamps	0–40 °C	Evacuable and bakable up to 200 °C	Akimoto et al. (1979a, b, c)
NIES, Tsukuba, Japan	Indoor	FEP	0.7	Fluorescent tubes	−5 to 40 °C		Deng et al. (2021)
KIST, Korea Institute of Science and Technology, Seoul, Korea	Indoor	FEP	5.8	Fluorescent tubes	Ambient	Dual	Lee et al. (2009)
CLOUD, CERN, Geneva, Switzerland	Indoor	Stainless steel	26	LED	Controlled		Duplissy et al. (2010)
Carnegie Mellon, Pittsburg, USA	Indoor	FEP	16	Fluorescent tubes	Controlled		Donahue et al. (2012)

(continued)

Table 1.4 (continued)

Chamber name and location	Indoor/Outdoor	Wall material	Volume (m³)	Light source	Temperature	Comment	References
GeorgiaTech, Atlanta, USA	Indoor	FEP	12	Fluorescent tubes	Controlled	Dual	Boyd et al. (2015)
CalTech, Pasadena, USA	Outdoor	FEP	28	Fluorescent tubes	18–50 °C		Cocker et al. (2001)
CalTech, Pasadena, USA	Outdoor	FEP	65	Sun	Ambient		Leone et al. (1985)
UC Irvine, Irvine, USA	Indoor	Stainless steel	0.5	Hg and flouorescent tubes	Controlled	−10 to 70 °C	De Haan et al. (1999)
UC Irvine, Irvine, USA	Indoor	FEP	5		Controlled		Nguyen et al. (2011)
UC Riverside, Riverside, USA	Indoor	FEP	90	Ar-Xe arc lamp	5–45 °C		Carter et al. (2005)
UC Riverside, Riverside, USA	Indoor	FEP	6.6	Fluorescent tubes	Ambient		Lim and Ziemann (2005)
UC Riverside, Riverside, USA	Indoor	Teflon coated aluminium	5.6	Xenon arc	Ambient	Evacuable	Beauchene et al. (1973), Winer et al. (1980)
UNC, Pitsboro, USA	Outdoor	FEP	135	Sun	Ambient	Half-cylinder-shaped dual chambers	Lee et al. (2004)

(continued)

Table 1.4 (continued)

Chamber name and location	Indoor/Outdoor	Wall material	Volume (m^3)	Light source	Temperature	Comment	References
Pacific Northwest National Laboratory, Richland, USA	Indoor	FEP	10.6	Fluorescent tubes	Controlled		Liu et al. (2012)
UNC, Chappel Hill, USA	Indoor	FEP	9	Fluorescent tubes	Ambient		Smith et al. (2019)
CU Boulder, Boulder, USA	Indoor	FEP	8	Fluorescent tubes	Ambient		Krechmer et al. (2016, 2017)
Harvard/HEC, USA	Indoor	FEP	4.7	Fluorescent tubes	Ambient		King et al. (2009)
NCAR, Boulder, USA	Indoor	FEP	10	Fluorescent tubes	Ambient		Fry et al. (2014)
University of Florida UF-APHOR, USA	Outdoor	FEP	52	Sun	Ambient	Half-cylinder-shaped dual chambers	Im et al. (2014)

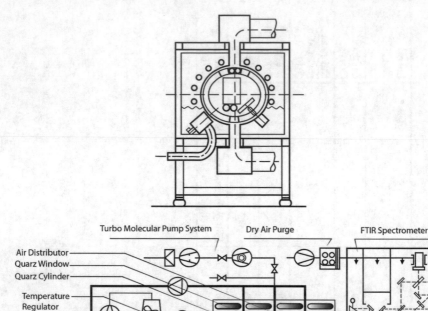

Fig. 1.4 Example of a 1 m³ indoor quartz chamber irradiated with UV fluorescent tube–the QUAREC chamber from the Bergische Universität Wuppertal–Germany. (© Bergische Universität Wuppertal)

1.2.2 Simulating Gas Phase Mechanism, Radical Cycles and Secondary Pollutant Formation

Studies on the formation of secondary pollutants are generally conducted in large outdoor chambers to avoid potential artefacts linked to a lack of realism in the irradiation and to minimize radical losses or conversion on the walls. Tropospheric ozone production studies were hence the first to benefit from chamber application. Nevertheless, for those studies to be of use for general modelling it is necessary to disentangle chamber effects from directly applicable results. Such an approach has led as early as the late 1970s to the first ozone isopleth diagrams, linking precursor levels to ozone production (Dodge 1977; Jeffries et al. 2013). Interestingly, because of the focus on photooxidants which is mostly driven by air quality legislation, operational model evaluation is often conducted by comparison with the results arising

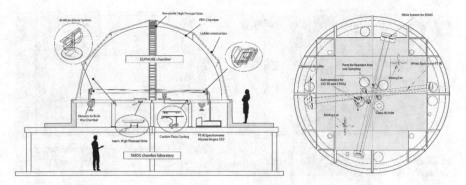

Fig. 1.5 Example of a large outdoor Teflon® chamber: the EUPHORE chamber–Valencia, Spain. (© EUPHORE)

from experiments conducted in these types of large chambers (Carter et al. 1979; Wagner et al. 2003; Bloss et al. 2005a, b; Carter 2008; Parikh et al. 2013).

Such chambers are made of FEP Teflon film, generally, several hundreds of cubic-meters in volume and are often installed on the roof of a dedicated laboratory (e.g. EUPHORE in Valencia, Spain Fig. 1.4 or Helios in Orleans, France) or in dedicated shelter structures (e.g. SAPHIR in Jülich, Germany Fig. 1.9 or UNC in North Carolina, USA). Because of their size and their outdoor installation, these facilities generally involve through-wall connections and inlets to connect the chamber with a measurement laboratory often located below. They also include devices such as a retractable roof to protect them from rain and wind. Temperature control cannot be achieved in such chambers and air inside the chamber may be heated by metal plates underneath the chamber when they are exposed to sunlight during the experiment. This effect is reduced if there is no direct contact of the metal plate with the chamber film and can be further reduced if the metal plate is cooled. Interestingly, even if their size is a significant advantage to minimize wall effects (on both gas phase and particulate phase), wind induced movements of the Teflon film lead to charge build-up that has the tendency to strongly reduce the physical lifetime of particle by drawing them to the wall (McMurry and Grosjean 1985) (Figs. 1.5 and 1.6).

1.2.3 Aerosol Processes

Originally considered as a technical problem during early smog simulation experiments, secondary organic aerosol (SOA) formation has since attracted very large interest from the scientific community. The availability of instruments such as Scanning Mobility Particle Sizers (SMPS), for the determination of particle number and size distribution with a time resolution of minutes, helped to promote the rapid development of experimental studies of SOA formation. This trend was further increased when mathematical formalisms were proposed to extrapolate the SOA yield from the

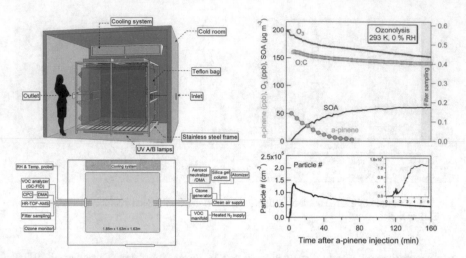

Fig. 1.6 Left: Example of a medium size indoor Teflon® chamber irradiated with UV fluorescent tubes/Right: Typical SOA production from terpene ozonolysis experiment (adapted from Kristensen et al. 2017)

high precursor concentrations used in chamber experiments to atmospheric conditions (Odum et al. 1996). The volatility basis set (VBS) formalism proposed by Donahue et al. (2006) was especially successful in providing a parameterization that could be inserted in models (3D included) and has triggered a renewed interest for chamber experiments from the modelling community. Both medium size and large chambers, as well as indoor and outdoor facilities, are regularly used for SOA experiments. Due to the multiphase nature of the processes studied and their even greater non-linearity, there is a general effort to reduce the starting concentration of the precursor to the ppb range (and sometimes below) in order to perform experiments at atmospherically relevant chemical conditions. These low concentrations make the results of these experiments very sensitive to wall effects on the gaseous species, such as wall loss of compounds that could normally participate in the aerosol mass or, on the other hand, the release of semi-volatile species. Further, physical wall losses of particles can also be significant. The quantitative characterization of these wall effects is still an open topic that requires a widely applicable formulation (see Chap. 2). It also depends highly on the properties of the wall (conductivity, permeability, reactivity, porosity…) in a context where the mechanisms involved are not yet well understood. Consequently, the combined use of several types of chambers, different in size but also made from different materials (Teflon film, glass, steel, aluminium…), is highly desirable for SOA experiments conducted at more realistic atmospheric concentrations of precursor gases. In parallel, a significant quantity of work has been conducted to better represent semi-volatile wall losses in this diversity of chambers (La et al. 2016; Krechmer et al. 2017; Lamkaddam 2017).

The contribution of simulation chambers to the understanding and quantification of SOA and related impacts is not limited to yield measurements. A wide body of work has focused on both online and offline chemical characterization with the aim of understanding the chemical composition of the SOA fraction but also the chemical processes that govern the formation and aging of organic aerosol. As a result of the amount of work carried out in medium size chambers, important breakthroughs have been made in these topics such as the identification of oligomerization processes in the aerosol phase (Kalberer et al. 2006), the chemical trends followed by oxidation during SOA aging (Jimenez et al. 2009; Ng et al. 2011a, b; Kourtchev et al. 2016), or the importance of auto-oxidation processes for the formation of SOA precursors (Ehn et al. 2014).

New particle formation was long considered as a barely controllable step in the formation of SOA during simulation chamber experiments. For reproducibility purposes, in most of the studies focusing on aerosol yield, it is hence recommended to use seed aerosol as a condensation medium in order to avoid nucleation. Nevertheless, dedicated chambers–often exhibiting a very low level of electrostatic charges on the wall–have been used to investigate this important process that is possibly controlling the number of cloud condensation nuclei in some parts of the atmosphere (Bonn et al. 2002; Kiendler-Scharr et al. 2009a, b; Kirkby et al. 2011, Boulon et al. 2012). One of the challenges in studying the early steps of nucleation in simulation chambers is, on the one hand, the ability to measure clusters and particles in the range of 1 to 3 nm and, on the other hand, the reduced lifetime of particles smaller than 20 nm in enclosed vessels (see Sect. 2.5 for particle wall losses analysis). Indeed, simulation chambers easily allow for aerosol lifetimes of several hours to a few days for particles in the range of a few hundreds of nanometers but due to their very high diffusivity, particles in the range of a few nanometer exhibit lifetimes in the range of a few minutes only.

Because of the importance of nucleation related processes, a dedicated facility was set-up at CERN: the CLOUD (Cosmics Leaving OUtdoor Droplets) experiment. The CLOUD chamber is a stainless steel atmospheric simulation chamber of 26.1 m^3 (Duplissy et al. 2010; Voightländer et al. 2012) operating under drastically clean conditions and installed in the T11 beamline at the CERN Proton Synchrotron. In order to study the effect of cosmic rays on nucleation, the chamber can be exposed to a 3.5 GeV/c positively-charged pion ($\pi+$) beam from a secondary target. The results from this atmospheric simulation chamber have led to significant advances in the understanding of nucleation including the elimination of the role of sulfuric acid alone as a nucleating agent, some insight on the effect of cosmic rays and the role of low volatility products from biogenic oxidation in initial cluster formation.

As aerosols refer to the particulate and gas phase, the investigation of aerosol processes in atmospheric simulation chambers also includes studies of heterogeneous processes. Prominent examples of systems studied include the chemical aging of aerosols and formation of brown carbon (e.g. Laskin et al. 2015) and the uptake of ozone on organic aerosol such as SOA formed from limonene ozonolysis (Leungsakul et al. 2005; Zhang et al. 2006). The N_2O_5 uptake coefficient on different particle types and the influence on gas phase oxidant levels were excessively studied in the Jülich indoor aerosol chamber (Mentel et al. 1996; Folkers et al. 2003; Anttila et al. 2006). More recently it was shown in atmospheric simulation chambers that levoglucosan, traditionally utilized as a source tracer for biomass burning aerosol, is reactive in the atmosphere (Hennigan et al. 2010, 2011; Sang et al. 2016; Bertrand et al. 2018; Pratap et al. 2019).

1.2.4   Cloud Processes

While "cloud chambers" have existed for a very long time, mostly to study the microphysics of fog and clouds, the past few decades have seen emerging chamber facilities which can generate clouds and fog under sufficiently clean conditions that multiphase chemistry, transformation at the droplet interface and cloud microphysical processes can be studied (Stehle et al. for the DRI chamber 1981; Hoppel et al. for the CALSPAN chamber 1994; Möhler et al. 2001 for the AIDA chamber; Duplissy et al. 2010 for the CLOUD chamber; Wang et al. 2011 for the CESAM chamber; Chang et al. 2016 for the Pi Chamber). All of these chambers are made of metal–mostly stainless steel (except for AIDA where the walls are made of aluminium)–because one of the most common protocols to generate a cloud is to perform a quasi-adiabatic expansion through a relatively fast decrease of the total pressure (from a few second to a few minutes) with or without controlling the wall temperature. For instance, the AIDA chamber allows for generating liquid droplets, mixed-phase (droplet and ice) and pure ice clouds. Further details can be found in Sect. 8.1 (Fig. 1.7).

These facilities have opened the door for realistic studies of cloud microphysics in the laboratory. The studies, which have been enabled due to careful control of the initial and boundary conditions, include investigations into the cloud condensation nuclei (CCN) and ice nucleation activity of various aerosol particles (Wagner et al. 2011; Henning et al. 2012; Hoose and Möhler 2012), homogeneous freezing of supercooled solution droplets (Möhler et al. 2003), scattering properties of ice crystals (Järvinen et al. 2014; Schnaiter et al. 2016), and the effects of non-precipitating water clouds on aerosol size distributions (Hoppel et al. 1994).

In parallel, a whole field of activity has been opened with the ability to study chemical transformations at the interface of droplets or even in the suspended aqueous phase. Using this approach, sulfate formation from the multiphase oxidation of SO_2 has clearly attracted the most attention (Stehle et al.1981; Miller et al. 1987; Lamb et al. 1987; Hoyle et al. 2016), but more recently, aqueous SOA formation from isoprene oxidation products (Brégonzio-Rozier et al. 2016) and brown carbon formation from fog processes of functionalized organics (De Haan et al. 2018) have also been investigated.

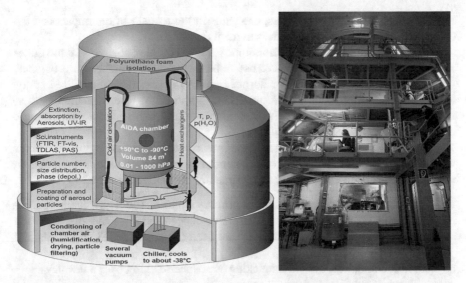

Fig. 1.7   Left: The AIDA facility at the Karlsruhe Institute of Technology with an 84 m³ aluminium chamber. Trace gas, aerosol and cloud experiments can be performed in a wide range of atmospheric temperatures (+60 °C to −90 °C), pressure (1–1000 hPa) and humidity (sub- and supersaturated with respect to liquid water and ice) conditions. Right: Typical evolution of pressure, temperature, relative humidity, and cloud droplet diameter for an adiabatic expansion experiment in AIDA

1.2.5   Characterization and Processing of Real-World Emissions

The development of atmospheric chemical mechanisms has been based on chamber studies of atmospheric oxidation of individual compounds. Hundreds of species have been studied following this approach and have contributed to the building of detailed chemical schemes, such as the Master Chemical Mechanism MCM (website: mcm. york.ac.uk). This effort is still ongoing to take into account new emissions and refine the chemical module of large-scale models. Nevertheless, in parallel, chamber studies that represent more realistic and more complex conditions are required to close the gap between well controlled but simplified laboratory experiments and observations in the real atmosphere.

Chamber studies, previously described here, have focused on chemical processes occurring in the gas and aerosol phases and have usually been limited to the simplified oxidation conditions and systems of selected precursors. More recent studies on real emissions from combustion sources such as engines and wood-burning stoves, or from natural emission sources such as plants or mineral dust, raise interesting possibilities for more relevant investigations of atmospheric processes.

In these studies, chambers are coupled to real emission sources (plant chambers, engines, wood burners, cooking stoves…) to study systems of real-world complexity. As much as one loses the ability to fully understand processes because of the complexity of the starting mixtures, one gains in the realism of the impact and the enhanced comparison with field measurements.

Experiments using real-world emissions involve complex sources that are either so intense that they need to be diluted before being added to chambers (e.g. engines, wood burners, cooking stoves) or do not require dilution (e.g. plants, sea spray, air fresheners and other household products). Approaches to ensure the quantitative transfer of all compounds of such complex emission blends into atmospheric chambers are described in detail in Chap. 5.

Concerning the first category, these experiments involve primary pollution sources whose aging is studied because of a potential formation of secondary pollution worsening their primary effect. The experimental challenges here are to

a. reproduce the atmospheric dilution of primary emission (both gaseous and particulate matter) while remaining in measurable concentrations: generally, a dilution factor ranging between 100 and 1000 are used (Platt et al. 2013, 2017; Gentner et al. 2017; Pereira et al. 2018)
b. establish a chemical system mimicking atmospheric aging over a few days.

Large and medium size chambers can be used for these studies. For example, Geiger et al. (2002) have connected a diesel engine fuelled with various diesel fuel formulations and mounted on a motor test bed directly to the EUPHORE chamber. In the dual outdoor simulation chambers, VOC mixtures containing a fixed ratio of *n*-butane, ethene and toluene were irradiated by natural sunlight in the presence and the absence of diesel exhaust. In this case, the large volume of the EUPHORE chamber (ca. 200 m^3) removed the need for a dilution system. For smaller simulation chambers (Chirico et al. 2010; Pereira et al. 2018; Platt et al. 2013) a conservative dilution system is needed to reduce the concentrations while keeping constant the various ratios between gaseous and particulate species, volatile and semi-volatile species. To do so, a specific aerosol diluter and heated lines are used. To preserve the efficiency of the atmospheric processes, prescribed VOC-to-NO$_x$ ratios are used which often require the addition of a VOC such as ethene, which is chosen for its ability not to add to the particulate mass during its oxidation. Aging is, for example, evaluated using the OH exposure index, defined as the cumulative OH concentration over the course of the experiment. The calculation of OH exposure requires the use of an OH tracer such as deuterated butanol-d$_9$ (Barmet et al. 2011) or the direct measurement of OH (e.g. Zhao et al. 2018) (Fig. 1.8).

These studies have demonstrated that, when considering car emission related fine particles, secondary pollution was as important as primary pollution and sometimes larger (Geiger et al. 2003; Bahreini et al. 2012; Platt et al. 2013, 2017; Gentner et al. 2017). In particular, the content of intermediate volatility organic compounds (IVOC) has been identified as critical in the ability to produce SOA (Pereira et al. 2018). The work in simulation chambers has allowed testing of the various types of vehicles, engines or fuel formulations that were already available on the market but, the interest that this methodology has raised among car manufacturers, allows one to hope for testing of future technology before its widespread deployment in vehicles.

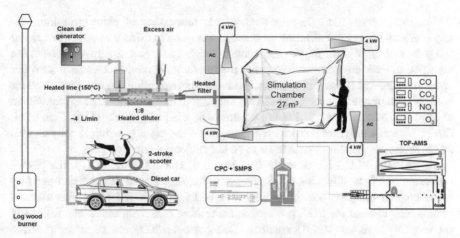

Fig. 1.8 Use of the PSI 27 m^3 Teflon chamber for investigating various real-world emissions transformation in the atmosphere. (Figure reused with permission from Heringa et al. (2012) Open access under a CC BY 3.0 license, https://creativecommons.org/licenses/by/3.0/)

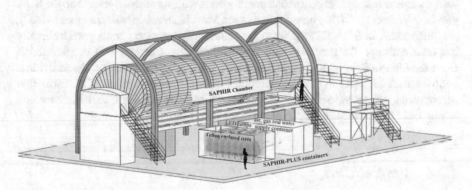

Fig. 1.9 SAPHIR-PLUS the combination of one of the largest outdoor simulation chamber (SAPHIR–Forschungszentrum Jülich, volume: ca. 270 m^3) with a controlled plant growing unit

A similar methodology can be applied to biomass combustion emissions. Considering the importance of this family of emissions, sources such as in-house open fires, agricultural burning, modern stoves or even barbecue emissions have been injected in a simulation chamber and aged in order to better quantify the extent of secondary pollution relatively to primary emission (Tiitta et al. 2016; Bertrand et al. 2017; Bhattu et al. 2019). Not only do these studies allow evaluation of the environmental impacts of combustion of various fuels (e.g. logwood, pellet, straw), types of combustion technology (e.g. stoves) and the various burning regimes (such as flaming or smouldering), but they also allow identification of molecular tracers and mass spectral signatures that can be monitored in the field to improve emissions inventories.

For experiments involving the atmospheric processing of plant emissions, the key challenge is not the dilution as these emissions are diffuse enough, but rather the preservation of their representativeness. Indeed, as living organisms, plants are sensitive to their environmental condition and any unwanted factors such as water stress, mechanical stress, biotic stress, oxidative stress or other abiotic stress from air composition may affect the composition and amount of their emissions (e.g. Kleist et al. 2012; Mentel et al. 2013; Wu et al. 2015; Yli-Pirilä et al. 2016; Zhao et al. 2017). Consequently, for studies involving plants, the plant growing facility as well as the emission transfer system have to be the subject of extreme care.

In SAPHIR-PLUS for example (see Fig. 1.9, Hohaus et al. 2016), the photo-oxidation of Pinus sylvestris L. (Scots pine) emissions were reacted and aged by ozonolysis in the presence of sunlight (Gkatzelis et al. 2018) which has allowed parameterization of the SOA production from these real plant emissions following the volatility basis set (VBS) formalism (Donahue et al. 2006). In a 9 m^3 temperature controlled Teflon simulation chamber, run in batch mode at the University of Eastern Finland, Failo et al. (2019) studied SOA formation from healthy Scots pine emissions and from the same plants infected with aphids. The aphid stressed pine were shown to emit more linear sesquiterpenes than healthy ones with significant effects on the SOA yields. Wyche et al. (2014) investigated in the Manchester Aerosol Chamber (MAC), the differences in SOA formed from predominantly terpene versus predominantly isoprene emitters. So far only very few studies have examined SOA production from the full range of VOCs made by plants. Since it was shown that the individual contributions of VOC in mixtures interact in non-linear ways in SOA formation mechanisms (Kiendler-Scharr et al. 2009a, b; McFiggans et al. 2019), there is a strong need for more studies exploring plant emissions.

1.2.6 Mineral Dust

aerosols are another key player in the atmospheric system. These particles contribute to the aerosol radiative effect and can act as cloud condensation nuclei (CCN) as well as ice nucleating particles (INPs). Mineral dust particles can deliver soluble elements needed for the development of oceanic life and eventually modify the CO_2 content of the atmosphere. Altogether, these kinds of aerosol particles affect Earth's weather and climate. Desert dust also affects human health, as an irritating agent at high concentrations causing respiratory diseases, as well as a vector for bacteria, viruses and possibly for severe infections like meningitis.

During transport, mineral dust can mix with air pollution and undergo chemical transformations that may affect their basic properties (composition, optical properties, CCN/IN activities, solubility...) and therefore their atmospheric impacts. Further, the multiphase chemistry occurring at their surface may also affect air composition. All these reasons have recently led a small number of research groups in the chamber community to apply the experimental simulation methodology to this science topic. This application implies solving various issues.

The first issue is the representativeness of the generated dust aerosol with respect to the atmosphere. Airborne mineral dust is a mixture of several minerals whose proportions change depending on the properties of the parent soil and wind speed. It forms an aerosol of an extended size distribution (extending from hundreds of nanometers to tenths of micrometers) that does not necessarily reflect the mineralogy of the soil due to the size-dependence fractionation between the soil and the aerosol phases that occurs at emission. There is hence a technological challenge in reproducing the dust generation from the soil process so that both the mineralogical composition and the size distribution are realistic (see Sect. 5.2). The global diversity of the mineralogical composition of natural parent soil is not reproduced by the commercially available minerals or standard mixtures. As much as possible, research tries to face this diversity by generating dust from natural soil collected across the world (Linke et al. 2006; Möhler et al. 2008a; Connolly et al. 2009; Wagner et al. 2012; Di Biagio et al. 2014, 2017a, b, 2019; Caponi et al. 2017), complementing and augmenting the many studies with model mineral dust such as Arizona Test Dust (Möhler et al. 2006, 2008a, b; Connolly et al. 2009; Vlasenko et al. 2006) or pure minerals such as illite (Möhler et al. 2008a, b) kaolinite (Tobo et al. 2012), hematite (Hiranuma et al. 2014) or Feldspar (Mogili et al. 2006, 2007; Atkinson et al. 2013).

Another critical issue for the study of mineral dust in simulation chambers is the reduced lifetime of these aerosols. Indeed, simulation chambers easily allow for aerosol lifetimes of several hours to a few days for particles in the range of a few hundreds of nanometers, but particles in the range of several micrometers undergo rapid sedimentation. As a consequence, in the absence of active resuspension processes, their lifetime in enclosed vessels is reduced to a few minutes only. This makes it difficult to study chemistry at the surface of the coarse fraction of mineral dust, but it is an advantage when one tries to reproduce the physical aging of dust plumes in the atmosphere. In fact, chamber experiments of a couple of hours duration can reproduce modifications to the size distribution of airborne dust that takes place over 2–3 days of transport (Di Biagio et al. 2017a, b). Chambers are therefore an emerging tool of choice to study the hygroscopicity and optical properties of mineral dust or the chemistry in the presence of the fine fraction only.

To date, most of the published results from chamber studies involving mineral dust have focused on their direct and indirect radiative effect. A large number of ice nucleation studies have been carried out at the AIDA chamber and LACIS (Leipzig Aerosol Cloud Interaction Simulator) on surrogate dust left bare (Möhler et al. 2006, 2008a, b; Tobo et al. 2012; Hiranuma et al. 2014; DeMott et al. 2015; Niedermeier et al. 2011, 2015; Hartmann et al. 2016) or covered with inorganic (Augustin-Bauditz et al. 2014; Niedermeier et al. 2011; Wex et al. 2014) and organic layers (Möhler et al. 2008a, b). In the CESAM chamber, most of the research to date has focused on optical properties and the derivation of complex refractive indexes in the long wave spectral ranges (Di Biagio et al. 2014, 2017a, b) and in the UV–visible (Di Biagio et al. 2019, Caponi et al. 2017).

To date, the number of studies of chemical reactivity at the surface of mineral dust in simulation chambers is rather limited due to the above-mentioned difficulties. They mostly involved ozone loss on the particles (Mogili et al. 2006) or SO_2 uptake and reactivity (Zhou et al. 2014).

1.3 Bridging the Gap Between Laboratory and Field Studies

Simulation chambers have been also used for the benefit of field experiments and long-term atmospheric monitoring (Kourtchev et al. 2016). These cross-community activities have first concerned instrumental development with a number of high technology new techniques being developed or tested at simulation chambers (see also 1.5). Prominent among these types of studies is the development of new techniques dedicated to atmospheric radical measurement (Schlosser et al. 2007; Onel et al. 2017), new techniques involving advanced optical setups such as optical cavities (Varma et al. 2009, 2013), the development of new advanced mass spectrometry instruments (Docherty et al. 2013) and chromatographic procedures for the elucidation of the aerosol organic fraction (Rossignol et al. 2012a, b).

1.3.1 Tracers and Sources of Fingerprint Studies

The use of simulation chambers for the benefit of field studies also includes the identification of specific signatures for emission sources (especially for aerosol mass spectrometry–see Aiken et al. 2008; Mohr et al. 2009; Kiendler-Scharr et al. 2009a, b; Zhang et al. 2011a, b; Schwartz et al. 2010). It also involves the identification of molecular tracers characteristic of specific processes. In this case, the ability of chambers to study specific processes is valuably used to separate the effect of the various potential oxidants or conditions. When well characterized, and found to be sufficiently unreactive in the atmosphere, these tracers are then searched for in the field to apply advanced apportionment procedures with the aims of not only elucidating the extent of primary sources but also of secondary processes (Jaoui et al. 2007; Kleindienst et al. 2007, 2012; Zhang et al. 2012).

In addition, important work has been carried out in characterizing the atmospheric tracers of primary sources such as levoglucosan or guaiacol (Hennigan et al. 2010; Bertrand et al. 2018; Pratap et al. 2019) that were initially thought fairly unreactive. This includes the use of stable isotopes as tracers for the extent of chemical processing (Sang et al. 2016; Gensch et al. 2014).

1.3.2 Instrument Comparison Campaigns

In addition to activities which involve generally one or only a few groups, large instrument comparison campaigns gather the wider atmospheric science community around chambers to characterize both established and emerging techniques using the ability of simulation chambers to precisely control the environmental conditions, while allowing different instruments to simultaneously sample from the same air mass. Suspected artefacts can hence be intentionally amplified and the sensitivity of the related techniques can be investigated and quantified. High precision water vapor measurement (Fahey et al. 2014), NO_x and NO_y measurements (Fuchs et al. 2010), oxygenated species measurements (Wisthaler et al. 2008; Apel et al. 2008; Thalman et al. 2015; Munoz et al. 2019), radical measurements (Schlosser et al. 2007; Fuchs et al. 2010; Fuchs et al. 2012a, b; Ródenas et al. 2013; Onel et al. 2017) or radical reactivity measurements (Fuchs et al. 2017) have been compared in large campaigns at chambers during the last 15 years.

1.3.3 Field Deployable Chamber

Recently a very innovative approach which combines the use of a simulation chamber with field studies has been developed both in Patras (Greece) and in Carnegie Mellon Institute (USA). It involves the use of portable simulation chambers directly in the field. This strategy is based upon a concept experiment: use ambient air as a starting point and allow the study of the evolution of atmospheric particulate matter at timescales longer than those achieved by traditional laboratory experiments (Kaltsonoudis et al. 2019).

This type of study can take place under more realistic environmental conditions but they could appear as being contrary to the whole simulation chamber experiment concept i.e. simplify and control the chemical system to better understand it. To solve this apparent contradiction, the group that is developing this new approach has developed a dual chamber strategy: after careful characterization of both chambers and so after verifying that they are producing comparable results, both are filled with the ambient being studied but one is "perturbed". The perturbation can consist of an additional oxidant injection such as ozone, addition of OH sources such as HONO or H_2O_2, or the addition of a compound potentially modifying the aerosol formation scheme such as α-pinene (Kaltsonoudis et al. 2019). The information on the chemical state of the sampled air is then deduced from the differential analysis of the results from the perturbed and control chambers (Fig. 1.10).

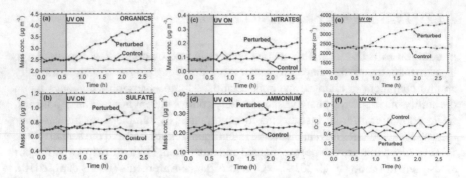

Fig. 1.10 Results from the operation of a dual field deployable simulation chamber during a campaign in Pittsburg (USA). One chamber is perturbed with the addition of HONO as an addition OH sources **a** submicronic aerosol mass **b** Sulfate content of sampled aerosol in both chambers as measured by an AMS **c** Nitrate aerosol content **d** Ammonium aerosol content **e** submicronic particle number concentration **f** Oxygen-to-carbon ratio in the organic fraction of the aerosol as measured by an AMS. (Reused with permission from Kaltsonoudis et al. 2019, open access under a CC BY 4.0 license, https://creativecommons.org/licenses/by/4.0/)

1.4 Emerging Applications

1.4.1 Air-Sea/Ice Sheet Interaction

Recently, even more specific installations have been developed across the simulation chamber community: a chambers dedicated to the elucidation of processes occurring at the air-sea interface. It consists of chambers that include a reservoir at their bottom where artificial or real sea water is kept under controlled conditions and in exchange with the atmosphere above. In Lyon (France) such a chamber has been developed and used to study the processes occurring in an organic film deposited at the water surface and potentially affecting the simulated atmosphere composition. From a modelled sea water containing, humic acid $(1-10$ mg $L^{-1})$ as a proxy for dissolved organic matter, and nonanoic acid $(0.1-10$ mM), a fatty acid proxy which formed an organic film at the air–water interface, this work has shown that a photosensitized production of marine secondary organic aerosol could occur (Bernard et al. 2016). These new results suggest that in addition to biogenic emissions, abiotic processes could be of importance for the marine boundary layer. In East Anglia (UK), the *Roland von Glasgow* Air-Sea-Ice Chamber (RvG-ASIC), named in honour of its late founder, allows users to simulate sea ice growth and decay in a controlled environment. The tank can be filled with artificial or natural seawater and can be capped with a Teflon sheet to reproduce an experimental atmosphere. Here the main challenge is to produce a realistic sea-ice from the cooling of the seawater tank (the whole facility can be temperature controlled from $+30$ to -55 °C). This new facility has allowed investigating the mechanisms governing the fate of persistent organic contaminants in sea ice. It has shown that sea ice formation results in the entrainment of chemicals from seawater, and concentration profiles in bulk ice generally showed the highest levels in both the upper (ice–atmosphere interface) and lower (ice–ocean interface) ice layers making them available from transit toward other compartments or interface reactivity.

1.4.2 Health Impacts

Even though the need to understand atmospheric chemistry has always been significantly motivated by public health issues and solving these issues has been part of the rationale for building many simulation chambers, until very recently, studies directly focused on health were rather scarce. In early investigations, the carcinogenicity and mutagenicity of chamber products were mostly evaluated after sampling of the contents and applying rather targeted offline in-vitro tests such as the Salmonella typhimurium plate-incorporation test (Claxton and Barnes 1981; Pitts 1983). In the past ten years, important progress has been made with the rise of surrogate indicators to qualify and quantify the potential health impact of particles such as the Reactive Oxygen Species content (ROS) (Fuller et al. 2014). The development of the corresponding instrumentation (Campbell et al. 2019) operating at high time resolution (on-line) now opens the way to building links between these indicators and the detailed chemical analysis often performed in the chamber. The goal is a better chemical characterization of the actual molecules or molecular functions involved in the oxidative stress.

In parallel, many groups have connected their simulation chambers with online samplers to expose living organisms such as lung cells or epithelial cells to the secondary pollutants produced in chambers (Savi et al. 2008; Mertes et al. 2013) in an attempt to understand the mechanisms that link cell toxicity with smog chemical and physical composition. This approach has led to important advances, especially when coupled with chamber experiments involving real world emissions (Künzi et al. 2013, 2015; Nordin et al. 2015). New directions have been explored by a few groups (Coll et al. 2018) which involve the use of simulation chambers for the long-term exposure (several days to several weeks) of living organisms such as murine models while complying with ethical standards. This new development requires overcoming substantial technical issues such as the stable and controlled production of secondary pollution over several days in a chamber. Their methodological research is pointing toward the use of indoor simulation chambers operated in batch mode. Development of such platforms in full cooperation with colleagues in the toxicology and medical communities may bring this health-related research to a better integration of the living body's functioning in the understanding of its response to air pollution.

1.4.3   Bioaerosols

Bioaerosols have been studied for over a decade in cloud chambers to investigate their potential ice nuclei activity (Möhler et al. 2008b). Given the public health problems associated with bioaerosol contamination and the many unknowns about the survival and transformation of bioaerosols, such as bacteria, in the atmospheric environment, innovative chamber work has recently started to address these issues (Amato et al. 2015; Brotto et al. 2015). These studies have led to the development of an indoor simulation chamber at the University of Genoa (Italy) where viable bioaerosol can be directly collected using Petri dishes without perturbing the course of the experiments while, in parallel, being online monitored by more classical techniques such as WIBS (Massabò et al. 2018). The goal is to derive parameterization of survival and activity of bioaerosols to eventually model the geographical extent of their contamination area.

1.4.4   Cultural Heritage

Works of art, with highly sensitive colours and materials, may be exposed to harmful levels of particulate matter in both indoor and ambient (i.e. outdoor) environments. Over time, these particles can deposit onto the surface of the artwork, which may influence the perceived colour. Reports over the concern of colour degradation to paintings, buildings, and other pieces of cultural heritage due to exposure to air pollution, acid rain, and other environmental factors have existed since at least the late 1800s due to London smog events (Brommelle 1964). However, the physical processes that connect exposure to particulate matter and the corresponding change in perceived colour are unknown, and first attempts to experimentally quantify the impact of particulate matter on painted works of art are only now emerging. The FORTH art exposure facility makes such an approach by developing protocols for the exposure of artwork to known levels of air pollutants and quantifying the effects of exposure using a portable colourimeter model WR-10 (FRU). Further developments in this emerging field will benefit from combining the expertise of exposure chamber approaches and atmospheric simulation chambers.

1.5   Considerations on the Design of an Atmospheric Simulation Chamber

The main objective of the guide is to serve as a reference for both new and current users of atmospheric simulation chambers. However, some readers may be considering the construction of a new chamber and this section is aimed at them. Additionally, it will provide to the new user, some insights into the design rationale of the chambers they will be working with.

This section mainly deals with the scientific issues and objectives that drive a particular chamber construction, but of course, practical limitations such as space, personnel and money will also influence chamber design. A particular focus is put on the requirements for the design of chambers dedicated to the exploration of atmospheric chemistry processes.

Atmospheric simulation chambers have several uses; firstly, they may be used to provide a controlled and realistic environment to simulate aspects of the real atmosphere or to test and compare field instrumentation. Secondly, chambers can be used as extended laboratory apparatus. For example, several hundreds of elementary reactions are involved in the complete oxidation of complex volatile organic compounds (VOC) such as isoprene (C_5H_8) or aromatic hydrocarbons. Some of these processes, particularly those occurring in the initial stages, can be studied individually by techniques such as laser flash photolysis or discharge flow, but many cannot. Atmospheric simulation chambers equipped with a wider range of instrumentation may either be able to directly measure rate coefficients, provide information on the yields of stable first-generation products, test entire chemical mechanisms or investigate aerosol chemistry. The main purpose of the experiments also strongly influences the design of the chamber.

1.5.1   Chemical Regime of Simulation Experiments

Whatever the objective of the chamber, the primary applications are to processes in the Earth's *troposphere* (extending from surface to the tropopause, where tropopause height varies with latitude from ~10 km in polar regions to ~18 km in the tropics). In the troposphere temperatures range from ~220–320 K and pressures of ~100–1000 mbar are found. In addition, we are often interested in the interactions of emissions (biogenic or anthropogenic) with the atmosphere and the interactions of atmospheric pollutants with humans, animals, plants and the ocean. Most of these interactions take place within the *boundary layer*, typically the first kilometre or so of the troposphere and therefore for many applications, operation at pressures close to 1000 mbar is appropriate. However, there is obviously still a wide range of temperature variation within the boundary layer and so temperature variation may be an important goal in chamber design. Relative humidity also varies over a wide range in the troposphere and affects many physical and chemical processes in the atmosphere. Therefore, depending on the application of the chamber, precise control of humidity is also vital.

Besides variations in physical parameters, there are also significant variations in the chemical composition desired in the simulation experiments that will influence the chamber design. Most studies focus on regions of the atmosphere with significant VOC emissions. The chemical oxidation of VOCs often includes the same initial reaction steps; the reaction of a radical species, X, (where X = OH, NO_3, Cl etc.) leads via abstraction or addition of the oxidant to an organic radical, R, which then rapidly adds O_2 to lead to an organic peroxy radical RO_2.

$$\text{e.g. OH} + \text{RH} \rightarrow \text{H}_2\text{O} + \text{R} \tag{R1}$$

$$\text{R} + \text{O}_2 \rightarrow \text{RO}_2 \tag{R2}$$

The atmospheric fate of the organic peroxy radicals depends on the relative abundance of concentrations of reaction partners such as nitric oxide ([NO]) and other peroxy radicals ([RO_2/HO_2]). In regions with high NO_x concentrations, the loss of RO_2 is typically dominated by the reaction with NO, generating an alkoxy radical (RO). The exact fate of the RO depends on its structure, but most often products are a carbonyl compound and hydroperoxyl radicals (HO_2). Further reaction of HO_2 with NO regenerates OH completing a reaction cycle (Fig. 1.10)

$$\text{RO}_2 + \text{NO} \rightarrow \text{RO} + \text{NO}_2 \tag{R3}$$

$$\text{e.g. RO} + \text{O}_2 \rightarrow \text{Carbonyl} + \text{HO}_2 \tag{R4}$$

$$\text{HO}_2 + \text{NO} \rightarrow \text{OH} + \text{NO}_2 \tag{R5}$$

The by-product of the NO to NO_2 conversion in reactions (R3) and (R5) is ozone, a significant secondary pollutant. This radical reaction chain is the only relevant chemical source for ozone in the troposphere.

However, in environments with low NO_x concentrations (typically [NO] < 50 pptv) such as the marine boundary layer or remote tropical or boreal forests, radical recombination reactions become the dominant RO_2 loss channel.

$$\text{RO}_2 + \text{RO}_2 \rightarrow \text{ROH} + \text{R}'\text{CHO} + \text{O}_2 \text{ or } 2\text{RO} \tag{R6}$$

$$\text{RO}_2 + \text{HO}_2 \rightarrow \text{ROOH} + \text{O}_2 \text{ or RO} + \text{OH} + \text{O}_2 \tag{R7}$$

These reactions terminate the radical chain. For specific RO_2 radicals, isomerization reactions can be competitive. Products can be again RO_2 radicals that may decompose and thereby form other radical species such as HO_2 or highly oxygenated molecules could be eventually formed. For example, significantly enhanced OH concentrations are observed in high isoprene and low NO_x environments that can be explained by radical production from isomerization reactions of isoprene derived RO_2 (Peeters et al. 2014; Novelli et al. 2020).

Due to the importance of the fate of RO_2 radicals for the chemical reaction system that should be investigated in the simulation experiments, considerations about the NO_x concentration that can be achieved in the chamber is important and can have implications on the chamber design (Fig. 1.11).

The chemical composition of the troposphere is also impacted by surface interactions such as bulk and aerosol surfaces. The interaction with bulk solid surfaces

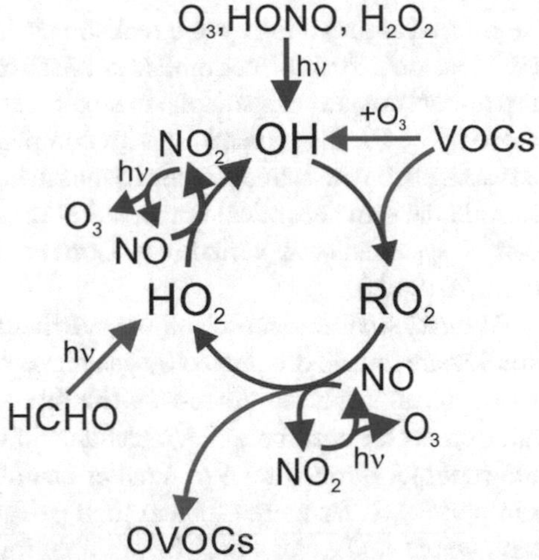

Fig. 1.11 Scheme of the radical reactions involved in atmospheric photochemical VOC oxidation and ozone production

O_3, HONO, H_2O_2

hv

NO_2 OH $\xleftarrow{+O_3}$ VOCs

hv

O_3

NO

HO_2 RO_2

hv

NO

HCHO

O_3

NO_2 hv

OVOCs

can be easily replicated in many chambers. Some chambers (e.g. ISAC) are specifically designed to investigate interactions with liquid surfaces and sea-ice like the Roland Van Glasow Air-Sea-Ice Chamber at the University of East Anglia. Aerosols, primary or secondary, organic or inorganic, are the other main surfaces in the troposphere and studies involving aerosols and gas/aerosol/cloud interactions may require specific design criteria and instrumentation.

1.5.2 Chamber Size

Whilst there may be specialized chambers for the investigation of interactions with bulk surfaces, often bulk surfaces and their associated heterogeneous chemistry are minimized to avoid that experiments are impacted by chamber wall effects. Minimizing the surface to volume ratio (S/V) helps and might be the only way to suppress chamber wall effects, if experiments are performed at atmospheric concentrations of trace gases. For example, the large chambers EUPHORE (200 m^3) and SAPHIR (270 m^3) have spherical and cylindrical shapes, respectively, to minimize the surface to volume ratio and are advantageous compared to cuboid structures. Cuboid shapes are commonly used for Teflon chambers as they can be easily mounted, illuminated and physically accessed.

Most chambers have capabilities to inject reagents and maintain a homogeneous mixture by operating fans. Clearly, the specifications of fans need to match the chamber size to ensure efficient operation. The practical issues concerning logistics are beyond the scope of this chapter, but it is worth highlighting that large chambers such as AIDA, EUPHORE and SAPHIR have significant numbers of dedicated personnel and additional infrastructure facilities for example for clean air generation and power requirements.

As well as providing a more realistic environment for simulations, large chambers are ideal tools for field instrument comparisons. The volumes of gas sampled by some instruments make comparisons in small chambers impossible and generally there is more space for instruments. In situ comparisons in the real atmosphere have their advantages, but instrument comparisons in large chambers ensure, that all instruments sample the same chemical composition in a controlled environment and conditions can be systematically varied (e.g. Dorn et al. 2013; Fuchs et al. 2010, 2017; Fuchs et al. 2012a, b).

Whilst a small surface to volume ratio helps in ensuring that the chemical processes studied are indeed dominated by gas phase chemistry and ensures the best representation of atmospheric processes, this may not be required for other purposes of environmental chambers. For mechanistic or relative rate reaction kinetic studies, the rapid turnaround time of smaller chambers, where several experiments can be run per day, is far more efficient than performing such experiments in large chambers where studies may only be possible for good weather conditions in the case of outdoor chambers and may be limited to one experiment per day. Smaller chambers (particularly if made from glass or metal) can be rapidly evacuated (and in some cases heated) to clean the surfaces or can be even physically cleaned. Surfaces can be coated to minimize wall effects. Furthermore, many small chambers are operated in steady state conditions contrary to the batch mode operation of large chambers.

1.5.3 Materials

In general, there are three types of materials used in chambers: Teflon (or equivalent), borosilicate glass, quartz, or stainless steel (see Table 1.4). All materials have their advantages and disadvantages with respect to surface properties and physical parameters (e.g. T, p) that can be regulated in the chamber. Depending on the purpose of the chamber, the possibility to simulate e.g. pseudo-adiabatic cloud expansion, ultra-clean air conditions, or photolytic conditions representative of the troposphere is a key driver of choices of material used.

Teflon (or equivalent). Due to their large size, all large ($> \sim 80$ m^3) chambers are constructed from fluoro-polymer plastics mounted on a metal frame. Such structures are light but fragile and need to be protected. Outdoor chambers like SAPHIR and EUPHORE have retractable protection, protecting the film from bad weather conditions, but also allow for experiments in the dark. The Helios chamber (~ 90 m^3) at CNRS-Orleans can be rapidly moved in and out of a permanent shelter. All of these chambers have a solid metal floor that can be used to place equipment such as FTIR mirrors and fans. In EUPHORE this forms part of the chamber surface and is cooled to prevent significant heating from solar radiation. In SAPHIR it is covered with Teflon and can be lowered for experiments such that the Teflon film does not have contact with the metal to avoid radiative transfer heating.

Teflon is also used in the construction of smaller chambers where glass or metal would be alternatives. Teflon has significant advantages in terms of cost. Additionally, as it is transparent, it is easy to fully illuminate the entire chamber with either solar or artificial light. Although Teflon is chemically inert, it is commonly observed that compounds can adhere to the wall and released in later experiments even if the chamber had been cleaned in between. For example, nitrous acid (HONO) is released, if humidified air is illuminated in Teflon chambers. The photolysis of HONO serves as a source of OH radicals, but also leads to an increase of nitrogen oxide species over the course of an experiment (Rohrer et al. 2005). The radical production from the chamber HONO source can be sufficiently high for performing OH oxidation experiments in large chambers as EUPHORE and SAPHIR (Fuchs et al. 2013). Smaller chambers can be manually cleaned, but this is not possible for larger chambers. As non-rigid structures, Teflon type chambers cannot be evacuated and are limited to operation at ambient pressures. Rather than evacuation, residual trace gases are removed by flowing clean gas through the chamber. For large chambers, this is typically done overnight. Smaller chambers can be enclosed in air-conditioned rooms to provide some degree of temperature control and variation.

Pyrex/Quartz Pyrex or quartz chambers are used for volumes of ~ 1 m^3 or less. Within EUROCHAMP, the chambers at Wuppertal and Iasi are of cylindrical shape (~ 0.5 m diameter) and have a volume of approximately 1 m^3. The end flanges of both chambers are metal allowing for easy access to instrumentation and provide a fixed framework for mounting FTIR mirrors (similar structures are also used in some Teflon type chambers too). Due to the fragility of glass, the chambers are mounted on a vibration resistant framework. The advantage of quartz is that it allows for the transmission of shorter wavelength UV radiation compared to Teflon (e.g. radiation from mercury lamps emitting at 254 nm) which can be useful for specific radical generation methods.

Whilst pyrex/quartz chambers are limited in size, their small size allows to uniformly distribute artificial light sources around the chamber. The rigid construction also allows to evacuate the chambers, so that the chamber can be cleaned within a short time between experiments and it can be operated at sub-ambient pressure. Smaller chambers such as those at the National Centre for Atmospheric Research (NCAR) in Boulder, US, are surrounded with air-conditioned liquid baths to perform studies in which the temperature is varied. Quartz and pyrex are well characterized and reasonably inert surfaces. Evacuation (in combination with heating if available), provides rapid and efficient cleaning, in extremis, the end flanges of large chambers can be removed to allow for physical cleaning.

Metal Chambers are typically of cylindrical shapes and have volumes of the order of 1–6 m^3, with the exception of the 84 m^3 large AIDA chamber at Karlsruhe Institute of Technology. Metal chambers are typically constructed from stainless steel and have significant advantages in their robustness compared to other materials allowing for rapid evacuation/operation at reduced pressure. Several systems are also equipped with a temperature control system. Temperature control can be useful for two main purposes; firstly, simulating the temperature variation both within the boundary layer at various latitudes/seasons and across the vertical extend of the troposphere; Secondly, elucidating the temperature dependence of chemical mechanisms.

For metal chambers, flanges with inlets for instruments are easy to install either in the main end flanges or elsewhere at the chamber. Although the end flanges of chambers can be large, they typically bow slightly if the chamber is operated at reduced pressure and therefore thought needs to be given on how to mount equipment requiring high spatial precision (e.g. multi-pass mirror) onto the end flanges.

The two significant disadvantages of metal as the construction material (besides the high S/V associated with the relatively small volume of most chambers) is the potential reactivity of the surface and the difficulty in generating a uniform light field. Surface effects can be accounted for (see Sects. 2.4 and 2.5) and efficient evacuation combined with overnight heating and/or oxidant exposure (e.g. O_3) ensures that the surface remains uniform over the course of an experimental campaign (see Chap. 3). Illumination issues are discussed in the next section.

1.5.4 Light Sources

Photochemistry is one of the main driving forces for atmospheric processes, so that whilst there are important dark reactions such as ozonolysis or nitrate radical (NO_3) initiated chemistry, light is required for most experiments.

The most obvious source, particularly if atmosphere-like conditions are simulated, is solar radiation and for large chambers such as Helios, EUPHORE and SAPHIR it is the only feasible option. Certain small/medium sized Teflon type chambers can be operated with either solar or artificial radiation.

The transmission of solar radiation by Teflon is good over the entire solar spectrum. Spectral radiometers inside the chamber can be used to measure the actinic flux (see also Sect. 2.3), both of the incoming solar radiation and of light reflected/emitted by the chamber floor. The disadvantage of outdoor chambers using sunlight is that experiments are dependent on the weather, because large chambers made of Teflon cannot be operated in windy conditions. Like in the atmosphere, the radiation field in the chamber changes over the course of a day-long experiment, both due to the change of the solar zenith angle and also due to short-term, transient variations caused by clouds.

Artificial radiation is used for a majority of smaller Teflon chambers and all glass and metal chambers. Depending on the main purpose of the chamber, light with a broad radiation distribution, including simulation of the solar spectrum, can be used or alternatively lamps with narrow outputs for example in the UV region (e.g. mercury lamps with emission lines at 254, 308, 365 nm) can be used. For many chambers it is possible to swap between different types of lamps.

For Teflon chambers lamps are often mounted on one side and the bank of lamps is directed into the chamber. The often cuboid nature of such chambers makes it easy to establish a uniform radiation field across the chamber. For glass chambers banks

Fig. 1.12 QUAREC
chamber, Wuppertal, the
lights are mounted outside
the chamber providing a
uniform radiation field in the
actual chamber

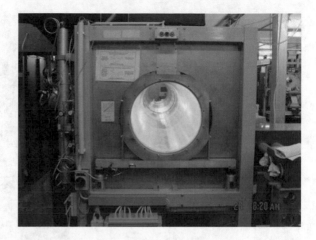

of tubular lamps surround the cylindrical chamber. Carefully arranged, the radiation
field inside the chamber can be very uniform.

The chamber construction determines the UV cut-off wavelength for example
quartz is transmissive for wavelengths higher than ~200 nm. Arranging lamps around
the chamber such that a uniform radiation field is obtained is clearly not possible for
a metal chamber. Two approaches are typically used. For example in the CESAM
chamber, radiations from xenon arc lamps are directed into the chamber through
windows, whereas in the HIRAC chamber quartz tubes mounted inside the chamber
are used as a light source (Figs. 1.12, 1.13, 1.14, and 1.15). Radiation fields in these
chambers are less uniform; variations can be measured with a spectral radiometer
(Sect. 2.3) and instruments can be designed to sample from various locations to test
for significant spatial variations of trace gas and radical concentrations.

1.5.5   Instrumentation

The type of instruments installed at the chamber depends on the primary purpose
of the individual study, for example, aerosol and gas phase experiments will require
different measurements. Table 1.5 summarizes typical instrumentation and measurement approaches utilized in chambers. Table 1.5 is structured into groups of instruments according to measurement parameters. Specialized and custom-built instrumentation may require significant technical support to ensure their operation. In some
cases, high costs for commercial systems can balance low, long-term running costs.

It is important to consider what instrumentation is going to be applied to the
chamber in advance of the construction, e.g. to allow for sufficient space and air conditioning. Although most commercial instruments and equipment that take samples for
later offline analysis can be easily placed at the chamber, some components that are
directly attached to the chamber (e.g. mirrors used for FTIR spectroscopy or special

Fig. 1.13 CESAM
Chamber, LISA, the chamber
is illuminated from above

Fig. 1.14 FORTH chamber,
a Teflon chamber with side
wall illumination

Fig. 1.15 HIRAC chamber
showing internal
illumination and modelling
of resultant radiation field
across the chamber.
Reproduced from Seakins
(2010)

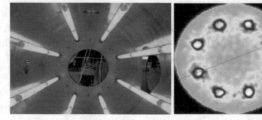

cavity ring-down systems) have to be considered in the early planning of the chamber
construction. Mirrors need to be mounted where they are unaffected by vibrations
from fans or pumps and the mounting needs to be rigid with respect to changes in
pressure or they need to be easily adjustable. Purge gas flows may be needed for

Table 1.5 Summary of typical instrumentation and measurement approaches utilized in chambers, structured in order to group the instrumentation according to measurement parameter

Gas phase concentration

Parameter	Instrument type	Online/offline/in situ	References
NO concentration	Chemiluminescence NO_x analyser Mo converter	Online	Dunlea et al. (2007)
NO and NO_2 concentration	Chemiluminescence NO_x analyser blue light Converter	Online	NO_2: Ryerson et al. (2000)
NO_2 concentration	NO_2 cavity-based absorption spectroscopy	Online	Fuchs et al. (2010), Kebabian et al. (2005)
Ozone concentration	UV absorption O_3 Analyser,	Online	Parrish and Fehsenfeld (2000)
Ozone concentration	Chemiluminescence		Ridley et al. (1992)
SO_2 concentration	fluorescence SO_2 Analyser	Online	Krechmer et al. (2016), Krechmer et al. (2017)
CO_2, CO, H_2O, CH_4 concentration	CRDS	Online	https://www.picarro.com/products/g2401_gas_concentration_analyzer
HCHO	Hantzsch reaction	Online	Junkermann and Burger (2006)
HONO	Chemical derivatization; HPLC or photometer absorption	Online	Kleffmann et al. (2002), Afif et al. (2016)
VOC, OVOC concentration	PTR-MS, CIMS	Online	DeGouw and Warneke (2007), Taipale et al. (2008), Barber et al. (2012), Cappellin et al. (2012), Duncianu et al. (2017)
VOC concentration	Sorbent cartridges + GC-MS	Offline	Rossignol et al. (2012a, b)
VOC concentration	Automatic GC- FID	Online	De Blas et al. (2011)
VOC, NO_x, NO_y, Ozone	long path FTIR	In situ	Barnes et al. (1985), Doussin et al. (1997)

(continued)

Table 1.5 (continued)

	Flash-photolysis combined with laser-induced fluorescence	Online	Fuchs et al. (2017) and references therein
OH Reactivity	Comparative Reactivity Method CRM		
NO$_3$ Reactivity	CRDS	Online	Dewald et al. (2020)
Water mixing ratio	Hygrometer gage, chilled mirror	In situ	Fahey et al. (2014)
Relative humidity	polymer sensors	In situ	Fahey et al. (2014)

Chamber physical properties

Parameter	Instrument type	Online/offline/in situ	References
Temperature	Thermocouple resistance thermometer	in situ	Dias et al. (2017)
Pressure	Capacitance manometer	in situ	Blado et al. (1970)
Irradiation spectrum/actinic flux	spectral radiometer	in situ	Bohn et al. (2008)

Radical measurements

Parameter	Instrument type	Online/offline/in situ	References
OH	Long Path DOAS Laser-induced fluorescence Chemical-ionization mass spectrometry	Online, in situ	Schlosser et al. (2009)
OH	Faraday Rotation Spectroscopy	Online	Zhao et al. (2018)

(continued)

Table 1.5 (continued)

HO$_2$	NIR-CRDS	Online, in situ	Onel et al. (2017)
HO$_2$	Chemical conversion and Laser-induced fluorescence	Online	Fuchs et al. (2010)
HO$_2$	CIMS(Br-)	Online	Albrecht et al. (2019)
CH$_3$O$_2$	NIR-CRDS	Online, in situ	Onel et al. (2020)
CH$_3$O$_2$	Laser-induced fluorescence	Online	Onel et al. (2017)
RO$_2$	CIMS	Online	Nozière and Hanson (2017)
RO$_2$	ROx-LIF	Online	Fuchs et al. (2008), Whalley et al. (2018)
Rox	ROx-MAS, PerCIMS	Online	Hanke et al. (2002), Edwards et al. (2003)
Rox	PERCA	Online	Cantrell et al. (1984)
Hydroperoxides, peroxides	HPLC with FLD detector	Offline	Lazrus et al. (1986)
NO$_3$	DOAS, cavity-based absorption spectroscopy	Online	Dorn et al. (2013), Fouqueau et al. (2020)
N$_2$O$_5$	FTIR, cavity-based absorption spectroscopy		Fuchs et al. (2012a, b)

Particulate matter physical properties

Parameter	Instrument type	Online/offline/in situ	References
Total concentration/Size distribution	Particle Size Magnifier (PSM)	Online	Vanhanen et al. (2011)
Total concentration	Diethylene Glycol–Condensation Particle Counter (DEG CPC)	Online	Jiang et al. (2011), Wimmer et al. (2013)

(continued)

Table 1.5 (continued)

Size distribution, Ion size distribution	Neutral and Air Ion Spectrometer (NAIS)	Online	Mirme and Mirme (2013)
Ion size distribution	Cluster Ion Counter (CIC)	Online	
Ion size distribution	Air Ion Spectrometer	Online	Mirme et al. (2007)
Total concentration	Condensation Particle Counter (CPC)	Online	Stolzenburg and McMurry (1991), Hering et al. (2005)
Total concentration/Size distribution	Condensation Particle Counter Battery (CPCb)	Online	Kulmala et al. (2007)
Size distribution	Differential Mobility Analyzer Train (DMA train)	Online	Stolzenburg et al. (2017)
Size distribution	Scanning Mobility Particle Sizer (SMPS/MPSS)	Online	Wang and Flagan (1990), Gaie-Levrel et al. (2020)
Size distribution	OPC–single wavelength (eg. Grimm)	Online	Heim et al. (2008)
Size distribution	OPC–white light	Online	Heim et al. (2008)
Size distribution	APS	Online	Pfeifer et al. (2016)
Aerosol spectral absorption	Photoacoustic spectrometer	Online	Arnott et al. (1999)
Aerosol spectral absorption	Aethalometer	Online	Arnott et al. (2005), Di Biagio et al. (2017a, b)
Aerosol spectral scattering	Nephelometer	Online	Andersoin and Ogren (1998)
Aerosol spectral extinction/scattering/SSA	External CRDS–Optical cavities	Online	Massoli et al. (2010)

(continued)

Table 1.5 (continued)

Parameter	Instrument type	Online/ offline/in situ	References
Aerosol spectral absorption	Filter sampling–Multi-Wavelength Absorbance Analyzer (MWAA)	Offline	Massabò et al. (2018)
Particle morphology/shape distribution	filter sampling–Electron microscopy	Offline	Chou et al. (2006)
Mass concentration	Online tapered element oscillating microbalance (TEOM)	Online	Tortajada-Genaro and Borrás (2011), Roberts and Nenes (2005)
Aerosol hygroscopicity	Hygroscopic Tandem Differential Mobility Analyser (HTDMA)	Online	Denjean et al. (2014), Good et al. (2010)
Cloud condensation Nucleous counter	CCN counter	Online	Wex et al. (2009)
Ice Nucleous Counter	IN counter	Online	DeMott et al. (2018), Mölher et al. (2021)
Particulate matter concentration and composition			
Parameter	**Instrument type**	**Online/ offline/in situ**	**References**
Organic aerosol chemical composition	Filter sampling - derivatization - GC–MS	Offline	Chiappini et al. (2006)
Elemental carbon and organic fraction of particle	EC/OC (Sunset)	Offline	Massabò et al. (2016)
Aerosol composition	ACSM-ToF Aerodyne		Ng et al. (2011a, b)
size segregated Aerosol composition	AMS		Kiendler-Scharr et al. (2009a, b)

(continued)

Table 1.5 (continued)

Surface state and composition	Filter sampling–X–ray induced photoelectron spectroscopy (XPS)	Offline	Denjean et al. (2015)
Refractory black carbon	Single Particle Soot Photometer (SP2)	Online	Laborde et al. (2012)
Inorganic/organic soluble fraction	Particle into Liquid Sampler (PILS)	Online	Pereira et al. (2015), Behera and Sharma (2012)
Inorganic soluble fraction	Filter sampling–Ion Chromatography (IC)	Offline	Chen and Jang (2012)
Organic aerosol chemical composition	filter sampling–UPLC-QTOF-MS	Offline	Nizkorodov et al. (2011), Hamilton et al. (2008)

optical systems to keep mirrors clean. Access to important equipment that may need regular cleaning or service must be assured. Some instruments will extract significant volumes of gas posing requirements on the chamber volume to ensure that dilution does not become a major loss term. A mechanism of regulating the replenishment flow to maintain a certain pressure or volume may be required.

There is a very strong synergy between chamber and field communities in terms of instrumentation, with chambers being used to design, develop, validate and compare field instruments. In general, instruments that work well in the field will be suitably sensitive and robust to ensure efficient use within chambers.

Homogeneous mixing within the chamber has to be ensured. This can be tested through a comparison of measurements that derive average concentrations across the pathlength of the system and point measurements at a single location, as well as through sampling from different locations. In addition, careful design of sampling systems (material, residence time, heating) and location of the instrumentation to minimize transfer distance limits the effect of sampling losses or transformation of reactive or instable species during the sampling process.

Making sensitive measurements of complex systems is a challenging task and even if carefully operated, systematic errors or inferences can occur. Having multiple, complementary methods (or regularly participating in inter-comparisons) can help identify these problems.

The above discussion focused on how scientific objectives and considerations influence the chamber design, construction and instrumentation. This section can only give a brief outline on considerations. This section can be used as an overall introduction, but details can be found in technical papers and reports. There is no perfect chamber design; each system has its own advantages in meeting particular objectives, but also disadvantages. In fact, having a variety of chamber designs and performing comparisons (i.e. reference experiments as detailed in Sects. 2.4 and 2.5, Donahue et al. 2012) highlights issues that would easily be missed in standardized approaches.

1.6   Conclusion

The original use of "smog chambers" for investigating chemical transformations in the atmosphere, for quantification of the rate, extent and relevance of the various possible pathways, for the identification of secondary pollutants remains just as relevant today as it did many decades ago. Indeed, the models that utilise chamber-derived data are still far from explicit, i.e. they do not include all of the processes that are required to represent and forecast the actual atmospheric composition, and there is still room for improvement, as well as the possibility of incorporating new chemistry to address future challenges. At the same time, the field of experimental atmospheric simulations has been extremely active over the past 15 years and considering the number of new facilities around the world, there is little doubt about its vitality over the next 15 years. A number of new methodologies and applications have risen, and they will bring the operational capacity of simulation chambers to a new level. This community effort will allow a much broader range of scientific and societal needs to be addressed, including the direct and indirect climate effect of atmospheric pollutants, the impact of air composition on health and cultural heritage, as well as on the various compartments of the Earth system. The application of simulation chambers in some of these areas is still in the early stages, but rapid progress is being made and already producing data that will help to open new ways of considering the complex interplays between atmospheric transformation and impacts.

References

Afif, C., Jambert, C., Michoud, V., Colomb, A., Eyglunent, G., Borbon, A., Daële, V., Doussin, J.-F., Perros, P.: NitroMAC: an instrument for the measurement of HONO and intercomparison with a long-path absorption photometer. J. Environ. Sci. **40**, 105–113 (2016). https://doi.org/10.1016/j.jes.2015.10.024

Aiken, A.C., DeCarlo, P.F., Kroll, J.H., Worsnop, D.R., Huffman, J.A., Docherty, K.S., Ulbrich, I.M., Mohr, C., Kimmel, J.R., Sueper, D., Sun, Y., Zhang, Q., Trimborn, A., Northway, M., Ziemann, P.J., Canagaratna, M.R., Onasch, T.B., Alfarra, M.R., Prevot, A.S.H., Dommen, J., Duplissy, J., Metzger, A., Baltensperger, U., Jimenez, J.L.: O/C and OM/OC ratios of primary, secondary, and ambient organic aerosols with high-resolution time-of-flight aerosol mass spectrometry. Environ. Sci. Technol. **42**, 4478–4485 (2008). https://doi.org/10.1021/es703009q

Akimoto, H., Hoshimo, M., Inoue, G., Sakamaki, F., Washida, N., Okuda, M.: Design and characterization of the evacuable and bakable photochemical smog chamber. Environ. Sci. Technol. **13**, 471–475 (1979a)

Akimoto, H., Hoshino, M., Inoue, G., Sakamaki, F., Washida, N., Okuda, M.: Design and characterization of the evacuable and bakable photochemical smog chamber. Environ. Sci. Technol. **13**, 471–475 (1979b)

Akimoto, H., Sakamaki, F., Hoshino, M., Inoue, G., Okuda, M.: Photochemical ozone formation in propylene-nitrogen oxide-dry air system. Environ. Sci. Technol. **13**, 53–58 (1979c). https://doi.org/10.1021/es60149a005

Albrecht, S.R., Novelli, A., Hofzumahaus, A., Kang, S., Baker, Y., Mentel, T., Wahner, A., Fuchs, H.: Measurements of hydroperoxy radicals (HO2) at atmospheric concentrations using bromide chemical ionisation mass spectrometry. Atmos. Meas. Tech. **12**, 891–902 (2019). https://doi.org/10.5194/amt-12-891-2019

Alfarra, M.R., Hamilton, J.F., Wyche, K.P., Good, N., Ward, M.W., Carr, T., Barley, M.H., Monks, P.S., Jenkin, M.E., Lewis, A.C., McFiggans, G.B.: The effect of photochemical ageing and initial precursor concentration on the composition and hygroscopic properties of β-caryophyllene secondary organic aerosol. Atmos. Chem. Phys. **12**, 6417–6436 (2012). https://doi.org/10.5194/acp-12-6417-2012

Amato, P., Joly, M., Schaupp, C., Attard, E., Möhler, O., Morris, C.E., Brunet, Y., Delort, A.M.: Survival and ice nucleation activity of bacteria as aerosols in a cloud simulation chamber. Atmos. Chem. Phys. **15**, 6455–6465 (2015). https://doi.org/10.5194/acp-15-6455-2015

Anderson, T.L., Ogren, J.A.: Determining aerosol radiative properties using the TSI 3563 integrating nephelometer. Aerosol Sci. Technol. **29**, 57–69 (1998). https://doi.org/10.1080/02786829808965551

Anttila, T., Kiendler-Scharr, A., Tillmann, R., Mentel, T.F.: On the reactive uptake of gaseous compounds by organic-coated aqueous aerosols: theoretical analysis and application to the heterogeneous hydrolysis of N2O5. J. Phys. Chem. A **110**, 10435–10443 (2006). https://doi.org/10.1021/jp062403c

Apel, E.C., Brauers, T., Koppmann, R., Bandowe, B., Boßmeyer, J., Holzke, C., Tillmann, R., Wahner, A., Wegener, R., Brunner, A., Jocher, M., Ruuskanen, T., Spirig, C., Steigner, D., Stein-brecher, R., Gomez Alvarez, E., Müller, K., Burrows, J.P., Schade, G., Solomon, S. J., Ladstätter-Weißenmayer, A., Simmonds, P., Young, D., Hopkins, J.R., Lewis, A.C., Legreid, G., Reimann, S., Hansel, A., Wisthaler, A., Blake, R.S., Ellis, A.M., Monks, P.S., Wyche, K.P.: Intercomparison of oxygenated volatile organic compound measurements at the SAPHIR atmosphere simulation chamber. J. Geophys. Res. Atmos. **113** (2008). https://doi.org/10.1029/2008jd009865

Arnott, P.W., Moosmüller, H., Rogers, F.C., Tianfeng, J., Reinhard, B.: Photoacoustic spectrometer for measuring light absorption by aerosol: instrument description. Atmos. Environ. **33**, 2845–2852 (1999). https://doi.org/10.1016/S1352-2310(98)00361-6

Arnott, W.P., Hamasha, K., Moosmüller, H., Sheridan, P.J., Ogren, J.A.: Towards aerosol light-absorption measurements with a 7-wavelength aethalometer: evaluation with a photoacoustic instrument and 3-wavelength nephelometer. Aerosol Sci. Technol. **39**, 17–29 (2005). https://doi.org/10.1080/027868290901972

Atkinson, R.: Kinetics and mechanisms of the gas-phase reactions of the hydroxyl radical with organic compounds under atmospheric conditions. Chem. Rev. **86**, 69–201 (1986). https://doi.org/10.1021/cr00071a004

Atkinson, R.: A structure-activity relationship for the estimation of rate constants for the gas-phase reactions of OH radicals with organic compounds. Int. J. Chem. Kinet. **19**, 799–828 (1987). https://doi.org/10.1002/kin.550190903

Atkinson, R.: Atmospheric chemistry of VOCs and NO_x. Atmos. Environ. **34**, 2063–2101 (2000)

Atkinson, J.D., Murray, B.J., Woodhouse, M.T., Whale, T.F., Baustian, K.J., Carslaw, K.S., Dobbie, S., O'Sullivan, D., Malkin, T.L.: The importance of feldspar for ice nucleation by mineral dust in mixed-phase clouds. Nature **498**, 355–358 (2013). https://doi.org/10.1038/nature12278

Augustin-Bauditz, S., Wex, H., Kanter, S., Ebert, M., Niedermeier, D., Stolz, F., Prager, A., Strat-mann, F.: The immersion mode ice nucleation behavior of mineral dusts: a comparison of different pure and surface modified dusts. Geophys. Res. Lett. **41**, 7375–7382 (2014). https://doi.org/10.1002/2014GL061317

Bahreini, R., Middlebrook, A.M., de Gouw, J. A., Warneke, C., Trainer, M., Brock, C.A., Stark, H., Brown, S.S., Dube, W.P., Gilman, J.B., Hall, K., Holloway, J.S., Kuster, W.C., Perring, A.E., Prevot, A.S.H., Schwarz, J.P., Spackman, J.R., Szidat, S., Wagner, N.L., Weber, R.J., Zotter, P., Parrish, D.D.: Gasoline emissions dominate over diesel in formation of secondary organic aerosol mass. Geophys. Res. Lett. **39** (2012). https://doi.org/10.1029/2011GL050718

Baltensperger, U., Kalberer, M., Dommen, J., Paulsen, D., Alfarra, M.R., Coe, H., Fisseha, R., Gascho, A., Gysel, M., Nyeki, S., Sax, M., Steinbacher, M., Prevot, A.S.H., Sjögren, S., Wein-gartner, E., Zenobi, R.: Secondary organic aerosols from anthropogenic and biogenic precursors. Faraday Discuss. **130**, 265–278 (2005). https://doi.org/10.1039/b417367h

Barber, S., Blake, R.S., White, I.R., Monks, P.S., Reich, F., Mullock, S., Ellis, A.M.: Increased sensitivity in proton transfer reaction mass spectrometry by incorporation of a radio frequency ion funnel. Anal. Chem. **84**, 5387–5391 (2012). https://doi.org/10.1021/ac300894t

Barmet, P., Dommen, J., DeCarlo, P., Tritscher, T., Praplan, A., Platt, S., Prevot, A., Donahue, N., Baltensperger, U.: OH clock determination by proton transfer reaction mass spectrometry at an environmental chamber. Atmos. Meas. Tech. Discuss. **4**, 7471–7498 (2011). https://doi.org/10.5194/amtd-4-7471-2011

Barnes, I., Becker, K.H., Fink, E.H., Kriesche, V., Wildt, J., Zabel, F.: Studies of atmospheric reaction systems in a temperature controlled reaction chamber using Fourier-Transform-spectroscopy. In: First European symposium on Physico-Chemical Behaviour of Atmospheric Pollutants, JRC-Ispra (Italy) (1979)

Barnes, I., Becker, K.H., Fink, E.H., Reimer, A., Zabel, F., Niki, H.: FTIR spectroscopic study of the gas-phase reaction of HO2 with H2CO. Chem. Phys. Lett. **115**, 1–8 (1985). https://doi.org/10.1016/0009-2614(85)80091-9

Barnes, I., Becker, K.H., Carlier, P., Mouvier, G.: FTIR study of the DMS/NO$_2$/I$_2$/N$_2$ photolysis system: the reaction of IO radicals with DMS. Int. J. Chem. Kinet. **19**, 489–501 (1987). https://doi.org/10.1002/kin.550190602

Barnes, I., Becker, K.H., Mihalopoulos, N.: An FTIR product study of the photooxidation of dimethyl disulfide. J. Atmos. Chem. **18**, 267–289 (1994). https://doi.org/10.1007/bf00696783

Barnes, I., Rudzinski, K.J.: Environmental Simulation Chambers: Application to Atmospheric Chemical Processes, Nato Science Series: IV, vol. 62, pp. 458. Springer Netherlands (2006)

Beauchene, J., Bekowies, P., Winer, A., McAfee, J., Zafonte, L., Pitts, J., Jr.: A Novel 20 KW Solar Simulator Designed for Air Pollution Research (1973)

Becker, K.H.: The European Photoreactor "EUPHORE", European Community, Final report (1996)

Becker, K.H.: Overview on the development of chambers for the study of atmospheric chemical processes. In: Environmental Simulation Chambers: Application to Atmospheric Chemical Processes, pp. 1–26. Dordrecht (2006)

Behera, S.N., Sharma, M.: Transformation of atmospheric ammonia and acid gases into components of PM2.5: an environmental chamber study. Environ. Sci. Pollut. Res. **19**, 1187–1197 (2012). https://doi.org/10.1007/s11356-011-0635-9

Behnke, W., Holländer, W., Koch, W., Nolting, F., Zetzsch, C.: A smog chamber for studies of the photochemical degradation of chemicals in the presence of aerosols. Atmos. Environ. **1967**(22), 1113–1120 (1988). https://doi.org/10.1016/0004-6981(88)90341-1

Bernard, F., Ciuraru, R., Boréave, A., George, C.: Photosensitized formation of secondary organic aerosols above the air/water interface. Environ. Sci. Technol. **50**, 8678–8686 (2016). https://doi.org/10.1021/acs.est.6b03520

Bertrand, A., Stefenelli, G., Bruns, E., Pieber, S., Temime-Roussel, B., Slowik, J., Prevot, A., Wortham, H., Haddad, I., Marchand, N.: Primary emissions and secondary aerosol production potential from woodstoves for residential heating: Influence of the stove technology and combustion efficiency. Atmos. Environ. **169** (2017). https://doi.org/10.1016/j.atmosenv.2017.09.005

Bertrand, A., Stefenelli, G., Pieber, S., Bruns, E., Temime-Roussel, B., Slowik, J., Wortham, H., Prevot, A., Haddad, I., Marchand, N.: Influence of the vapor wall loss on the degradation rate constants in chamber experiments of levoglucosan and other biomass burning markers. Atmos. Chem. Phys. **18**, 10915–10930 (2018). https://doi.org/10.5194/acp-18-10915-2018

Bhattu, D., Zotter, P., Zhou, J., Stefenelli, G., Klein, F., Bertrand, A., Temime-Roussel, B., Marchand, N., Slowik, J.G., Baltensperger, U., Prévôt, A.S.H., Nussbaumer, T., El Haddad, I., Dommen, J.: Effect of stove technology and combustion conditions on gas and particulate emissions from residential biomass combustion. Environ. Sci. Technol. **53**, 2209–2219 (2019). https://doi.org/10.1021/acs.est.8b05020

Blado, L.D., Lilienkamp, R.H.: In Situ Vacuum Gauge Calibration by the Reference Transfer Method. In: ASTM/IES/AIAA Space Simulation Conference (1970)

Bloss, C., Wagner, V., Bonzanini, A., Jenkin, M.E., Wirtz, K., Martin-Reviejo, M., Pilling, M.J.: Evaluation of detailed aromatic mechanisms (MCMv3 and MCMv3.1) against environmental chamber data. Atmos. Chem. Phys. **5**, 623–639 (2005a). https://doi.org/10.5194/acp-5-623-2005a

Bloss, C., Wagner, V., Jenkin, M.E., Volkamer, R., Bloss, W.J., Lee, J.D., Heard, D.E., Wirtz, K., Martin-Reviejo, M., Rea, G., Wenger, J.C., Pilling, M.J.: Development of a detailed chemical mechanism (MCMv3.1) for the atmospheric oxidation of aromatic hydrocarbons. Atmos. Chem. Phys. **5**, 641–664 (2005b). https://doi.org/10.5194/acp-5-641-2005b

Bohn, B., Corlett, G.K., Gillmann, M., Sanghavi, S., Stange, G., Tensing, E., Vrekoussis, M., Bloss, W.J., Clapp, L.J., Kortner, M., Dorn, H.P., Monks, P.S., Platt, U., Plass-Dülmer, C., Mihalopoulos, N., Heard, D.E., Clemitshaw, K.C., Meixner, F.X., Prevot, A.S.H., Schmitt, R.: Photolysis frequency measurement techniques: results of a comparison within the ACCENT project. Atmos. Chem. Phys. **8**, 5373–5391 (2008). https://doi.org/10.5194/acp-8-5373-2008

Bonn, B., Schuster, G., Moortgat, G.K.: Influence of water vapor on the process of new particle formation during monoterpene ozonolysis. J. Phys. Chem. A **106**, 2869–2881 (2002). https://doi.org/10.1021/jp012713p

Boulon, J., Sellegri, K., Katrib, Y., Wang, J., Miet, K., Langmann, B., Laj, P., Doussin, J.F.: Sub-3 nm particles detection in a large photoreactor background: possible implications for new particles formation studies in a smog chamber. Aerosol Sci. Technol. **47**, 153–157 (2012). https://doi.org/10.1080/02786826.2012.733040

Boyd, C.M., Sanchez, J., Xu, L., Eugene, A.J., Nah, T., Tuet, W.Y., Guzman, M.I., Ng, N.L.: Secondary organic aerosol formation from the β-pinene+NO$_3$ system: effect of humidity and peroxy radical fate. Atmos. Chem. Phys. **15**, 7497–7522 (2015). https://doi.org/10.5194/acp-15-7497-2015

Brauers, T., Finlayson-Pitts, B.J.: Analysis of relative rate measurements. Int. J. Chem. Kinet. **29**, 665–672 (1997). https://doi.org/10.1002/(SICI)1097-4601(1997)29:9%3c665::AID-KIN3%3e3.0.CO;2-S

Brauers, T., Bohn, B., Johnen, F.-J., Rohrer, R., Rodriguez Bares, S., Tillmann, R., Wahner, A.: The atmosphere simulation chamber SAPHIR: a tool for the investigation of photochemistry (2003)

Brégonzio-Rozier, L., Giorio, C., Siekmann, F., Pangui, E., Morales, S.B., Temime-Roussel, B., Gratien, A., Michoud, V., Cazaunau, M., DeWitt, H.L., Tapparo, A., Monod, A., Doussin, J.F.: Secondary organic aerosol formation from isoprene photooxidation during cloud condensation–evaporation cycles. Atmos. Chem. Phys. **16**, 1747–1760 (2016). https://doi.org/10.5194/acp-16-1747-2016

Brommelle, N.S.: The Russell and Abney report on the action of light on water colours. Stud. Conserv. **9**, 140–152 (1964). https://doi.org/10.2307/1505213

Brotto, P., Repetto, B., Formenti, P., Pangui, E., Livet, A., Bousserrhine, N., Martini, I., Varnier, O., Doussin, J.F., Prati, P.: Use of an atmospheric simulation chamber for bioaerosol investigation: a feasibility study. Aerobiologia **31**, 445–455 (2015). https://doi.org/10.1007/s10453-015-9378-2

Campbell, S.J., Utinger, B., Lienhard, D.M., Paulson, S.E., Shen, J., Griffiths, P.T., Stell, A.C., Kalberer, M.: Development of a physiologically relevant online chemical assay to quantify aerosol oxidative potential. Anal. Chem. **91**, 13088–13095 (2019). https://doi.org/10.1021/acs.analchem.9b03282

Cantrell, C.A., Stedman, D.H., Wendel, G.J.: Measurement of atmospheric peroxy radicals by chemical amplification. Anal. Chem. **56**, 1496–1502 (1984). https://doi.org/10.1021/ac00272a065

Caponi, L., Formenti, P., Massabo, D., Di Biagio, C., Cazaunau, M., Pangui, E., Chevaillier, S., Landrot, G., Andreae, M.O., Kandler, K., Piketh, S., Saeed, T., Seibert, D., Williams, E., Balkanski, Y., Prati, P., Doussin, J.F.: Spectral- and size-resolved mass absorption efficiency of mineral dust aerosols in the shortwave spectrum: a simulation chamber study. Atmos. Chem. Phys. **17**, 7175–7191 (2017). https://doi.org/10.5194/acp-17-7175-2017

Cappellin, L., Karl, T., Probst, M., Ismailova, O., Winkler, P.M., Soukoulis, C., Aprea, E., Märk, T.D., Gasperi, F., Biasioli, F.: On quantitative determination of volatile organic compound concentrations using proton transfer reaction time-of-flight mass spectrometry. Environ. Sci. Technol. **46**, 2283–2290 (2012). https://doi.org/10.1021/es203985t

Carter, W.P.L., Lloyd, A.C., Sprung, J.L., Pitts, J.N., Jr.: Computer modeling of smog chamber data: progress in validation of a detailed mechanism for the photooxidation of propene and n-butane in photochemical smog. Int. J. Chem. Kinet. **11**, 45–101 (1979). https://doi.org/10.1002/kin.550110105

Carter, W.P.L., Lurmann, F.W.: Evaluation of a detailed gas-phase atmospheric reaction mechanism using environmental chamber data. Atmos. Environ. Part A Gen. Top. **25**, 2771–2806 (1991). https://doi.org/10.1016/0960-1686(91)90206-M

Carter, W.P.L.: Development of ozone reactivity scales for volatile organic compounds. Air Waste **44**, 881–899 (1994). https://doi.org/10.1080/1073161x.1994.10467290

Carter, W.P.L., Cocker, D.R., III., Fitz, D.R., Malkina, I.L., Bumiller, K., Sauer, C.G., Pisano, J.T., Bufalino, C., Song, C.: A new environmental chamber for evaluation of gas-phase chemical mechanisms and secondary aerosol formation. Atmos. Environ. **39**, 7768–7788 (2005)

Carter, W.P.L.: Development of the SAPRC-07 chemical mechanism. Atmos. Environ. **44**, 5324–5335 (2010). https://doi.org/10.1016/j.atmosenv.2010.01.026

Carter W.P.L.: Development of the SAPRC-07 chemical mechanism and updated ozone reactivity scales. Final Report to the California Air Resources Board Contract No. 03-318 (2008)

Chang, K., Bench, J., Brege, M., Cantrell, W., Chandrakar, K., Ciochetto, D., Mazzoleni, C., Mazzoleni, L.R., Niedermeier, D., Shaw, R.A.: A laboratory facility to study gas–aerosol–cloud interactions in a turbulent environment: the Π chamber. Bull. Am. Meteor. Soc. **97**, 2343–2358 (2016). https://doi.org/10.1175/bams-d-15-00203.1

Chen, T., Jang, M.: Chamber simulation of photooxidation of dimethyl sulfide and isoprene in the presence of NO_x. Atmos. Chem. Phys. **12**, 10257–10269 (2012). https://doi.org/10.5194/acp-12-10257-2012

Chen, T., Liu, Y., Chu, B., Liu, C., Liu, J., Ge, Y., Ma, Q., Ma, J., He, H.: Differences of the oxidation process and secondary organic aerosol formation at low and high precursor concentrations. J. Environ. Sci. **79**, 256–263 (2019). https://doi.org/10.1016/j.jes.2018.11.011

Chiappini, L., Perraudin, E., Durand-Jolibois, R., Doussin, J.F.: Development of a supercritical fluid extraction-gas chromatography-mass spectrometry method for the identification of highly polar compounds in secondary organic aerosols formed from biogenic hydrocarbons in smog chamber experiments. Anal. Bioanal. Chem. **386**, 1749–1759 (2006)

Chirico, R., DeCarlo, P.F., Heringa, M.F., Tritscher, T., Richter, R., Prévôt, A.S.H., Dommen, J., Weingartner, E., Wehrle, G., Gysel, M., Laborde, M., Baltensperger, U.: Impact of aftertreatment devices on primary emissions and secondary organic aerosol formation potential from in-use diesel vehicles: results from smog chamber experiments. Atmos. Chem. Phys. **10**, 11545–11563 (2010). https://doi.org/10.5194/acp-10-11545-2010

Chou, C., Formenti, P., Maille, M., Ausset, P., Helas, G., Harrison, M., Osborne, S.: Size distribution, shape, and composition of mineral dust aerosols collected during the African Monsoon Multidisciplinary Analysis Special Observation Period 0: Dust and Biomass-Burning Experiment field campaign in Niger, January 2006. J. Geophys. Res. Atmos. **113** (2008). https://doi.org/10.1029/2008JD009897

Ciarelli, G., Aksoyoglu, S., El Haddad, I., Bruns, E.A., Crippa, M., Poulain, L., Äijälä, M., Carbone, S., Freney, E., O'Dowd, C., Baltensperger, U., Prévôt, A.S.H.: Modelling winter organic aerosol at the European scale with CAMx: evaluation and source apportionment with a VBS parameterization based on novel wood burning smog chamber experiments. Atmos. Chem. Phys. **17**, 7653–7669 (2017). https://doi.org/10.5194/acp-17-7653-2017

Claxton, L.D., Barnes, H.M.: The mutagenicity of diesel-exhaust particle extracts collected under smog-chamber conditions using the Salmonella typhimurium test system. Mutat. Res. Genet. Toxicol. **88**, 255–272 (1981). https://doi.org/10.1016/0165-1218(81)90037-9

Cocker, D.R., Flagan, R.C., Seinfeld, J.H.: State of the art chamber facility for studying atmospheric aerosol chemistry. Environ. Sci. Technol. **35**, 2594–2601 (2001)

Coll, P., Cazaunau, M., Boczkowski, J., Zysman, M., Doussin, J.-F., Gratien, A., Derumeaux, G., Pini, M., Di Biagio, C., Pangui, É., Gaimoz, C., HÜE, S., Relaix, F., Der Vartanian, A., Coll, I., Michoud, V., Formenti, P., Foret, G., Thavaratnasingam, L., Lanone, S.: Pollurisk: an innovative experimental platform to investigate health impacts of air quality, pp. 557–565 (2018)

Connolly, P.J., Möhler, O., Field, P.R., Saathoff, H., Burgess, R., Choularton, T., Gallagher, M.: Studies of heterogeneous freezing by three different desert dust samples. Atmos. Chem. Phys. **9**, 2805–2824 (2009). https://doi.org/10.5194/acp-9-2805-2009

Crutzen, P.J., Howard, C.J.: The effect of the HO2+NO reaction rate constant on one-dimensional model calculations of stratospheric ozone perturbations. Pure Appl. Geophys. **116**, 497–510 (1978). https://doi.org/10.1007/bf01636903

D'Ambro, E.L., Møller, K.H., Lopez-Hilfiker, F.D., Schobesberger, S., Liu, J., Shilling, J.E., Lee, B.H., Kjaergaard, H.G., Thornton, J.A.: Isomerization of second-generation isoprene peroxy radicals: epoxide formation and implications for secondary organic aerosol yields. Environ. Sci. Technol. **51**, 4978–4987 (2017). https://doi.org/10.1021/acs.est.7b00460

de Blas, M., Navazo, M., Alonso, L., Durana, N., Iza, J.: Automatic on-line monitoring of atmospheric volatile organic compounds: Gas chromatography–mass spectrometry and gas chromatography–flame ionization detection as complementary systems. Sci. Total Environ. **409**, 5459–5469 (2011). https://doi.org/10.1016/j.scitotenv.2011.08.072

de Gouw, J., Warneke, C.: Measurements of volatile organic compounds in the earth's atmosphere using proton-transfer-reaction mass spectrometry. Mass Spectrom. Rev. **26**, 223–257 (2007). https://doi.org/10.1002/mas.20119

De Haan, D. O., Brauers, T., Oum, K., Stutz, J., Nordmeyer, T., Finlayson-Pitts, B.J.: Heterogeneous chemistry in the troposphere: experimental approaches and applications to the chemistry of sea salt particles. Int. Rev. Phys. Chem. 343–385 (1999)

De Haan, D.O., Tapavicza, E., Riva, M., Cui, T., Surratt, J.D., Smith, A.C., Jordan, M.-C., Nilakantan, S., Almodovar, M., Stewart, T.N., de Loera, A., De Haan, A.C., Cazaunau, M., Gratien, A., Pangui, E., Doussin, J.-F.: Nitrogen-containing, light-absorbing oligomers produced in aerosol particles exposed to methylglyoxal, photolysis, and cloud cycling. Environ. Sci. Technol. **52**, 4061–4071 (2018). https://doi.org/10.1021/acs.est.7b06105

DeMott, P.J., Prenni, A.J., McMeeking, G.R., Sullivan, R.C., Petters, M.D., Tobo, Y., Niemand, M., Möhler, O., Snider, J.R., Wang, Z., Kreidenweis, S.M.: Integrating laboratory and field data to quantify the immersion freezing ice nucleation activity of mineral dust particles. Atmos. Chem. Phys. **15**, 393–409 (2015). https://doi.org/10.5194/acp-15-393-2015

DeMott, P.J., Möhler, O., Cziczo, D.J., Hiranuma, N., Petters, M.D., Petters, S.S., Belosi, F., Bingemer, H.G., Brooks, S.D., Budke, C., Burkert-Kohn, M., Collier, K.N., Danielczok, A., Eppers, O., Felgitsch, L., Garimella, S., Grothe, H., Herenz, P., Hill, T.C.J., Höhler, K., Kanji, Z.A., Kiselev, A., Koop, T., Kristensen, T.B., Krüger, K., Kulkarni, G., Levin, E.J.T., Murray, B.J., Nicosia, A., O'Sullivan, D., Peckhaus, A., Polen, M.J., Price, H.C., Reicher, N., Rothenberg, D.A., Rudich, Y., Santachiara, G., Schiebel, T., Schrod, J., Seifried, T.M., Stratmann, F., Sullivan, R.C., Suski, K.J., Szakáll, M., Taylor, H.P., Ullrich, R., Vergara-Temprado, J., Wagner, R., Whale, T.F., Weber, D., Welti, A., Wilson, T.W., Wolf, M.J., Zenker, J.: The fifth international workshop on ice nucleation phase 2 (FIN-02): laboratory intercomparison of ice nucleation measurements. Atmos. Meas. Tech. **11**, 6231–6257 (2018). https://doi.org/10.5194/amt-11-6231-2018

Deng, Y., Inomata, S., Sato, K., Ramasamy, S., Morino, Y., Enami, S., Tanimoto, H.: Temperature and acidity dependence of secondary organic aerosol formation from α-pinene ozonolysis with a compact chamber system. Atmos. Chem. Phys. **21**, 5983–6003 (2021). https://doi.org/10.5194/acp-21-5983-2021

Denjean, C., Formenti, P., Picquet-Varrault, B., Katrib, Y., Pangui, E., Zapf, P., Doussin, J.F.: A new experimental approach to study the hygroscopic and optical properties of aerosols: application to ammonium sulfate particles. Atmos. Meas. Tech. **7**, 183–197 (2014). https://doi.org/10.5194/amt-7-183-2014

Denjean, C., Formenti, P., Picquet-Varrault, B., Pangui, E., Zapf, P., Katrib, Y., Giorio, C., Tapparo, A., Monod, A., Temime-Roussel, B., Decorse, P., Mangeney, C., Doussin, J.F.: Relating hygroscopicity and optical properties to chemical composition and structure of secondary organic aerosol particles generated from the ozonolysis of α-pinene. Atmos. Chem. Phys. **15**, 3339–3358 (2015). https://doi.org/10.5194/acp-15-3339-2015

Dewald, P., Liebmann, J.M., Friedrich, N., Shenolikar, J., Schuladen, J., Rohrer, F., Reimer, D., Tillmann, R., Novelli, A., Cho, C., Xu, K., Holzinger, R., Bernard, F., Zhou, L., Mellouki, W., Brown, S.S., Fuchs, H., Lelieveld, J., Crowley, J.N.: Evolution of NO_3 reactivity during the oxidation of isoprene. Atmos. Chem. Phys. **20**, 10459–10475 (2020). https://doi.org/10.5194/acp-20-10459-2020

Di Biagio, C., Formenti, P., Styler, S.A., Pangui, E., Doussin, J.-F.: Laboratory chamber measurements of the longwave extinction spectra and complex refractive indices of African and Asian mineral dusts. Geophys. Res. Lett. **41**, 6289–6297 (2014). https://doi.org/10.1002/2014gl060213

Di Biagio, C., Formenti, P., Balkanski, Y., Caponi, L., Cazaunau, M., Pangui, E., Journet, E., Nowak, S., Caquineau, S., Andreae, M.O., Kandler, K., Saeed, T., Piketh, S., Seibert, D., Williams, E., Doussin, J.F.: Global scale variability of the mineral dust long-wave refractive index: a new dataset of in situ measurements for climate modeling and remote sensing. Atmos. Chem. Phys. **17**, 1901–1929 (2017a). https://doi.org/10.5194/acp-17-1901-2017

Di Biagio, C., Formenti, P., Cazaunau, M., Pangui, E., Marchand, N., Doussin, J.F.: Aethalometer multiple scattering correction Cref for mineral dust aerosols. Atmos. Meas. Tech. **10**, 2923–2939 (2017b). https://doi.org/10.5194/amt-10-2923-2017

Di Biagio, C., Formenti, P., Balkanski, Y., Caponi, L., Cazaunau, M., Pangui, E., Journet, E., Nowak, S., Andreae, M.O., Kandler, K., Saeed, T., Piketh, S., Seibert, D., Williams, E., Doussin, J.-F.: Complex refractive indices and single-scattering albedo of global dust aerosols in the shortwave spectrum and relationship to size and iron content. Atmos. Chem. Phys. **19**, 15503–15531 (2019). https://doi.org/10.5194/acp-19-15503-2019

Dias, A., Ehrhart, S., Vogel, A., Williamson, C., Almeida, J., Kirkby, J., Mathot, S., Mumford, S., Onnela, A.: Temperature uniformity in the CERN CLOUD chamber. Atmos. Meas. Tech. **10**, 5075–5088 (2017). https://doi.org/10.5194/amt-10-5075-2017

Docherty, K.S., Jaoui, M., Corse, E., Jimenez, J.L., Offenberg, J.H., Lewandowski, M., Kleindienst, T.E.: Collection efficiency of the aerosol mass spectrometer for chamber-generated secondary organic aerosols. Aerosol Sci. Technol. **47**, 294–309 (2013). https://doi.org/10.1080/02786826.2012.752572

Dodge, M.C.: Chemical oxidant mechanisms for air quality modeling: critical review. Atmos. Environ. **34**, 2103–2130 (2000)

Dodge M.: Combined use of modeling techniques and smog chamber data to derive ozone–precursor relationships. In: International Conference on Photochemical Oxidant Pollution and its Control Proceedings, Raleigh, NC (1977)

Donahue, N.M., Hartz, K.E.H., Chuong, B., Presto, A.A., Stanier, C.O., Rosenhorn, T., Robinson, A.L., Pandis, S.N.: Critical factors determining the variation in SOA yields from terpene ozonolysis: a combined experimental and computational study. Faraday Discuss. **130**, 295–309 (2005)

Donahue, N.M., Robinson, A.L., Stanier, C.O., Pandis, S.N.: Coupled partitioning, dilution, and chemical aging of semivolatile organics. Environ. Sci. Technol. **40**, 2635–2643 (2006). https://doi.org/10.1021/es052297c

Donahue, N.M., Henry, K.M., Mentel, T.F., Kiendler-Scharr, A., Spindler, C., Bohn, B., Brauers, T., Dorn, H.P., Fuchs, H., Tillmann, R., Wahner, A., Saathoff, H., Naumann, K.-H., Möhler, O., Leisner, T., Müller, L., Reinnig, M.-C., Hoffmann, T., Salo, K., Hallquist, M., Frosch, M., Bilde, M., Tritscher, T., Barmet, P., Praplan, A.P., DeCarlo, P.F., Dommen, J., Prévôt, A.S.H., Baltensperger, U.: Aging of biogenic secondary organic aerosol via gas-phase OH radical reactions. Proc. Natl. Acad. Sci. **109**, 13503–13508 (2012). https://doi.org/10.1073/pnas.1115186109

Dorn, H.P., Apodaca, R.L., Ball, S.M., Brauers, T., Brown, S.S., Crowley, J.N., Dubé, W.P., Fuchs, H., Häseler, R., Heitmann, U., Jones, R.L., Kiendler-Scharr, A., Labazan, I., Langridge, J.M., Meinen, J., Mentel, T.F., Platt, U., Pöhler, D., Rohrer, F., Ruth, A.A., Schlosser, E., Schuster, G., Shillings, A.J.L., Simpson, W.R., Thieser, J., Tillmann, R., Varma, R., Venables, D.S., Wahner, A.: Intercomparison of NO3 radical detection instruments in the atmosphere simulation chamber SAPHIR. Atmos. Meas. Tech. **6**, 1111–1140 (2013). https://doi.org/10.5194/amt-6-1111-2013

Doussin, J.F., Ritz, D., DurandJolibois, R., Monod, A., Carlier, P.: Design of an environmental chamber for the study of atmospheric chemistry: new developments in the analytical device. Analusis **25**, 236–242 (1997)

Doyle, G.J., Lloyd, A.C., Darnall, K.R., Winer, A.M., Pitts, J.N.: Gas phase kinetic study of relative rates of reaction of selected aromatic compounds with hydroxyl radicals in an environmental chamber. Environ. Sci. Technol. **9**, 237–241 (1975). https://doi.org/10.1021/es60101a002

Duncianu, M., David, M., Kartigueyane, S., Cirtog, M., Doussin, J.F., Picquet-Varrault, B.: Measurement of alkyl and multifunctional organic nitrates by proton-transfer-reaction mass spectrometry. Atmos. Meas. Tech. **10**, 1445–1463 (2017). https://doi.org/10.5194/amt-10-1445-2017

Dunlea, E.J., Herndon, S.C., Nelson, D.D., Volkamer, R.M., San Martini, F., Sheehy, P.M., Zahniser, M.S., Shorter, J.H., Wormhoudt, J.C., Lamb, B.K., Allwine, E.J., Gaffney, J.S., Marley, N.A.,

Grutter, M., Marquez, C., Blanco, S., Cardenas, B., Retama, A., Ramos Villegas, C.R., Kolb, C.E., Molina, L.T., Molina, M.J.: Evaluation of nitrogen dioxide chemiluminescence monitors in a polluted urban environment. Atmos. Chem. Phys. **7**, 2691–2704 (2007). https://doi.org/10. 5194/acp-7-2691-2007

Duplissy, J., Enghoff, M.B., Aplin, K.L., Arnold, F., Aufmhoff, H., Avngaard, M., Baltensperger, U., Bondo, T., Bingham, R., Carslaw, K., Curtius, J., David, A., Fastrup, B., Gagné, S., Hahn, F., Harrison, R.G., Kellett, B., Kirkby, J., Kulmala, M., Laakso, L., Laaksonen, A., Lillestol, E., Lockwood, M., Mäkelä, J., Makhmutov, V., Marsh, N.D., Nieminen, T., Onnela, A., Pedersen, E., Pedersen, J.O.P., Polny, J., Reichl, U., Seinfeld, J.H., Sipilä, M., Stozhkov, Y., Stratmann, F., Svensmark, H., Svensmark, J., Veenhof, R., Verheggen, B., Viisanen, Y., Wagner, P.E., Wehrle, G., Weingartner, E., Wex, H., Wilhelmsson, M., Winkler, P.M.: Results from the CERN pilot CLOUD experiment. Atmos. Chem. Phys. **10**, 1635–1647 (2010). https://doi.org/10.5194/acp-10-1635-2010

Edwards, G.D., Cantrell, C.A., Stephens, S., Hill, B., Goyea, O., Shetter, R.E., Mauldin, R.L., Kosciuch, E., Tanner, D.J., Eisele, F.L.: Chemical ionization mass spectrometer instrument for the measurement of tropospheric HO2 and RO2. Anal. Chem. **75**, 5317–5327 (2003). https://doi. org/10.1021/ac034402b

Ehn, M., Thornton, J.A., Kleist, E., Sipilä, M., Junninen, H., Pullinen, I., Springer, M., Rubach, F., Tillmann, R., Lee, B., Lopez-Hilfiker, F., Andres, S., Acir, I.-H., Rissanen, M., Jokinen, T., Schobesberger, S., Kangasluoma, J., Kontkanen, J., Nieminen, T., Kurtén, T., Nielsen, L.B., Jørgensen, S., Kjaergaard, H.G., Canagaratna, M., Maso, M.D., Berndt, T., Petäjä, T., Wahner, A., Kerminen, V.-M., Kulmala, M., Worsnop, D.R., Wildt, J., Mentel, T.F.: A large source of low-volatility secondary organic aerosol. Nature **506**, 476–479 (2014). https://doi.org/10.1038/ nature13032

Etzkorn, T., Klotz, B., Sorensen, S., Patroescu, I., Barnes, I., Becker, K.H., Platt, U.: Gas-phase absorption cross sections of 24 monocyclic aromatic hydrocarbons in the UV and IR spectral ranges. Atmos. Environ. **33**, 525–540 (1999)

Evans, L.F., Weeks, I.A., Eccleston, A.J.: A chamber study of photochemical smog in Melbourne, Australia—present and future. Atmos. Environ. **1967**(20), 1355–1368 (1986). https://doi.org/10. 1016/0004-6981(86)90006-5

Fahey, D.W., Gao, R.S., Möhler, O., Saathoff, H., Schiller, C., Ebert, V., Krämer, M., Peter, T., Amarouche, N., Avallone, L.M., Bauer, R., Bozóki, Z., Christensen, L.E., Davis, S.M., Durry, G., Dyroff, C., Herman, R.L., Hunsmann, S., Khaykin, S.M., Mackrodt, P., Meyer, J., Smith, J.B., Spelten, N., Troy, R.F., Vömel, H., Wagner, S., Wienhold, F.G.: The AquaVIT-1 intercomparison of atmospheric water vapor measurement techniques. Atmos. Meas. Tech. **7**, 3177–3213 (2014). https://doi.org/10.5194/amt-7-3177-2014

Faiola, C.L., Pullinen, I., Buchholz, A., Khalaj, F., Ylisirniö, A., Kari, E., Miettinen, P., Holopainen, J.K., Kivimäenpää, M., Schobesberger, S., Yli-Juuti, T., Virtanen, A.: Secondary organic aerosol formation from healthy and aphid-stressed scots pine emissions. ACS Earth Space Chem. **3**, 1756–1772 (2019). https://doi.org/10.1021/acsearthspacechem.9b00118

Field, P.R., Möhler, O., Connolly, P., Krämer, M., Cotton, R., Heymsfield, A.J., Saathoff, H., Schnaiter, M.: Some ice nucleation characteristics of Asian and Saharan desert dust. Atmos. Chem. Phys. **6**, 2991–3006 (2006). https://doi.org/10.5194/acp-6-2991-2006

Finlayson-Pitts, B.J., Pitts, J.N., Jr.: Atmospheric chemistry: fundamentals and experimental techniques. John Wiley and Sons, New-York (1986). (edited by: Publication, W. I.)

Finlayson-Pitts, B.J., Pitts, J.N., Jr.: The Chemistry of the Lower and Upper Atmosphere: Theory, Experiments and Applications. Academic Press, New-York (2000)

Fitz, D.R., Dodd, M.C., Winer, A.M.: Photooxidation of apinene at near-ambient concentrations under simulated atmospheric conditions. In: 74th Annual Meeting, Air Pollution Control Association, Philadeplphia, PA, 21–26 Juen (1981)

Folkers, M., Mentel, T.F., Wahner, A.: Influence of an organic coating on the reactivity of aqueous aerosols probed by the heterogeneous hydrolysis of N2O5. Geophys. Res. Lett. **30** (2003). https:// doi.org/10.1029/2003GL017168

Fouqueau, A., Cirtog, M., Cazaunau, M., Pangui, E., Zapf, P., Siour, G., Landsheere, X., Méjean, G., Romanini, D., Picquet-Varrault, B.: Implementation of an incoherent broadband cavity-enhanced absorption spectroscopy technique in an atmospheric simulation chamber for in situ NO_3 monitoring: characterization and validation for kinetic studies. Atmos. Meas. Tech. **13**, 6311–6323 (2020). https://doi.org/10.5194/amt-13-6311-2020

Fox, D.L., Kamens, R., Jeffries, H.E.: Photochemical smog systems: effect of dilution on ozone formation. Science **188**, 1113–1114 (1975). https://doi.org/10.1126/science.188.4193.1113

Fry, J.L., Draper, D.C., Barsanti, K.C., Smith, J.N., Ortega, J., Winkler, P.M., Lawler, M.J., Brown, S.S., Edwards, P.M., Cohen, R.C., Lee, L.: Secondary organic aerosol formation and organic nitrate yield from NO_3 oxidation of biogenic hydrocarbons. Environ. Sci. Technol. **48**, 11944–11953 (2014). https://doi.org/10.1021/es502204x

Fuchs, H., Ball, S.M., Bohn, B., Brauers, T., Cohen, R.C., Dorn, H.P., Dubé, W.P., Fry, J.L., Häseler, R., Heitmann, U., Jones, R.L., Kleffmann, J., Mentel, T.F., Müsgen, P., Rohrer, F., Rollins, A.W., Ruth, A.A., Kiendler-Scharr, A., Schlosser, E., Shillings, A.J.L., Tillmann, R., Varma, R.M., Venables, D.S., Villena Tapia, G., Wahner, A., Wegener, R., Wooldridge, P.J., Brown, S.S.: Intercomparison of measurements of NO_2 concentrations in the atmosphere simulation chamber SAPHIR during the NO_3Comp campaign. Atmos. Meas. Tech. **3**, 21–37 (2010). https://doi.org/10.5194/amt-3-21-2010

Fuchs, H., Dorn, H.P., Bachner, M., Bohn, B., Brauers, T., Gomm, S., Hofzumahaus, A., Holland, F., Nehr, S., Rohrer, F., Tillmann, R., Wahner, A.: Comparison of OH concentration measurements by DOAS and LIF during SAPHIR chamber experiments at high OH reactivity and low NO concentration. Atmos. Meas. Tech. **5**, 1611–1626 (2012a). https://doi.org/10.5194/amt-5-1611-2012

Fuchs, H., Simpson, W.R., Apodaca, R.L., Brauers, T., Cohen, R.C., Crowley, J.N., Dorn, H.P., Dubé, W.P., Fry, J.L., Häseler, R., Kajii, Y., Kiendler-Scharr, A., Labazan, I., Matsumoto, J., Mentel, T.F., Nakashima, Y., Rohrer, F., Rollins, A.W., Schuster, G., Tillmann, R., Wahner, A., Wooldridge, P.J., Brown, S.S.: Comparison of N_2O_5 mixing ratios during NO_3Comp 2007 in SAPHIR. Atmos. Meas. Tech. **5**, 2763–2777 (2012b). https://doi.org/10.5194/amt-5-2763-2012

Fuchs, H., Hofzumahaus, A., Rohrer, F., Bohn, B., Brauers, T., Dorn, H.P., Häseler, R., Holland, F., Kaminski, M., Li, X., Lu, K., Nehr, S., Tillmann, R., Wegener, R., Wahner, A.: Experimental evidence for efficient hydroxyl radical regeneration in isoprene oxidation. Nat. Geosci. **6**, 1023–1026 (2013). https://doi.org/10.1038/ngeo1964

Fuchs, H., Novelli, A., Rolletter, M., Hofzumahaus, A., Pfannerstill, E.Y., Kessel, S., Edtbauer, A., Williams, J., Michoud, V., Dusanter, S., Locoge, N., Zannoni, N., Gros, V., Truong, F., Sarda-Esteve, R., Cryer, D.R., Brumby, C.A., Whalley, L.K., Stone, D., Seakins, P.W., Heard, D.E., Schoemaecker, C., Blocquet, M., Coudert, S., Batut, S., Fittschen, C., Thames, A.B., Brune, W.H., Ernest, C., Harder, H., Muller, J.B.A., Elste, T., Kubistin, D., Andres, S., Bohn, B., Hohaus, T., Holland, F., Li, X., Rohrer, F., Kiendler-Scharr, A., Tillmann, R., Wegener, R., Yu, Z., Zou, Q., Wahner, A.: Comparison of OH reactivity measurements in the atmospheric simulation chamber SAPHIR. Atmos. Meas. Tech. **10**, 4023–4053 (2017). https://doi.org/10.5194/amt-10-4023-2017

Fuller, S.J., Wragg, F.P.H., Nutter, J., Kalberer, M.: Comparison of on-line and off-line methods to quantify reactive oxygen species (ROS) in atmospheric aerosols. Atmos. Environ. **92**, 97–103 (2014). https://doi.org/10.1016/j.atmosenv.2014.04.006

Gaie-Levrel, F., Bau, S., Bregonzio-Rozier, L., Payet, R., Artous, S., Jacquinot, S., Guiot, A., Ouf, F.X., Bourrous, S., Marpillat, A., Foulquier, C., Smith, G., Crenn, V., Feltin, N.: An intercomparison exercise of good laboratory practices for nano-aerosol size measurements by mobility spectrometers. J. Nanopart. Res. **22**, 103 (2020). https://doi.org/10.1007/s11051-020-04820-y

Geiger, H., Kleffmann, J., Wiesen, P.: Smog chamber studies on the influence of diesel exhaust on photosmog formation. Atmos. Environ. **36**, 1737–1747 (2002). https://doi.org/10.1016/S1352-2310(02)00175-9

Geiger, H., Becker, K., Wiesen, P.: Effect of gasoline formulation on the formation of photosmog: a box model study. J. Air Waste Manag. Assoc. **1995**(53), 425–433 (2003). https://doi.org/10.1080/10473289.2003.10466173

Gensch, I., Kiendler-Scharr, A., Rudolph, J.: Isotope ratio studies of atmospheric organic compounds: principles, methods, applications and potential. Int. J. Mass Spectrom. **365–366**, 206–221 (2014). https://doi.org/10.1016/j.ijms.2014.02.004

Gentner, D.R., Jathar, S.H., Gordon, T.D., Bahreini, R., Day, D.A., El Haddad, I., Hayes, P.L., Pieber, S.M., Platt, S.M., de Gouw, J., Goldstein, A.H., Harley, R.A., Jimenez, J.L., Prévôt, A.S.H., Robinson, A.L.: Review of urban secondary organic aerosol formation from gasoline and diesel motor vehicle emissions. Environ. Sci. Technol. **51**, 1074–1093 (2017). https://doi.org/10.1021/acs.est.6b04509

Gery, M.W., Whitten, G.Z., Killus, J.P., Dodge, M.C.: A photochemical kinetics mechanism for urban and regional scale computer modeling. J. Geophys. Res. Atmos. **94**, 12925–12956 (1989). https://doi.org/10.1029/JD094iD10p12925

Gkatzelis, G.I., Hohaus, T., Tillmann, R., Gensch, I., Müller, M., Eichler, P., Xu, K.M., Schlag, P., Schmitt, S.H., Yu, Z., Wegener, R., Kaminski, M., Holzinger, R., Wisthaler, A., Kiendler-Scharr, A.: Gas-to-particle partitioning of major biogenic oxidation products: a study on freshly formed and aged biogenic SOA. Atmos. Chem. Phys. **18**, 12969–12989 (2018). https://doi.org/10.5194/acp-18-12969-2018

Glaschick-Schimpf, I., Leiss, A., Monkhouse, P.B., Schurath, U., Becker, K.H., Fink, E.H.: A kinetic study of the reactions of HO2/DO2 radicals with nitric oxide using near-infrared chemiluminescence detection. Chem. Phys. Lett. **67**, 318–323 (1979). https://doi.org/10.1016/0009-2614(79)85170-2

Glowacki, D., Goddard, A., Hemavibool, K., Malkin, T., Commane, R., Anderson, F., Bloss, W., Heard, D., Ingham, T., Pilling, M., Seakins, P.: Design of and initial results from a highly instrumented reactor for atmospheric chemistry (HIRAC). Atmos. Chem. Phys. **7**, 5371–5390 (2007). https://doi.org/10.5194/acpd-7-10687-2007

Goliff, W.S., Stockwell, W.R., Lawson, C.V.: The regional atmospheric chemistry mechanism, version 2. Atmos. Environ. **68**, 174–185 (2013). https://doi.org/10.1016/j.atmosenv.2012.11.038

Good, N., Coe, H., McFiggans, G.: Instrumentational operation and analytical methodology for the reconciliation of aerosol water uptake under sub- and supersaturated conditions. Atmos. Meas. Tech. **3**, 1241–1254 (2010). https://doi.org/10.5194/amt-3-1241-2010

Greiner, N.R.: Hydroxyl-radical kinetics by kinetic spectroscopy. I. Reactions with H2, CO, and CH4 at 300°K. J. Chem. Phys. **46**, 2795–2799 (1967). https://doi.org/10.1063/1.1841115

Groth, W., Becker, K.H., Comsa, G.H., Elzer, A., Fink, E., Jud, W., Kley, D., Schurath, U., Thran, D.: Untersuchungen in der "Großen Bonner Kugel." Naturwissenschaften **59**, 379–387 (1972). https://doi.org/10.1007/bf00623126

Haagen-Smit, A.J.: Chemistry and physiology of Los Angeles smog. Ind. Eng. Chem. **44**, 1342–1346 (1952). https://doi.org/10.1021/ie50510a045

Hack, W., Preuss, A.W., Temps, F., Wagner, H.G., Hoyermann, K.: Direct determination of the rate constant of the reaction NO + HO₂ → NO₂ + OH with the LMR. Int. J. Chem. Kinet. **12**, 851–860 (1980). https://doi.org/10.1002/kin.550121104

Hallquist, M., Wenger, J.C., Baltensperger, U., Rudich, Y., Simpson, D., Claeys, M., Dommen, J., Donahue, N.M., George, C., Goldstein, A.H., Hamilton, J.F., Herrmann, H., Hoffmann, T., Iinuma, Y., Jang, M., Jenkin, M.E., Jimenez, J.L., Kiendler-Scharr, A., Maenhaut, W., McFiggans, G., Mentel, T.F., Monod, A., Prévôt, A.S.H., Seinfeld, J.H., Surratt, J.D., Szmigielski, R., Wildt, J.: The formation, properties and impact of secondary organic aerosol: current and emerging issues. Atmos. Chem. Phys. **9**, 5155–5236 (2009). https://doi.org/10.5194/acp-9-5155-2009

Hamilton, J.F., Lewis, A.C., Carey, T.J., Wenger, J.C.: Characterization of polar compounds and oligomers in secondary organic aerosol using liquid chromatography coupled to mass spectrometry. Anal. Chem. **80**, 474–480 (2008). https://doi.org/10.1021/ac701852t

Hanke, M., Uecker, J., Reiner, T., Arnold, F.: Atmospheric peroxy radicals: ROXMAS, a new mass-spectrometric methodology for speciated measurements of HO2 and $\sum$RO2 and first results. Int. J. Mass Spectrom. **213**, 91–99 (2002). https://doi.org/10.1016/S1387-3806(01)00548-6

Hanst, P.L.: Spectroscopic methods of air pollution measurement. Adv. Environ. Sci. Technol. **2**, 91–213 (1971)

Hartmann, S., Wex, H., Clauss, T., Augustin-Bauditz, S., Niedermeier, D., Rösch, M., Stratmann, F.: Immersion freezing of kaolinite: scaling with particle surface area. J. Atmos. Sci. **73**, 263–278 (2016). https://doi.org/10.1175/jas-d-15-0057.1

Hatakeyama, S., Akimoto, H., Washida, N.: Effect of temperature on the formation of photochemical ozone in a propene-NO(x)-air-irradiation system. Environ. Sci. Technol. **25**, 1884–1890 (2002). https://doi.org/10.1021/es00023a007

Heicklen, J., Westberg, K., Cohen, N.: Conversion of nitrogen oxide to nitrogen dioxide in polluted atmospheres, pp. 115–169. Pennsylvania State University, University Park, Penn (1969)

Heim, M., Mullins, B.J., Umhauer, H., Kasper, G.: Performance evaluation of three optical particle counters with an efficient "multimodal" calibration method. J. Aerosol Sci. **39**, 1019–1031 (2008). https://doi.org/10.1016/j.jaerosci.2008.07.006

Hennigan, C.J., Sullivan, A.P., Collett, J.L. Jr., Robinson, A.L.: Levoglucosan stability in biomass burning particles exposed to hydroxyl radicals. Geophys. Res. Lett. **37** (2010). https://doi.org/10.1029/2010gl043088

Hennigan, C.J., Miracolo, M.A., Engelhart, G.J., May, A.A., Presto, A.A., Lee, T., Sullivan, A.P., McMeeking, G.R., Coe, H., Wold, C.E., Hao, W.M., Gilman, J.B., Kuster, W.C., de Gouw, J., Schichtel, B.A., Collett, J.L., Jr., Kreidenweis, S.M., Robinson, A.L.: Chemical and physical transformations of organic aerosol from the photo-oxidation of open biomass burning emissions in an environmental chamber. Atmos. Chem. Phys. **11**, 7669–7686 (2011). https://doi.org/10.5194/acp-11-7669-2011

Henning, S., Ziese, M., Kiselev, A., Saathoff, H., Möhler, O., Mentel, T.F., Buchholz, A., Spindler, C., Michaud, V., Monier, M., Sellegri, K., Stratmann, F.: Hygroscopic growth and droplet activation of soot particles: uncoated, succinic or sulfuric acid coated. Atmos. Chem. Phys. **12**, 4525–4537 (2012). https://doi.org/10.5194/acp-12-4525-2012

Hering, S.V., Stolzenburg, M.R., Quant, F.R., Oberreit, D.R., Keady, P.B.: A laminar-flow, water-based condensation particle counter (WCPC). Aerosol Sci. Technol. **39**, 659–672 (2005). https://doi.org/10.1080/02786820500182123

Heringa, M.F., DeCarlo, P.F., Chirico, R., Tritscher, T., Clairotte, M., Mohr, C., Crippa, M., Slowik, J.G., Pfaffenberger, L., Dommen, J., Weingartner, E., Prévôt, A.S.H., Baltensperger, U.: A new method to discriminate secondary organic aerosols from different sources using high-resolution aerosol mass spectra. Atmos. Chem. Phys. **12**, 2189–2203 (2012). https://doi.org/10.5194/acp-12-2189-2012

Herriott, D.R., Kogelnik, H., Kompfner, R.: Off-axis paths in spherical mirror interferometers. Appl. Opt. **3**, 523–526 (1964). https://doi.org/10.1364/ao.3.000523

Herriott, D.R., Schulte, H.J.: Folded optical delay lines. Appl. Opt. **4**, 883–889 (1965). https://doi.org/10.1364/ao.4.000883

Hiranuma, N., Hoffmann, N., Kiselev, A., Dreyer, A., Zhang, K., Kulkarni, G., Koop, T., Möhler, O.: Influence of surface morphology on the immersion mode ice nucleation efficiency of hematite particles. Atmos. Chem. Phys. **14**, 2315–2324 (2014). https://doi.org/10.5194/acp-14-2315-2014

Hofzumahaus, A., Rohrer, F., Lu, K., Bohn, B., Brauers, T., Chang, C.-C., Fuchs, H., Holland, F., Kita, K., Kondo, Y., Li, X., Lou, S., Shao, M., Zeng, L., Wahner, A., Zhang, Y.: Amplified trace gas removal in the troposphere. Science **324**, 1702–1704 (2009). https://doi.org/10.1126/science.1164566

Hohaus, T., Kuhn, U., Andres, S., Kaminski, M., Rohrer, F., Tillmann, R., Wahner, A., Wegener, R., Yu, Z., Kiendler-Scharr, A.: A new plant chamber facility, PLUS, coupled to the atmosphere simulation chamber SAPHIR. Atmos. Meas. Tech. **9**, 1247–1259 (2016). https://doi.org/10.5194/amt-9-1247-2016

Hoose, C., Möhler, O.: Heterogeneous ice nucleation on atmospheric aerosols: a review of results from laboratory experiments. Atmos. Chem. Phys. **12**, 9817–9854 (2012). https://doi.org/10.5194/acp-12-9817-2012

Hoppel, W.A., Frick, G.M., Fitzgerald, J.W., Wattle, B.J.: A cloud chamber study of the effect that nonprecipitating water clouds have on the aerosol size distribution. Aerosol Sci. Technol. **20**, 1 (1994). https://doi.org/10.1080/02786829408959660

Howard, C.J., Evenson, K.M.: Kinetics of the reaction of HO2 with NO. Geophys. Res. Lett. **4**, 437–440 (1977). https://doi.org/10.1029/GL004i010p00437

Howard, C.J.: Temperature dependence of the reaction HO2+NO→OH+NO$_2$. J. Chem. Phys. **71**, 2352–2359 (1979). https://doi.org/10.1063/1.438639

Howard, C. J.: Kinetic study of the equilibrium HO2 + NO .dblarw. OH + NO$_2$ and the thermochemistry of HO2. J. Am. Chem. Soc. **102**, 6937–6941 (1980). https://doi.org/10.1021/ja0054 3a006

Hoyle, C.R., Fuchs, C., Järvinen, E., Saathoff, H., Dias, A., El Haddad, I., Gysel, M., Coburn, S.C., Tröstl, J., Bernhammer, A.K., Bianchi, F., Breitenlechner, M., Corbin, J.C., Craven, J., Donahue, N.M., Duplissy, J., Ehrhart, S., Frege, C., Gordon, H., Höppel, N., Heinritzi, M., Kristensen, T.B., Molteni, U., Nichman, L., Pinterich, T., Prévôt, A.S.H., Simon, M., Slowik, J.G., Steiner, G., Tomé, A., Vogel, A.L., Volkamer, R., Wagner, A.C., Wagner, R., Wexler, A.S., Williamson, C., Winkler, P.M., Yan, C., Amorim, A., Dommen, J., Curtius, J., Gallagher, M.W., Flagan, R.C., Hansel, A., Kirkby, J., Kulmala, M., Möhler, O., Stratmann, F., Worsnop, D.R., Baltensperger, U.: Aqueous phase oxidation of sulphur dioxide by ozone in cloud droplets. Atmos. Chem. Phys. **16**, 1693–1712 (2016). https://doi.org/10.5194/acp-16-1693-2016

Huang, W., Saathoff, H., Pajunoja, A., Shen, X., Naumann, K.H., Wagner, R., Virtanen, A., Leisner, T., Mohr, C.: α-Pinene secondary organic aerosol at low temperature: chemical composition and implications for particle viscosity. Atmos. Chem. Phys. **18**, 2883–2898 (2018). https://doi.org/ 10.5194/acp-18-2883-2018

Hynes, R.G., Angove, D., Saunders, S., Haverd, V., and Azzi, M.: Evaluation of two MCM v3.1 alkene mechanisms using indoor environmental chamber data. Atmos. Environ. **39**, 7251–7262 (2005). https://doi.org/10.1016/j.atmosenv.2005.09.005

Im, Y., Jang, M., Beardsley, R.L.: Simulation of aromatic SOA formation using the lumping model integrated with explicit gas-phase kinetic mechanisms and aerosol-phase reactions. Atmos. Chem. Phys. **14**, 4013–4027 (2014). https://doi.org/10.5194/acp-14-4013-2014

Jaoui, M., Lewandowski, M., Kleindienst, T., Offenberg, J., and Edney, E.: β-caryophyllinic acid: An atmospheric tracer for β-caryophyllene secondary organic aerosol. Geophys. Res. Lett. **34** (2007). https://doi.org/10.1029/2006gl028827

Järvinen, E., Vochezer, P., Möhler, O., Schnaiter, M.: Laboratory study of microphysical and scattering properties of corona-producing cirrus clouds. Appl. Opt. **53**, 7566–7575 (2014). https:// doi.org/10.1364/ao.53.007566

Jeffries, H., Fox, D., Kamens, R.: Outdoor smog chamber studies: light effects relative to indoor chambers. Environ. Sci. Technol. **10**, 1006–1011 (1976). https://doi.org/10.1021/es60121a016

Jeffries, H.E., Kamens, R.M., Sexton, K.: Early history and rationale for outdoor chamber work at the University of North Carolina. Environ. Chem. **10**, 349–364 (2013). https://doi.org/10.1071/ EN13901

Jenkin, M.E., Wyche, K.P., Evans, C.J., Carr, T., Monks, P.S., Alfarra, M.R., Barley, M.H., McFiggans, G.B., Young, J.C., Rickard, A.R.: Development and chamber evaluation of the MCM v3.2 degradation scheme for β-caryophyllene. Atmos. Chem. Phys. **12**, 5275–5308 (2012). https://doi. org/10.5194/acp-12-5275-2012.

Jiang, J., Chen, M., Kuang, C., Attoui, M., McMurry, P.H.: Electrical mobility spectrometer using a diethylene glycol condensation particle counter for measurement of aerosol size distributions down to 1 nm. Aerosol Sci. Technol. **45**, 510–521 (2011). https://doi.org/10.1080/02786826. 2010.547538

Jimenez, J.L., Canagaratna, M.R., Donahue, N.M., Prevot, A.S.H., Zhang, Q., Kroll, J.H., DeCarlo, P.F., Allan, J.D., Coe, H., Ng, N.L., Aiken, A.C., Docherty, K.S., Ulbrich, I.M., Grieshop, A.P., Robinson, A.L., Duplissy, J., Smith, J.D., Wilson, K.R., Lanz, V.A., Hueglin, C., Sun, Y.L., Tian, J., Laaksonen, A., Raatikainen, T., Rautiainen, J., Vaattovaara, P., Ehn, M., Kulmala, M., Tomlinson, J.M., Collins, D.R., Cubison, M.J., Dunlea, J., Huffman, J.A., Onasch, T.B., Alfarra, M.R., Williams, P.I., Bower, K., Kondo, Y., Schneider, J., Drewnick, F., Borrmann, S., Weimer, S., Demerjian, K., Salcedo, D., Cottrell, L., Griffin, R., Takami, A., Miyoshi, T., Hatakeyama, S., Shimono, A., Sun, J.Y., Zhang, Y.M., Dzepina, K., Kimmel, J.R., Sueper, D., Jayne, J.T.,

Herndon, S.C., Trimborn, A.M., Williams, L.R., Wood, E.C., Middlebrook, A.M., Kolb, C.E., Baltensperger, U., Worsnop, D.R.: Evolution of organic aerosols in the atmosphere. Science **326**, 1525–1529 (2009). https://doi.org/10.1126/science.1180353

Junkermann, W., Burger, J.M.: A new portable instrument for continuous measurement of formaldehyde in ambient air. J. Atmos. Oceanic Tech. **23**, 38–45 (2006). https://doi.org/10.1175/jtech1831.1

Kaduwela, A., Luecken, D., Carter, W., Derwent, R.: New directions: atmospheric chemical mechanisms for the future. Atmos. Environ. **122**, 609–610 (2015). https://doi.org/10.1016/j.atmosenv.2015.10.031

Kalberer, M., Sax, M., Samburova, V.: Molecular size evolution of oligomers in organic aerosols collected in urban atmospheres and generated in a smog chamber. Environ. Sci. Technol. **40**, 5917–5922 (2006). https://doi.org/10.1021/es0525760

Kaltsonoudis, C., Kostenidou, E., Louvaris, E., Psichoudaki, M., Tsiligiannis, E., Florou, K., Liangou, A., Pandis, S.N.: Characterization of fresh and aged organic aerosol emissions from meat charbroiling. Atmos. Chem. Phys. **17**, 7143–7155 (2017). https://doi.org/10.5194/acp-17-7143-2017

Kaltsonoudis, C., Jorga, S.D., Louvaris, E., Florou, K., Pandis, S.N.: A portable dual-smog-chamber system for atmospheric aerosol field studies. Atmos. Meas. Tech. **12**, 2733–2743 (2019). https://doi.org/10.5194/amt-12-2733-2019

Kanakidou, M., Seinfeld, J.H., Pandis, S.N., Barnes, I., Dentener, F.J., Facchini, M.C., Van Dingenen, R., Ervens, B., Nenes, A., Nielsen, C.J., Swietlicki, E., Putaud, J.P., Balkanski, Y., Fuzzi, S., Horth, J., Moortgat, J., Winterhalter, R., Myhre, C.E.L., Tsigaridis, K., Vignati, E., Stephanou, E.G., Wilson, J.: Organic aerosol and global climate modelling: a review. Atmos. Chem. Phys. **4**, 5855–6024 (2005)

Kebabian, P.L., Herndon, S.C., Freedman, A.: Detection of nitrogen dioxide by cavity attenuated phase shift spectroscopy. Anal. Chem. **77**, 724–728 (2005). https://doi.org/10.1021/ac048715y

Kelly, N.A.: Characterization of fluorocarbon-film bags as smog chambers. Environ. Sci. Technol. **16**, 763–770 (1982). https://doi.org/10.1021/es00105a007

Kiendler-Scharr, A., Wildt, J., Maso, M.D., Hohaus, T., Kleist, E., Mentel, T.F., Tillmann, R., Uerlings, R., Schurr, U., Wahner, A.: New particle formation in forests inhibited by isoprene emissions. Nature **461**, 381–384 (2009a). https://doi.org/10.1038/nature08292

Kiendler-Scharr, A., Zhang, Q., Hohaus, T., Kleist, E., Mensah, A., Mentel, T.F., Spindler, C., Uerlings, R., Tillmann, R., Wildt, J.: Aerosol mass spectrometric features of biogenic SOA: observations from a plant chamber and in rural atmospheric environments. Environ. Sci. Technol. **43**, 8166–8172 (2009b). https://doi.org/10.1021/es901420b

King, S.M., Rosenoern, T., Shilling, J.E., Chen, Q., Martin, S.T.: Increased cloud activation potential of secondary organic aerosol for atmospheric mass loadings. Atmos. Chem. Phys. **9**, 2959–2971 (2009). https://doi.org/10.5194/acp-9-2959-2009

Kirkby, J., Curtius, J., Almeida, J., Dunne, E., Duplissy, J., Ehrhart, S., Franchin, A., Gagné, S., Ickes, L., Kürten, A., Kupc, A., Metzger, A., Riccobono, F., Rondo, L., Schobesberger, S., Tsagkogeorgas, G., Wimmer, D., Amorim, A., Bianchi, F., Breitenlechner, M., David, A., Dommen, J., Downard, A., Ehn, M., Flagan, R. C., Haider, S., Hansel, A., Hauser, D., Jud, W., Junninen, H., Kreissl, F., Kvashin, A., Laaksonen, A., Lehtipalo, K., Lima, J., Lovejoy, E.R., Makhmutov, V., Mathot, S., Mikkilä, J., Minginette, P., Mogo, S., Nieminen, T., Onnela, A., Pereira, P., Petäjä, T., Schnitzhofer, R., Seinfeld, J.H., Sipilä, M., Stozhkov, Y., Stratmann, F., Tomé, A., Vanhanen, J., Viisanen, Y., Vrtala, A., Wagner, P.E., Walther, H., Weingartner, E., Wex, H., Winkler, P.M., Carslaw, K.S., Worsnop, D.R., Baltensperger, U., Kulmala, M.: Role of sulphuric acid, ammonia and galactic cosmic rays in atmospheric aerosol nucleation. Nature **476**, 429 (2011). https://doi.org/10.1038/nature10343. https://www.nature.com/articles/nature10343#supplementary-information

Kleffmann, J., Heland, J., Kurtenbach, R., Lörzer, J.C., Wiesen, P.: A new instrument (LOPAP) for the detection of nitrous acid (HONO). Environ. Sci. Pollut. Res. **9**, 48–54 (2002)

Kleindienst, T.E., Jaoui, M., Lewandowski, M., Offenberg, J.H., Lewis, C.W., Bhave, P.V., Edney, E.O.: Estimates of the contributions of biogenic and anthropogenic hydrocarbons to secondary organic aerosol at a southeastern US location. Atmos. Environ. **41**, 8288–8300 (2007)

Kleindienst, T.E., Jaoui, M., Lewandowski, M., Offenberg, J.H., Docherty, K.S.: The formation of SOA and chemical tracer compounds from the photooxidation of naphthalene and its methyl analogs in the presence and absence of nitrogen oxides. Atmos. Chem. Phys. **12**, 8711–8726 (2012). https://doi.org/10.5194/acp-12-8711-2012

Kleist, E., Mentel, T.F., Andres, S., Bohne, A., Folkers, A., Kiendler-Scharr, A., Rudich, Y., Springer, M., Tillmann, R., Wildt, J.: Irreversible impacts of heat on the emissions of monoterpenes, sesquiterpenes, phenolic BVOC and green leaf volatiles from several tree species. Biogeosciences **9**, 5111–5123 (2012). https://doi.org/10.5194/bg-9-5111-2012

Kostenidou, E., Kaltsonoudis, C., Tsiflikiotou, M., Louvaris, E., Russell, L.M., Pandis, S.N.: Burning of olive tree branches: a major organic aerosol source in the Mediterranean. Atmos. Chem. Phys. **13**, 8797–8811 (2013). https://doi.org/10.5194/acp-13-8797-2013

Kourtchev, I., Giorio, C., Manninen, A., Wilson, E., Mahon, B., Aalto, J., Kajos, M., Venables, D., Ruuskanen, T., Levula, J., Loponen, M., Connors, S., Harris, N., Zhao, D., Kiendler-Scharr, A., Mentel, T., Rudich, Y., Hallquist, M., Doussin, J.-F., Maenhaut, W., Bäck, J., Petäjä, T., Wenger, J., Kulmala, M., Kalberer, M.: Enhanced Volatile Organic Compounds emissions and organic aerosol mass increase the oligomer content of atmospheric aerosols. Sci. Rep. **6**, 35038 (2016). https://doi.org/10.1038/srep35038

Krechmer, J.E., Pagonis, D., Ziemann, P.J., Jimenez, J.L.: Quantification of gas-wall partitioning in teflon environmental chambers using rapid bursts of low-volatility oxidized species generated in situ. Environ. Sci. Technol. **50**, 5757–5765 (2016). https://doi.org/10.1021/acs.est.6b00606

Krechmer, J.E., Day, D.A., Ziemann, P.J., Jimenez, J.L.: Direct measurements of gas/particle partitioning and mass accommodation coefficients in environmental chambers. Environ. Sci. Technol. **51**, 11867–11875 (2017). https://doi.org/10.1021/acs.cst.7b02144

Kristensen, K., Jensen, L.N., Glasius, M., Bilde, M.: The effect of sub-zero temperature on the formation and composition of secondary organic aerosol from ozonolysis of alpha-pinene. Environ. Sci. Process Impacts **19**, 1220–1234 (2017). https://doi.org/10.1039/c7em00231a

Kulmala, M., Mordas, G., Petäjä, T., Grönholm, T., Aalto, P.P., Vehkamäki, H., Hienola, A.I., Herrmann, E., Sipilä, M., Riipinen, I., Manninen, H.E., Hämeri, K., Stratmann, F., Bilde, M., Winkler, P.M., Birmili, W., Wagner, P.E.: The condensation particle counter battery (CPCB): a new tool to investigate the activation properties of nanoparticles. J. Aerosol Sci. **38**, 289–304 (2007). https://doi.org/10.1016/j.jaerosci.2006.11.008

Künzi, L., Mertes, P., Schneider, S., Jeannet, N., Menzi, C., Dommen, J., Baltensperger, U., Prévôt, A.S.H., Salathe, M., Kalberer, M., Geiser, M.: Responses of lung cells to realistic exposure of primary and aged carbonaceous aerosols. Atmos. Environ. **68**, 143–150 (2013). https://doi.org/10.1016/j.atmosenv.2012.11.055

Künzi, L., Krapf, M., Daher, N., Dommen, J., Jeannet, N., Schneider, S., Platt, S., Slowik, J.G., Baumlin, N., Salathe, M., Prévôt, A.S.H., Kalberer, M., Strähl, C., Dümbgen, L., Sioutas, C., Baltensperger, U., Geiser, M.: Toxicity of aged gasoline exhaust particles to normal and diseased airway epithelia. Sci. Rep. **5**, 11801 (2015). https://doi.org/10.1038/srep11801. https://www.nature.com/articles/srep11801#supplementary-information

Kwok, E.S.C., Atkinson, R.: Estimation of hydroxyl radical reaction rate constants for gas-phase organic compounds using a structure-reactivity relationship: an update. Atmos. Environ. **29**, 1685–1695 (1995). https://doi.org/10.1016/1352-2310(95)00069-B

La, Y.S., Camredon, M., Ziemann, P.J., Valorso, R., Matsunaga, A., Lannuque, V., Lee-Taylor, J., Hodzic, A., Madronich, S., Aumont, B.: Impact of chamber wall loss of gaseous organic compounds on secondary organic aerosol formation: explicit modeling of SOA formation from alkane and alkene oxidation. Atmos. Chem. Phys. **16**, 1417–1431 (2016). https://doi.org/10.5194/acp-16-1417-2016

Laborde, M., Schnaiter, M., Linke, C., Saathoff, H., Naumann, K.H., Möhler, O., Berlenz, S., Wagner, U., Taylor, J.W., Liu, D., Flynn, M., Allan, J.D., Coe, H., Heimerl, K., Dahlkötter, F., Weinzierl, B., Wollny, A.G., Zanatta, M., Cozic, J., Laj, P., Hitzenberger, R., Schwarz, J.P., Gysel,

M.: Single particle soot photometer intercomparison at the AIDA chamber. Atmos. Meas. Tech. **5**, 3077–3097 (2012). https://doi.org/10.5194/amt-5-3077-2012

Lamb, D., Miller, D.F., Robinson, N.F., Gertler, A.W.: The importance of liquid water concentration in the atmospheric oxidation of SO2. Atmos. Environ. **1967**(21), 2333–2344 (1987). https://doi.org/10.1016/0004-6981(87)90369-6

Lamkaddam, H.: Study under simulated condition of the secondary organic aerosol from the photooxydation of n-dodecane: impact of the physical-chemical processes. Université Paris-Est (2017)

Laskin, A., Laskin, J., Nizkorodov, S.A.: Chemistry of atmospheric brown carbon. Chem. Rev. **115**, 4335–4382 (2015). https://doi.org/10.1021/cr5006167

Lazrus, A.L., Kok, G.L., Lind, J.A., Gitlin, S.N., Heikes, B.G., Shetter, R.E.: Automated fluorometric method for hydrogen peroxide in air. Anal. Chem. **58**, 594–597 (1986). https://doi.org/10.1021/ac00294a024

Lee, S., Jang, M., Kamens, R.M.: SOA formation from the photooxidation of α-pinene in the presence of freshly emitted diesel soot exhaust. Atmos. Environ. **38**, 2597–2605 (2004). https://doi.org/10.1016/j.atmosenv.2003.12.041

Lee, S.-B., Bae, G.-N., and Moon, K.-C.: Smog chamber measurements. In: Kim, Y.J., Platt, U., Gu, M.B., Iwahashi, H. (eds.) Atmospheric and Biological Environmental Monitoring, pp. 105–136. Springer Netherlands, Dordrecht (2009)

Leone, J.A., Flagan, R.C., Grosjean, D., Seinfeld, J.H.: An outdoor smog chamber and modeling study of toluene–NO_x photooxidation. Int. J. Chem. Kinet. **17**, 177–216 (1985). https://doi.org/10.1002/kin.550170206

Leskinen, A., Yli-Pirilä, P., Kuuspalo, K., Sippula, O., Jalava, P., Hirvonen, M.R., Jokiniemi, J., Virtanen, A., Komppula, M., Lehtinen, K.E.J.: Characterization and testing of a new environmental chamber. Atmos. Meas. Tech. **8**, 2267–2278 (2015). https://doi.org/10.5194/amt-8-2267-2015

Leu, M.T.: Rate constant for the reaction $HO_2+NO \rightarrow OH+NO_2$. J. Chem. Phys. **70**, 1662–1666 (1979). https://doi.org/10.1063/1.437680

Leungsakul, S., Jaoui, M., Kamens, R.M.: Kinetic mechanism for predicting secondary organic aerosol formation from the reaction of d-limonene with ozone. Environ. Sci. Technol. **39**, 9583–9594 (2005). https://doi.org/10.1021/es0492687

Levy, H.: Normal atmosphere: large radical and formaldehyde concentrations predicted. Science **173**, 141–143 (1971). https://doi.org/10.1126/science.173.3992.141

Li, C., Ma, Z., Chen, J., Wang, X., Ye, X., Wang, L., Yang, X., Kan, H., Donaldson, D.J., Mellouki, A.: Evolution of biomass burning smoke particles in the dark. Atmos. Environ. **120**, 244–252 (2015). https://doi.org/10.1016/j.atmosenv.2015.09.003

Li, C., Hu, Y., Zhang, F., Chen, J., Ma, Z., Ye, X., Yang, X., Wang, L., Tang, X., Zhang, R., Mu, M., Wang, G., Kan, H., Wang, X., Mellouki, A.: Multi-pollutant emissions from the burning of major agricultural residues in China and the related health-economic effects. Atmos. Chem. Phys. **17**, 4957–4988 (2017). https://doi.org/10.5194/acp-17-4957-2017

Li, K., Lin, C., Geng, C., White, S., Chen, L., Bao, Z., Zhang, X., Zhao, Y., Han, L., Yang, W., Azzi, M.: Characterization of a new smog chamber for evaluating SAPRC gas-phase chemical mechanism. J. Environ. Sci. **95**, 14–22 (2020). https://doi.org/10.1016/j.jes.2020.03.028

Li, J., Li, H., Wang, X., Wang, W., Ge, M., Zhang, H., Zhang, X., Li, K., Chen, Y., Wu, Z., Chai, F., Meng, F., Mu, Y., Mellouki, A., Bi, F., Zhang, Y., Wu, L., Liu, Y.: A large-scale outdoor atmospheric simulation smog chamber for studying atmospheric photochemical processes: characterization and preliminary application. J. Environ. Sci. **102**, 185–197 (2021). https://doi.org/10.1016/j.jes.2020.09.015

Lim, Y.B., Ziemann, P.J.: Products and mechanism of secondary organic aerosol formation from reactions of n-alkanes with OH radicals in the presence of NO_x. Environ. Sci. Technol. **39**, 9229–9236 (2005). https://doi.org/10.1021/es051447g

Linke, C., Möhler, O., Veres, A., Mohácsi, A., Bozóki, Z., Szabó, G., Schnaiter, M.: Optical properties and mineralogical composition of different Saharan mineral dust samples: a laboratory study. Atmos. Chem. Phys. **6**, 3315–3323 (2006)

Liu, S., Shilling, J.E., Song, C., Hiranuma, N., Zaveri, R.A., Russell, L.M.: Hydrolysis of organonitrate functional groups in aerosol particles. Aerosol Sci. Technol. **46**, 1359–1369 (2012). https://doi.org/10.1080/02786826.2012.716175

Luo, H., Li, G., Chen, J., Wang, Y., An, T.: Reactor characterization and primary application of a state of art dual-reactor chamber in the investigation of atmospheric photochemical processes. J. Environ. Sci. **98**, 161–168 (2020). https://doi.org/10.1016/j.jes.2020.05.021

Martín-Reviejo, M., Wirtz, K.: Is benzene a precursor for secondary organic aerosol? Environ. Sci. Technol. **39**, 1045–1054 (2005). https://doi.org/10.1021/es049802a

Massabò, D., Caponi, L., Bove, M.C., Prati, P.: Brown carbon and thermal–optical analysis: a correction based on optical multi-wavelength apportionment of atmospheric aerosols. Atmos. Environ. **125**, 119–125 (2016). https://doi.org/10.1016/j.atmosenv.2015.11.011

Massabò, D., Danelli, S.G., Brotto, P., Comite, A., Costa, C., Di Cesare, A., Doussin, J.F., Ferraro, F., Formenti, P., Gatta, E., Negretti, L., Oliva, M., Parodi, F., Vezzulli, L., Prati, P.: ChAMBRe: a new atmospheric simulation chamber for aerosol modelling and bio-aerosol research. Atmos. Meas. Tech. **11**, 5885–5900 (2018). https://doi.org/10.5194/amt-11-5885-2018

Massoli, P., Kebabian, P.L., Onasch, T.B., Hills, F.B., Freedman, A.: Aerosol light extinction measurements by cavity attenuated phase shift (CAPS) spectroscopy: laboratory validation and field deployment of a compact aerosol particle extinction monitor. Aerosol Sci. Technol. **44**, 428–435 (2010). https://doi.org/10.1080/02786821003716599

McFiggans, G., Mentel, T.F., Wildt, J., Pullinen, I., Kang, S., Kleist, E., Schmitt, S., Springer, M., Tillmann, R., Wu, C., Zhao, D., Hallquist, M., Faxon, C., Le Breton, M., Hallquist, Å.M., Simpson, D., Bergström, R., Jenkin, M.E., Ehn, M., Thornton, J.A., Alfarra, M.R., Bannan, T.J., Percival, C.J., Priestley, M., Topping, D., Kiendler-Scharr, A.: Secondary organic aerosol reduced by mixture of atmospheric vapours. Nature **565**, 587–593 (2019). https://doi.org/10.1038/s41586-018-0871-y

McMurry, P.H., Grosjean, D.: Gas and aerosol wall losses in Teflon film smog chambers. Environ. Sci. Technol. **19**, 1176–1182 (1985)

McVay, R.C., Zhang, X., Aumont, B., Valorso, R., Camredon, M., La, Y.S., Wennberg, P.O., Seinfeld, J.H.: SOA formation from the photooxidation of α-pinene: systematic exploration of the simulation of chamber data. Atmos. Chem. Phys. **16**, 2785–2802 (2016). https://doi.org/10.5194/acp-16-2785-2016

Mentel, T., Bleilebens, D., Wahner, A.: A study of nighttime nitrogen oxide oxidation in a large reaction chamber-the fate of NO_2, N_2O_5, HNO_3 and O_3 at different humidities. Atmos. Environ. **30**, 4007–4020 (1996)

Mentel, T.F., Kleist, E., Andres, S., Dal Maso, M., Hohaus, T., Kiendler-Scharr, A., Rudich, Y., Springer, M., Tillmann, R., Uerlings, R., Wahner, A., Wildt, J.: Secondary aerosol formation from stress-induced biogenic emissions and possible climate feedbacks. Atmos. Chem. Phys. **13**, 8755–8770 (2013). https://doi.org/10.5194/acp-13-8755-2013

Mertes, P., Praplan, A.P., Künzi, L., Dommen, J., Baltensperger, U., Geiser, M., Weingartner, E., Ricka, J., Fierz, M., Kalberer, M.: A compact and portable deposition chamber to study nanoparticles in air-exposed tissue. J. Aerosol Med. Pulm. Drug Deliv. **26**, 228–235 (2013). https://doi.org/10.1089/jamp.2012.0985

Metzger, A., Dommen, J., Gaeggeler, K., Duplissy, J., Prevot, A.S.H., Kleffmann, J., Elshorbany, Y., Wisthaler, A., Baltensperger, U.: Evaluation of 1,3,5 trimethylbenzene degradation in the detailed tropospheric chemistry mechanism, MCMv3.1, using environmental chamber data. Atmos. Chem. Phys. **8**, 6453–6468 (2008). https://doi.org/10.5194/acp-8-6453-2008

Meyer, N.K., Duplissy, J., Gysel, M., Metzger, A., Dommen, J., Weingartner, E., Alfarra, M.R., Prevot, A.S.H., Fletcher, C., Good, N., McFiggans, G., Jonsson, Å.M., Hallquist, M.,

Baltensperger, U., Ristovski, Z.D.: Analysis of the hygroscopic and volatile properties of ammonium sulphate seeded and unseeded SOA particles. Atmos. Chem. Phys. **9**, 721–732 (2009). https://doi.org/10.5194/acp-9-721-2009

Miller, D.F., Lamb, D., Gertler, A.W.: SO2 oxidation in cloud drops containing NaCl or sea salt as condensation nuclei. Atmos. Environ. **1967**(21), 991–993 (1987). https://doi.org/10.1016/0004-6981(87)90096-5

Mirme, A., Tamm, E., Mordas, G., Vana, M., Uin, J., Mirme, S., Bernotas, T., Laakso, L., Hirsikko, A., Kulmala, M.: A wide-range multi-channel air ion spectrometer. Boreal Env. Res. **12**, 247–264 (2007)

Mirme, S., Mirme, A.: The mathematical principles and design of the NAIS–a spectrometer for the measurement of cluster ion and nanometer aerosol size distributions. Atmos. Meas. Tech. **6**, 1061–1071 (2013). https://doi.org/10.5194/amt-6-1061-2013

Mogili, P.K., Kleiber, P.D., Young, M.A., Grassian, V.H.: Heterogeneous uptake of ozone on reactive components of mineral dust aerosol: an environmental aerosol reaction chamber study. J. Phys. Chem. A **110**, 13799–13807 (2006). https://doi.org/10.1021/jp063620g

Mogili, P.K., Yang, K.H., Young, M.A., Kleiber, P.D., Grassian, V.H.: Environmental aerosol chamber studies of extinction spectra of mineral dust aerosol components: broadband IR-UV extinction spectra. J. Geophys. Res. **112**(D21) (2007). https://doi.org/10.1029/2007JD008890

Möhler, O., Nink, A., Saathoff, H., Schaefers, S., Schnaiter, M., Schöck, W., Schurath, U.: The Karlsruhe aerosol chamber facility AIDA: technical description and first results of homogeneous and heterogeneous ice nucleation experiments, pp. 163–168 (2001)

Möhler, O., Stetzer, O., Schaefers, S., Linke, C., Schnaiter, M., Tiede, R., Saathoff, H., Krämer, M., Mangold, A., Budz, P., Zink, P., Schreiner, J., Mauersberger, K., Haag, W., Kärcher, B., Schurath, U.: Experimental investigation of homogeneous freezing of sulphuric acid particles in the aerosol chamber AIDA. Atmos. Chem. Phys. **3**, 211–223 (2003). https://doi.org/10.5194/acp-3-211-2003

Möhler, O., Field, P.R., Connolly, P., Benz, S., Saathoff, H., Schnaiter, M., Wagner, R., Cotton, R., Krämer, M., Mangold, A., Heymsfield, A.J.: Efficiency of the deposition mode ice nucleation on mineral dust particles. Atmos. Chem. Phys. **6**, 3007–3021 (2006). https://doi.org/10.5194/acp-6-3007-2006

Möhler, O., Benz, S., Saathoff, H., Schnaiter, M., Wagner, R., Schneider, J., Walter, S., Ebert, V., Wagner, S.: The effect of organic coating on the heterogeneous ice nucleation efficiency of mineral dust aerosols. Environ. Res. Lett. **3**, 025007 (2008a). https://doi.org/10.1088/1748-9326/3/2/025007

Möhler, O., Georgakopoulos, D.G., Morris, C.E., Benz, S., Ebert, V., Hunsmann, S., Saathoff, H., Schnaiter, M., Wagner, R.: Heterogeneous ice nucleation activity of bacteria: new laboratory experiments at simulated cloud conditions. Biogeosciences **5**, 1425–1435 (2008b). https://doi.org/10.5194/bg-5-1425-2008

Möhler, O., Adams, M., Lacher, L., Vogel, F., Nadolny, J., Ullrich, R., Boffo, C., Pfeuffer, T., Hobl, A., Weiß, M., Vepuri, H.S.K., Hiranuma, N., Murray, B.J.: The Portable Ice Nucleation Experiment (PINE): a new online instrument for laboratory studies and automated long-term field observations of ice-nucleating particles. Atmos. Meas. Tech. **14**, 1143–1166 (2021). https://doi.org/10.5194/amt-14-1143-2021

Mohr, C., Huffman, J.A., Cubison, M.J., Aiken, A.C., Docherty, K.S., Kimmel, J.R., Ulbrich, I.M., Hannigan, M., Jimenez, J.L.: Characterization of primary organic aerosol emissions from meat cooking, trash burning, and motor vehicles with high-resolution aerosol mass spectrometry and comparison with ambient and chamber observations. Environ. Sci. Technol. **43**, 2443–2449 (2009). https://doi.org/10.1021/es8011518

Munoz, A., Ródenas, M., Borras, E., Brenan, A., Dellen, J., Escalante, J.M., Gratien, A., Gomez, T., Herrmann, H., Kari, E., Michoud, V., Mutzel, A., Olariu, R., Seakins, P., Tillmann, R., Vera, T., Virtanen, A.: Intercomparison of instruments to measure OVOCs: assessment of performance under different relevant controlled conditions (EUPHORE chambers), EGU General Assembly 2019, Vienna, Austria (2019)

Ng, N.L., Canagaratna, M.R., Jimenez, J.L., Chhabra, P.S., Seinfeld, J.H., Worsnop, D.R.: Changes in organic aerosol composition with aging inferred from aerosol mass spectra. Atmos. Chem. Phys. **11**, 6465–6474 (2011a). https://doi.org/10.5194/acp-11-6465-2011

Ng, N.L., Herndon, S.C., Trimborn, A., Canagaratna, M.R., Croteau, P.L., Onasch, T.B., Sueper, D., Worsnop, D.R., Zhang, Q., Sun, Y.L., Jayne, J.T.: An aerosol chemical speciation monitor (ACSM) for routine monitoring of the composition and mass concentrations of ambient aerosol. Aerosol Sci. Technol. **45**, 780–794 (2011b). https://doi.org/10.1080/02786826.2011.560211

Nguyen, T.B., Roach, P.J., Laskin, J., Laskin, A., Nizkorodov, S.A.: Effect of humidity on the composition of isoprene photooxidation secondary organic aerosol. Atmos. Chem. Phys. **11**, 6931–6944 (2011). https://doi.org/10.5194/acp-11-6931-2011

Niedermeier, D., Hartmann, S., Clauss, T., Wex, H., Kiselev, A., Sullivan, R.C., DeMott, P.J., Petters, M.D., Reitz, P., Schneider, J., Mikhailov, E., Sierau, B., Stetzer, O., Reimann, B., Bundke, U., Shaw, R.A., Buchholz, A., Mentel, T.F., Stratmann, F.: Experimental study of the role of physicochemical surface processing on the IN ability of mineral dust particles. Atmos. Chem. Phys. **11**, 11131–11144 (2011). https://doi.org/10.5194/acp-11-11131-2011

Niedermeier, D., Augustin-Bauditz, S., Hartmann, S., Wex, H., Ignatius, K., Stratmann, F.: Can we define an asymptotic value for the ice active surface site density for heterogeneous ice nucleation? J. Geophys. Res. Atmos. **120**, 5036–5046 (2015). https://doi.org/10.1002/2014JD022814

Niki, H., Daby, E.E., Weinstock, B.: Mechanisms of smog reactions. In: Photochemical Smog and Ozone Reactions, Advances in Chemistry, vol. 113, pp. 16–57. American Chemical Society (1972)

Niki, H., Maker, P., Savage, C., Breitenbach, L.: An FTIR study of mechanisms for the HO radical initiated oxidation of C2H4 in the presence of NO: detection of glycolaldehyde. Chem. Phys. Lett. **80**, 499–503 (1981)

Nizkorodov, S.A., Laskin, J., Laskin, A.: Molecular chemistry of organic aerosols through the application of high resolution mass spectrometry. Phys. Chem. Chem. Phys. **13**, 3612–3629 (2011). https://doi.org/10.1039/c0cp02032j

Nordin, E.Z., Uski, O., Nyström, R., Jalava, P., Eriksson, A.C., Genberg, J., Roldin, P., Bergvall, C., Westerholm, R., Jokiniemi, J., Pagels, J.H., Boman, C., Hirvonen, M. R.: Influence of ozone initiated processing on the toxicity of aerosol particles from small scale wood combustion. Atmos. Environ. **102**, 282–289 (2015). https://doi.org/10.1016/j.atmosenv.2014.11.068

Novelli, A., Vereecken, L., Bohn, B., Dorn, H.P., Gkatzelis, G.I., Hofzumahaus, A., Holland, F., Reimer, D., Rohrer, F., Rosanka, S., Taraborrelli, D., Tillmann, R., Wegener, R., Yu, Z., Kiendler-Scharr, A., Wahner, A., Fuchs, H.: Importance of isomerization reactions for OH radical regeneration from the photo-oxidation of isoprene investigated in the atmospheric simulation chamber SAPHIR. Atmos. Chem. Phys. **20**, 3333–3355 (2020). https://doi.org/10.5194/acp-20-3333-2020

Nozière, B., Hanson, D.R.: Speciated monitoring of gas-phase organic peroxy radicals by chemical ionization mass spectrometry: cross-reactions between CH3O2, CH3(CO)O2, (CH3)3CO2, and c-C6H11O2. J. Phys. Chem. A **121**, 8453–8464 (2017). https://doi.org/10.1021/acs.jpca.7b06456

Odum, J.R., Hoffmann, T., Bowman, F., Collins, D., Flagan, R.C., Seinfeld, J.H.: Gas/particle partitioning and secondary organic aerosol yields. Environ. Sci. Technol. **30**, 2580–2585 (1996)

Onel, L., Brennan, A., Seakins, P.W., Whalley, L., Heard, D.E.: A new method for atmospheric detection of the CH3O2 radical. Atmos. Meas. Tech. **10**, 3985–4000 (2017). https://doi.org/10.5194/amt-10-3985-2017

Onel, L., Brennan, A., Gianella, M., Hooper, J., Ng, N., Hancock, G., Whalley, L., Seakins, P.W., Ritchie, G.A.D., Heard, D.E.: An intercomparison of CH3O2 measurements by fluorescence assay by gas expansion and cavity ring-down spectroscopy within HIRAC (Highly Instrumented Reactor for Atmospheric Chemistry). Atmos. Meas. Tech. **13**, 2441–2456 (2020). https://doi.org/10.5194/amt-13-2441-2020

Pandis, S.N., Paulson, S.E., Seinfeld, J.H., Flagan, R.C.: Aerosol formation in the photooxidation of isoprene and β-pinene. Atmos. Environ. Part A Gen. Top. **25**, 997–1008 (1991). https://doi.org/10.1016/0960-1686(91)90141-S

Pankow, J.F.: An absorption model of the gas/aerosol partitioning involved in the formation of secondary organic aerosol. Atmos. Environ. **28**, 189–193 (1994)

Parikh, H., Jeffries, H., Sexton, K., Luecken, D., Kamens, R., Vizuete, W.: Evaluation of aromatic oxidation reactions in seven chemical mechanisms with an outdoor chamber. Environ. Chem. **10**, 245 (2013). https://doi.org/10.1071/en13039

Parrish, D.D., Fehsenfeld, F.C.: Methods for gas-phase measurements of ozone, ozone precursors and aerosol precursors. Atmos. Environ. **34**, 1921–1957 (2000). https://doi.org/10.1016/S1352-2310(99)00454-9

Pathak, R.K., Presto, A.A., Lane, T.E., Stanier, C.O., Donahue, N.M., Pandis, S.N.: Ozonolysis of-pinene: parameterization of secondary organic aerosol mass fraction. Atmos. Chem. Phys. **7**, 3811–3821 (2007)

Paulsen, D., Dommen, J., Kalberer, M., Prévôt, A.S.H., Richter, R., Sax, M., Steinbacher, M., Weingartner, E., Baltensperger, U.: Secondary organic aerosol formation by irradiation of 1,3,5-trimethylbenzene−NO_x−H_2O in a new reaction chamber for atmospheric chemistry and physics. Environ. Sci. Technol. **39**, 2668–2678 (2005). https://doi.org/10.1021/es0489137

Peeters, J., Müller, J.-F., Stavrakou, T., Nguyen, V.S.: Hydroxyl radical recycling in isoprene oxidation driven by hydrogen bonding and hydrogen tunneling: the upgraded LIM1 mechanism. J. Phys. Chem. A **118**, 8625–8643 (2014). https://doi.org/10.1021/jp5033146

Pereira, K.L., Hamilton, J.F., Rickard, A.R., Bloss, W.J., Alam, M.S., Camredon, M., Ward, M.W., Wyche, K.P., Muñoz, A., Vera, T., Vázquez, M., Borrás, E., Ródenas, M.: Insights into the formation and evolution of individual compounds in the particulate phase during aromatic photo-oxidation. Environ. Sci. Technol. **49**, 13168–13178 (2015). https://doi.org/10.1021/acs.est.5b0 3377

Pereira, K.L., Dunmore, R., Whitehead, J., Alfarra, M.R., Allan, J.D., Alam, M.S., Harrison, R.M., McFiggans, G., Hamilton, J.F.: Technical note: use of an atmospheric simulation chamber to investigate the effect of different engine conditions on unregulated VOC-IVOC diesel exhaust emissions. Atmos. Chem. Phys. **18**, 11073–11096 (2018). https://doi.org/10.5194/acp-18-11073-2018

Person, A., Eyglunent, G., Daële, V., Mellouki, A., Mu, Y.: The near UV absorption cross-sections and the rate coefficients for the ozonolysis of a series of styrene-like compounds. J. Photochem. Photobiol. A Chem. **195**, 54–63 (2008)

Pfeifer, S., Müller, T., Weinhold, K., Zikova, N., Martins dos Santos, S., Marinoni, A., Bischof, O.F., Kykal, C., Ries, L., Meinhardt, F., Aalto, P., Mihalopoulos, N., Wiedensohler, A.: Intercomparison of 15 aerodynamic particle size spectrometers (APS 3321): uncertainties in particle sizing and number size distribution. Atmos. Meas. Tech. **9**, 1545–1551 (2016). https://doi.org/10.5194/amt-9-1545-2016

Picquet-Varrault, B., Doussin, J.F., Durand-Jolibois, R., Carlier, P.: FTIR spectroscopic study of the OH-induced oxidation of two linear acetates: ethyl and n-propyl acetates. Phys. Chem. Chem. Phys. **3**, 2595–2606 (2001)

Pitts, J.N.: Formation and fate of gaseous and particulate mutagens and carcinogens in real and simulated atmospheres. Environ. Health Perspect. **47**, 115–140 (1983)

Platt, S.M., El Haddad, I., Zardini, A.A., Clairotte, M., Astorga, C., Wolf, R., Slowik, J.G., Temime-Roussel, B., Marchand, N., Ježek, I., Drinovec, L., Močnik, G., Möhler, O., Richter, R., Barmet, P., Bianchi, F., Baltensperger, U., Prévôt, A.S.H.: Secondary organic aerosol formation from gasoline vehicle emissions in a new mobile environmental reaction chamber. Atmos. Chem. Phys. **13**, 9141–9158 (2013). https://doi.org/10.5194/acp-13-9141-2013

Platt, S.M., El Haddad, I., Pieber, S.M., Zardini, A.A., Suarez-Bertoa, R., Clairotte, M., Daellenbach, K.R., Huang, R.J., Slowik, J.G., Hellebust, S., Temime-Roussel, B., Marchand, N., de Gouw, J., Jimenez, J.L., Hayes, P.L., Robinson, A.L., Baltensperger, U., Astorga, C., Prévôt, A.S.H.: Gasoline cars produce more carbonaceous particulate matter than modern filter-equipped diesel cars. Sci. Rep. **7**, 4926 (2017). https://doi.org/10.1038/s41598-017-03714-9

Pratap, V., Bian, Q., Kiran, S.A., Hopke, P.K., Pierce, J.R., Nakao, S.: Investigation of levoglucosan decay in wood smoke smog-chamber experiments: the importance of aerosol loading, temperature, and vapor wall losses in interpreting results. Atmos. Environ. **199**, 224–232 (2019). https://doi.org/10.1016/j.atmosenv.2018.11.020

Pun, B.K., Wu, S.Y., Seigneur, C., Seinfeld, J.H., Griffin, R.J., Pandis, S.: Uncertainties in modelling secondary organic aerosols: three-dimensional modeling studies in Nashville/Western Tennessee. Environ. Sci. Technol. **37**, 3647–3661 (2003)

Qi, X., Zhu, S., Zhu, C., Hu, J., Lou, S., Xu, L., Dong, J., Cheng, P.: Smog chamber study of the effects of NO_x and NH_3 on the formation of secondary organic aerosols and optical properties from photo-oxidation of toluene. Sci. Total Environ. **727**, 138632 (2020). https://doi.org/10.1016/j.scitotenv.2020.138632

Ren, Y., Grosselin, B., Daële, V., Mellouki, A.: Investigation of the reaction of ozone with isoprene, methacrolein and methyl vinyl ketone using the HELIOS chamber. Faraday Discuss. **200**, 289–311 (2017). https://doi.org/10.1039/c7fd00014f

Ridley, B.A., Grahek, F.E., Walega, J.G.: A small high-sensitivity, medium-response ozone detector suitable for measurements from light aircraft. J. Atmos. Oceanic Tech. **9**, 142–148 (1992). https://doi.org/10.1175/1520-0426(1992)009%3c0142:ashsmr%3e2.0.co;2

Roberts, G.C., Nenes, A.: A continuous-flow streamwise thermal-gradient CCN chamber for atmospheric measurements. Aerosol Sci. Technol. **39**, 206–221 (2005). https://doi.org/10.1080/027868290913988

Ródenas, M., Muñoz, A., Alacreu, F., Brauers, T., Dorn, H.-P., Kleffmann, J., Bloss, W.: Assessment of HONO measurements: The FIONA campaign at EUPHORE, pp. 45–58 (2013)

Rohrer, F., Bohn, B., Brauers, T., Brüning, D., Johnen, F.J., Wahner, A., Kleffmann, J.: Characterisation of the photolytic HONO-source in the atmosphere simulation chamber SAPHIR. Atmos. Chem. Phys. **5**, (2005a). https://doi.org/10.5194/acp-5-2189-2005

Rohrer, F., Bohn, B., Brauers, T., Brüning, D., Johnen, F.J., Wahner, A., Kleffmann, J.: Characterisation of the photolytic HONO-source in the atmosphere simulation chamber SAPHIR. Atmos. Chem. Phys. **5**, 2189–2201 (2005b). https://doi.org/10.5194/acp-5-2189-2005

Rossignol, S., Chiappini, L., Perraudin, E., Rio, C., Fable, S., Valorso, R., Doussin, J.F.: Development of parallel sampling and analysis for the elucidation of gas/particle partitioning of oxygenated semi-volatile organics: a limonene ozonolysis study. Atmos. Meas. Techn. Discuss. **5**, 1153–1231 (2012a)

Rossignol, S., Chiappini, L., Perraudin, E., Rio, C., Fable, S., Valorso, R., Doussin, J.F.: Development of a parallel sampling and analysis method for the elucidation of gas/particle partitioning of oxygenated semi-volatile organics: a limonene ozonolysis study. Atmos. Meas. Tech. **5**, 1459–1489 (2012b). https://doi.org/10.5194/amt-5-1459-2012

Ryerson, T.B., Williams, E.J., Fehsenfeld, F.C.: An efficient photolysis system for fast-response NO_2 measurements. J. Geophys. Res. Atmos. **105**, 26447–26461 (2000). https://doi.org/10.1029/2000JD900389

Sang, X.F., Gensch, I., Kammer, B., Khan, A., Kleist, E., Laumer, W., Schlag, P., Schmitt, S.H., Wildt, J., Zhao, R., Mungall, E.L., Abbatt, J.P.D., Kiendler-Scharr, A.: Chemical stability of levoglucosan: an isotopic perspective. Geophys. Res. Lett. **43**, 5419–5424 (2016). https://doi.org/10.1002/2016GL069179

Saunders, S.M., Jenkin, M.E., Derwent, R.G., Pilling, M.J.: Protocol for the development of the master chemical mechanism, MCM v3 (Part A): tropospheric degradation of non-aromatic volatile organic compounds. Atmos. Chem. Phys. **3**, 161–180 (2003). https://doi.org/10.5194/acp-3-161-2003

Savi, M., Kalberer, M., Lang, D., Ryser, M., Fierz, M., Gaschen, A., Rička, J., Geiser, M.: A novel exposure system for the efficient and controlled deposition of aerosol particles onto cell cultures. Environ. Sci. Technol. **42**, 5667–5674 (2008). https://doi.org/10.1021/es703075q

Schlosser, E., Bohn, B., Brauers, T., Dorn, H.-P., Fuchs, H., Häseler, R., Hofzumahaus, A., Holland, F., Rohrer, F., Rupp, L.O., Siese, M., Tillmann, R., Wahner, A.: Intercomparison of two hydroxyl radical measurement techniques at the atmosphere simulation chamber SAPHIR. J. Atmos. Chem. **56**, 187–205 (2007). https://doi.org/10.1007/s10874-006-9049-3

Schlosser, E., Brauers, T., Dorn, H.-P., Fuchs, H., Häseler, R., Hofzumahaus, A., Holland, F., Wahner, A., Kanaya, Y., Kajii, Y., Miyamoto, K., Nishida, S., Watanabe, K., Yoshino, A., Kubistin, D., Martinez, M., Rudolf, M., Harder, H., Berresheim, H., Elste, T., Plass-Dülmer, C., Stange, G.,

Schurath, U.: Formal blind intercomparison of OH measurements: results from the international campaign HOxComp. Atmos. Chem. Phys. **9**, 7923–7948 (2009). https://doi.org/10.5194/acp-9-7923-2009

Schnaiter, M., Järvinen, E., Vochezer, P., Abdelmonem, A., Wagner, R., Jourdan, O., Mioche, G., Shcherbakov, V.N., Schmitt, C.G., Tricoli, U., Ulanowski, Z., Heymsfield, A.J.: Cloud chamber experiments on the origin of ice crystal complexity in cirrus clouds. Atmos. Chem. Phys. **16**, 5091–5110 (2016). https://doi.org/10.5194/acp-16-5091-2016

Schwartz, R.E., Russell, L.M., Sjostedt, S.J., Vlasenko, A., Slowik, J.G., Abbatt, J.P.D., Macdonald, A.M., Li, S.M., Liggio, J., Toom-Sauntry, D., Leaitch, W.R.: Biogenic oxidized organic functional groups in aerosol particles from a mountain forest site and their similarities to laboratory chamber products. Atmos. Chem. Phys. **10**, 5075–5088 (2010). https://doi.org/10.5194/acp-10-5075-2010

Seakins, P.W.: A brief review of the use of environmental chambers for gas phase studies of kinetics, chemical mechanisms and characterisation of field instruments. In: Boutron, C. (ed.) Erca 9: From the Global Mercury Cycle to the Discoveries of Kuiper Belt Objects. EPJ Web of Conferences, E D P Sciences, Cedex A, pp. 143–163 (2010).

Smith, D.M., Fiddler, M.N., Sexton, K.G., Bililign, S.: Construction and characterization of an indoor smog chamber for measuring the optical and physicochemical properties of aging biomass burning aerosols. Aerosol Air Qual. Res. **19**, 467–483 (2019). https://doi.org/10.4209/aaqr.2018.06.0243

Spicer, C.W.: Smog chamber studies of nitrogen oxide (NO_x) transformation rate and nitrate precursor relationships. Environ. Sci. Technol. **17**, 112–120 (1983). https://doi.org/10.1021/es00108a010

Stedman, D.H.M., Morris, E.D., Jr., Daby, E.E., Niki, H., Weinstock, B.: The role of OH radicals in photochemical smog reactions. In: 160th National Meeting of the American Chemical Society, Chicago, Illinois (1970)

Stehle, R.L., Gertler, A.W., Katz, U., Lamb, D., Miller, D.F.: Cloud chamber studies of dark transformations of sulfur dioxide in cloud droplets. Atmos. Environ. **1967**(15), 2341–2352 (1981). https://doi.org/10.1016/0004-6981(81)90264-X

Stephens, E.R.: Long-path infrared spectrocopy for air pollution research. Appl. Spectrosc. **12**, 80–84 (1958)

Stockwell, W.R., Saunders, E., Goliff, W.S., Fitzgerald, R.M.: A perspective on the development of gas-phase chemical mechanisms for Eulerian air quality models. J. Air Waste Manag. Assoc. **70**, 44–70 (2020). https://doi.org/10.1080/10962247.2019.1694605

Stolzenburg, M.R., McMurry, P.H.: An ultrafine aerosol condensation nucleus counter. Aerosol Sci. Technol. **14**, 48–65 (1991). https://doi.org/10.1080/02786829108959470

Stolzenburg, D., Steiner, G., Winkler, P.M.: A DMA-train for precision measurement of sub-10 nm aerosol dynamics. Atmos. Meas. Tech. **10**, 1639–1651 (2017). https://doi.org/10.5194/amt-10-1639-2017

Storelvmo, T., Tan, I.: The Wegener-Bergeron-Findeisen process-its discovery and vital importance for weather and climate. Meteorol. Z. **24**, 455–461 (2015). https://doi.org/10.1127/metz/2015/0626

Taipale, R., Ruuskanen, T.M., Rinne, J., Kajos, M.K., Hakola, H., Pohja, T., Kulmala, M.: Technical note: quantitative long-term measurements of VOC concentrations by PTR-MS–measurement, calibration, and volume mixing ratio calculation methods. Atmos. Chem. Phys. **8**, 6681–6698 (2008). https://doi.org/10.5194/acp-8-6681-2008

Takekawa, H., Minoura, H., Yamazaki, S.: Temperature dependence of secondary organic aerosol formation by photo-oxidation of hydrocarbons. Atmos. Environ. **37**, 3413–3424 (2003)

Thalman, R., Baeza-Romero, M.T., Ball, S.M., Borrás, E., Daniels, M.J.S., Goodall, I.C.A., Henry, S.B., Karl, T., Keutsch, F.N., Kim, S., Mak, J., Monks, P.S., Muñoz, A., Orlando, J., Peppe, S., Rickard, A.R., Ródenas, M., Sánchez, P., Seco, R., Su, L., Tyndall, G., Vázquez, M., Vera, T., Waxman, E., Volkamer, R.: Instrument intercomparison of glyoxal, methyl glyoxal and NO_2 under simulated atmospheric conditions. Atmos. Meas. Tech. **8**, 1835–1862 (2015). https://doi.org/10.5194/amt-8-1835-2015

Thomas, M., France, J., Crabeck, O., Hall, B., Hof, V., Notz, D., Rampai, T., Riemenschneider, L., Tooth, O.J., Tranter, M., Kaiser, J.: The roland von glasow air-sea-ice chamber (RvG-ASIC): an experimental facility for studying ocean–sea-ice–atmosphere interactions. Atmos. Meas. Tech. **14**, 1833–1849 (2021). https://doi.org/10.5194/amt-14-1833-2021

Thrush, B.A., Wilkinson, J.P.T.: The rate of reaction of HO2 radicals with HO and with NO. Chem. Phys. Lett. **81**, 1–3 (1981). https://doi.org/10.1016/0009-2614(81)85314-6

Tiitta, P., Leskinen, A., Hao, L., Yli-Pirilä, P., Kortelainen, M., Grigonyte, J., Tissari, J., Lamberg, H., Hartikainen, A., Kuuspalo, K., Kortelainen, A.M., Virtanen, A., Lehtinen, K.E.J., Komppula, M., Pieber, S., Prévôt, A.S.H., Onasch, T.B., Worsnop, D.R., Czech, H., Zimmermann, R., Jokiniemi, J., Sippula, O.: Transformation of logwood combustion emissions in a smog chamber: formation of secondary organic aerosol and changes in the primary organic aerosol upon daytime and nighttime aging. Atmos. Chem. Phys. **16**, 13251–13269 (2016). https://doi.org/10.5194/acp-16-13251-2016

Tobo, Y., DeMott, P.J., Raddatz, M., Niedermeier, D., Hartmann, S., Kreidenweis, S.M., Stratmann, F., Wex, H.: Impacts of chemical reactivity on ice nucleation of kaolinite particles: a case study of levoglucosan and sulfuric acid. Geophys. Res. Lett. **39** (2012). https://doi.org/10.1029/2012GL053007

Tortajada-Genaro, L.-A., Borrás, E.: Temperature effect of tapered element oscillating microbalance (TEOM) system measuring semi-volatile organic particulate matter. J. Environ. Monit. **13**, 1017–1026 (2011). https://doi.org/10.1039/c0em00451k

Vanhanen, J., Mikkilä, J., Lehtipalo, K., Sipilä, M., Manninen, H.E., Siivola, E., Petäjä, T., Kulmala, M.: Particle size magnifier for nano-CN detection. Aerosol Sci. Technol. **45**, 533–542 (2011). https://doi.org/10.1080/02786826.2010.547889

Varma, R., Venables, D., Ruth, A., Heitmann, U., Schlosser, E., Dixneuf, S.: Long optical cavities for open-path monitoring of atmospheric trace gases and aerosol extinction. Appl. Opt. **48**, B159-171 (2009). https://doi.org/10.1364/ao.48.00b159

Varma, R.M., Ball, S.M., Brauers, T., Dorn, H.P., Heitmann, U., Jones, R.L., Platt, U., Pöhler, D., Ruth, A.A., Shillings, A.J.L., Thieser, J., Wahner, A., Venables, D.S.: Light extinction by secondary organic aerosol: an intercomparison of three broadband cavity spectrometers. Atmos. Meas. Tech. **6**, 3115–3130 (2013). https://doi.org/10.5194/amt-6-3115-2013

Vlasenko, A., Sjogren, S., Weingartner, E., Stemmler, K., Gäggeler, H.W., Ammann, M.: Effect of humidity on nitric acid uptake to mineral dust aerosol particles. Atmos. Chem. Phys. **6**, 2147–2160 (2006). https://doi.org/10.5194/acp-6-2147-2006

Voigtländer, J., Duplissy, J., Rondo, L., Kürten, A., Stratmann, F.: Numerical simulations of mixing conditions and aerosol dynamics in the CERN CLOUD chamber. Atmos. Chem. Phys. **12**, 2205–2214 (2012). https://doi.org/10.5194/acp-12-2205-2012

Wagner, V., Jenkin, M.E., Saunders, S.M., Stanton, J., Wirtz, K., Pilling, M.J.: Modelling of the photooxidation of toluene: conceptual ideas for validating detailed mechanisms. Atmos. Chem. Phys. **3**, 89–106 (2003). https://doi.org/10.5194/acp-3-89-2003

Wagner, R., Bunz, H., Linke, C., Möhler, O., Naumann, K.-H., Saathoff, H., Schnaiter, M., Schurath, U.: Chamber Simulations of Cloud Chemistry: The AIDA Chamber, Environmental Simulation Chambers: Application to Atmospheric Chemical Processes, Dordrecht, pp. 67–82 (2006)

Wagner, R., Möhler, O., Saathoff, H., Schnaiter, M., Leisner, T.: New cloud chamber experiments on the heterogeneous ice nucleation ability of oxalic acid in the immersion mode. Atmos. Chem. Phys. **11**, 2083–2110 (2011). https://doi.org/10.5194/acp-11-2083-2011

Wagner, R., Ajtai, T., Kandler, K., Lieke, K., Linke, C., Müller, T., Schnaiter, M., Vragel, M.: Complex refractive indices of Saharan dust samples at visible and near UV wavelengths: a laboratory study. Atmos. Chem. Phys. **12**, 2491–2512 (2012). https://doi.org/10.5194/acp-12-2491-2012

Wahner, A., Mentel, T.F., Sohn, M., Stier, J.: Heterogeneous reaction of N2O5 on sodium nitrate aerosol. J. Geophys. Res. Atmos. **103**, 31103–31112 (1998). https://doi.org/10.1029/1998JD100022

Wahner, A.: SAPHIR: simulation of atmospheric photochemistry in a large reaction chamber: a novel instrument. American Chemical Society 2002, U306–U306.

Wang, S.C., Flagan, R.C.: Scanning electrical mobility spectrometer. Aerosol Sci. Technol. **13**, 230–240 (1990). https://doi.org/10.1080/02786829008959441

Wang, J., Doussin, J.F., Perrier, S., Perraudin, E., Katrib, Y., Pangui, E., Picquet-Varrault, B.: Design of a new multi-phase experimental simulation chamber for atmospheric photosmog, aerosol and cloud chemistry research. Atmos. Meas. Tech. **4**, 2465–2494 (2011). https://doi.org/10.5194/amt-4-2465-2011

Wang, X., Liu, T., Bernard, F., Ding, X., Wen, S., Zhang, Y., Zhang, Z., He, Q., Lü, S., Chen, J., Saunders, S., Yu, J.: Design and characterization of a smog chamber for studying gas-phase chemical mechanisms and aerosol formation. Atmos. Meas. Tech. **7**, 301–313 (2014). https://doi.org/10.5194/amt-7-301-2014

Weinstock, B.: Carbon monoxide: residence time in the atmosphere. Science **166**, 224–225 (1969). https://doi.org/10.1126/science.166.3902.224

Wex, H., Petters, M.D., Carrico, C.M., Hallbauer, E., Massling, A., McMeeking, G.R., Poulain, L., Wu, Z., Kreidenweis, S.M., Stratmann, F.: Towards closing the gap between hygroscopic growth and activation for secondary organic aerosol: Part 1–evidence from measurements. Atmos. Chem. Phys. **9**, 3987–3997 (2009). https://doi.org/10.5194/acp-9-3987-2009

Wex, H., DeMott, P.J., Tobo, Y., Hartmann, S., Rösch, M., Clauss, T., Tomsche, L., Niedermeier, D., Stratmann, F.: Kaolinite particles as ice nuclei: learning from the use of different kaolinite samples and different coatings. Atmos. Chem. Phys. **14**, 5529–5546 (2014). https://doi.org/10.5194/acp-14-5529-2014

Whalley, L.K., Stone, D., Dunmore, R., Hamilton, J., Hopkins, J.R., Lee, J.D., Lewis, A.C., Williams, P., Kleffmann, J., Laufs, S., Woodward-Massey, R., Heard, D.E.: Understanding in situ ozone production in the summertime through radical observations and modelling studies during the Clean air for London project (ClearfLo). Atmos. Chem. Phys. **18**, 2547–2571 (2018). https://doi.org/10.5194/acp-18-2547-2018

White, J.U.: Long optical paths of large aperture. J. Opt. Soc. Am. **32**, 285–288 (1942). https://doi.org/10.1364/josa.32.000285

White, J.U.: Very long optical paths in air. J. Opt. Soc. Am. **66**, 411–416 (1976). https://doi.org/10.1364/josa.66.000411

Wimmer, D., Lehtipalo, K., Franchin, A., Kangasluoma, J., Kreissl, F., Kürten, A., Kupc, A., Metzger, A., Mikkilä, J., Petäjä, T., Riccobono, F., Vanhanen, J., Kulmala, M., Curtius, J.: Performance of diethylene glycol-based particle counters in the sub-3 nm size range. Atmos. Meas. Tech. **6**, 1793–1804 (2013). https://doi.org/10.5194/amt-6-1793-2013

Winer, A.M., Graham, R.A., Doyle, J.G., Bekowies, P.J., Mac Affee, J.M., Pitts, J.N.: An evacuable environmental chamber and solar simulator facility for the study of atmospheric photochemistry. In: Pitts, J.N., Metcalf, R.L., Grosjean, D. (eds.) Advances in Environmental Science and Technology, pp. 461–511. Wiley, New-York (1980)

Wisthaler, A., Apel, E.C., Bossmeyer, J., Hansel, A., Junkermann, W., Koppmann, R., Meier, R., Müller, K., Solomon, S.J., Steinbrecher, R., Tillmann, R., Brauers, T.: Technical note: intercomparison of formaldehyde measurements at the atmosphere simulation chamber SAPHIR. Atmos. Chem. Phys. **8**, 2189–2200 (2008). https://doi.org/10.5194/acp-8-2189-2008

Wu, C.H., Japar, S.M., Niki, H.: Relative reactivities of ho-hydrocarbon reactions from smog reactor studies. J. Environ. Sci. Health Part A Environ. Sci. Eng. **11**, 191–200 (1976). https://doi.org/10.1080/10934527609385765

Wu, S., Lü, Z., Hao, J., Zhao, Z., Li, J., Takekawa, H., Minoura, H., Yasuda, A.: Construction and characterization of an atmospheric simulation smog chamber. Adv. Atmos. Sci. **24**, 250–258 (2007). https://doi.org/10.1007/s00376-007-0250-3

Wu, C., Pullinen, I., Andres, S., Carriero, G., Fares, S., Goldbach, H., Hacker, L., Kasal, T., Kiendler-Scharr, A., Kleist, E., Paoletti, E., Wahner, A., Wildt, J., Mentel, T.F.: Impacts of soil moisture on de novo monoterpene emissions from European beech, Holm Oal, Scots Pine, and Norway Spruce. Biogeosciences **12**, 177–191 (2015). https://doi.org/10.5194/bg-12-177-2015

Wyche, K.P., Ryan, A.C., Hewitt, C.N., Alfarra, M.R., McFiggans, G., Carr, T., Monks, P.S., Smallbone, K.L., Capes, G., Hamilton, J.F., Pugh, T.A.M., MacKenzie, A.R.: Emissions of biogenic volatile organic compounds and subsequent photochemical production of secondary organic aerosol in mesocosm studies of temperate and tropical plant species. Atmos. Chem. Phys. **14**, 12781–12801 (2014). https://doi.org/10.5194/acp-14-12781-2014

Yli-Pirilä, P., Copolovici, L., Kännaste, A., Noe, S., Blande, J.D., Mikkonen, S., Klemola, T., Pulkkinen, J., Virtanen, A., Laaksonen, A., Joutsensaari, J., Niinemets, Ü., Holopainen, J.K.: Herbivory by an outbreaking moth increases emissions of biogenic volatiles and leads to enhanced secondary organic aerosol formation capacity. Environ. Sci. Technol. **50**, 11501–11510 (2016). https://doi.org/10.1021/acs.est.6b02800

Zádor, J., Turányi, T., Wirtz, K., Pilling, M.J.: Measurement and investigation of chamber radical sources in the European Photoreactor (EUPHORE). J. Atmos. Chem. **55**, 147–166 (2006). https://doi.org/10.1007/s10874-006-9033-y

Zhang, J., Huff Hartz, K.E., Pandis, S.N., Donahue, N.M.: Secondary organic aerosol formation from limonene ozonolysis: homogeneous and heterogeneous influences as a function of NO_x. J. Phys. Chem. A **110**, 11053–11063 (2006). https://doi.org/10.1021/jp062836f

Zhang, H., Hu, D., Chen, J., Ye, X., Wang, S.X., Hao, J.M., Wang, L., Zhang, R., An, Z.: Particle size distribution and polycyclic aromatic hydrocarbons emissions from agricultural crop residue burning. Environ. Sci. Technol. **45**, 5477–5482 (2011a). https://doi.org/10.1021/es1037904

Zhang, Q., Jimenez, J.L., Canagaratna, M.R., Ulbrich, I.M., Ng, N.L., Worsnop, D.R., Sun, Y.: Understanding atmospheric organic aerosols via factor analysis of aerosol mass spectrometry: a review. Anal. Bioanal. Chem. **401**, 3045–3067 (2011b). https://doi.org/10.1007/s00216-011-5355-y

Zhang, H., Worton, D.R., Lewandowski, M., Ortega, J., Rubitschun, C.L., Park, J.-H., Kristensen, K., Campuzano-Jost, P., Day, D.A., Jimenez, J.L., Jaoui, M., Offenberg, J.H., Kleindienst, T.E., Gilman, J., Kuster, W.C., de Gouw, J., Park, C., Schade, G.W., Frossard, A.A., Russell, L., Kaser, L., Jud, W., Hansel, A., Cappellin, L., Karl, T., Glasius, M., Guenther, A., Goldstein, A.H., Seinfeld, J.H., Gold, A., Kamens, R.M., Surratt, J.D.: Organosulfates as tracers for secondary organic aerosol (SOA) formation from 2-methyl-3-buten-2-ol (MBO) in the atmosphere. Environ. Sci. Technol. **46**, 9437–9446 (2012). https://doi.org/10.1021/es301648z

Zhang, Q., Xu, Y., Jia, L.: Secondary organic aerosol formation from OH-initiated oxidation of m-xylene: effects of relative humidity on yield and chemical composition. Atmos. Chem. Phys. **19**, 15007–15021 (2019). https://doi.org/10.5194/acp-19-15007-2019

Zhao, D.F., Buchholz, A., Mentel, T.F., Müller, K.P., Borchardt, J., Kiendler-Scharr, A., Spindler, C., Tillmann, R., Trimborn, A., Zhu, T., Wahner, A.: Novel method of generation of $Ca(HCO_3)_2$ and $CaCO_3$ aerosols and first determination of hygroscopic and cloud condensation nuclei activation properties. Atmos. Chem. Phys. **10**, 8601–8616 (2010). https://doi.org/10.5194/acp-10-8601-2010

Zhao, D.F., Buchholz, A., Tillmann, R., Kleist, E., Wu, C., Rubach, F., Kiendler-Scharr, A., Rudich, Y., Wildt, J., Mentel, T.F.: Environmental conditions regulate the impact of plants on cloud formation. Nat. Commun. **8**, 14067 (2017). https://doi.org/10.1038/ncomms14067

Zhao, D., Schmitt, S.H., Wang, M., Acir, I.H., Tillmann, R., Tan, Z., Novelli, A., Fuchs, H., Pullinen, I., Wegener, R., Rohrer, F., Wildt, J., Kiendler-Scharr, A., Wahner, A., Mentel, T.F.: Effects of NO_x and SO_2 on the secondary organic aerosol formation from photooxidation of α-pinene and limonene. Atmos. Chem. Phys. **18**, 1611–1628 (2018). https://doi.org/10.5194/acp-18-1611-2018

Zhou, L., Wang, W., Gai, Y., Ge, M.: Knudsen cell and smog chamber study of the heterogeneous uptake of sulfur dioxide on Chinese mineral dust. J. Environ. Sci. **26**, 2423–2433 (2014). https://doi.org/10.1016/j.jes.2014.04.005

Chapter 2
Physical and Chemical Characterization of the Chamber

Rami Alfarra, Marie Camredon, Mathieu Cazaunau,
Jean-François Doussin, Hendrik Fuchs, Spiro Jorga, Gordon McFiggans,
Mike J. Newland, Spyros Pandis, Andrew R. Rickard, and Harald Saathoff

Abstract In order to perform experiments in the chamber, characterization of physical properties is essential for the evaluation and interpretation of experiments. In this chapter, recommendations are given how to measure physical parameters such as temperature and pressure. For photochemistry experiments, knowledge of the radiation either provided by the sun or lamps is key to calculate photolysis frequencies. Standard protocols are described how to validate the calculation of the radiation inside the chamber using actinometry experiments. In addition, the characterization of loss processes for gas-phase species as well as for aerosol is discussed. Reference experiments can be used to test the state of the chamber. Different types of reference experiments focusing on gas-phase photo-oxidation experiments are recommended and described in detail in this chapter.

R. Alfarra · G. McFiggans
University of Manchester, Manchester, UK
e-mail: rami.alfarra@manchester.ac.uk

G. McFiggans
e-mail: g.mcfiggans@manchester.ac.uk

M. Camredon
Université Paris Est Créteil, Créteil, France
e-mail: marie.camredon@lisa.ipsl.fr

M. Cazaunau · J.-F. Doussin
Centre National de la Recherche Scientifique, Paris, France
e-mail: mathieu.cazaunau@lisa.ipsl.fr

J.-F. Doussin
e-mail: jean-francois.doussin@lisa.ipsl.fr

H. Fuchs (✉)
Forschungszentrum Jülich, Jülich, Germany
e-mail: h.fuchs@fz-juelich.de

S. Jorga
Carnegie Mellon Institute, Pittsburgh, USA
e-mail: sjorga@andrew.cmu.edu

M. J. Newland
University of York, Heslington, UK
e-mail: mike.newland@york.ac.uk

© The Author(s) 2023
J.-F. Doussin et al. (eds.), *A Practical Guide to Atmospheric Simulation Chambers*,
https://doi.org/10.1007/978-3-031-22277-1_2

2.1   Measurements of Temperature, Pressure, and Humidity

Temperature, pressure, and humidity are basic parameters required for the interpretation of almost any experiment carried out in an atmospheric simulation chamber. This is obvious, for example, in cloud studies where small changes in temperature and corresponding relative humidity can lead to cloud activation of aerosol particles, but also for chemical reaction kinetics for which reaction rates can have strong pressure and temperature dependencies. Therefore, we briefly summarize some recommendations on how to measure these parameters in atmospheric simulation chambers. The quality and traceability of such parameters are becoming increasingly important not only allowing for better comparability of experimental results, but especially if data will be used in atmospheric measurement networks like ACTRIS where all data require traceable quality standards. Recently, the European metrological institutions have addressed the issue of traceability. A consortium of national laboratory developed metrological methods for improving atmospheric measurements of pressure, temperature, humidity and airspeed has been carried out in the EURAMET project METEOMET. These methods include corresponding laboratory methods and traceability chains, which are also useful for simulation chambers and are summarized in the METEOMET project report (METEOMET 2020). Measurement procedures, standard operating procedures, good laboratory practices or definitions of traceability chains have been defined by the World Meteorological Organisation (WMO 2018) and the National Institute of Standards (NIST 2019).

   Measuring temperatures can be achieved by placing thermocouples (e.g., type J or K), resistant sensors (e.g., PT100 with four wire technique), ultrasonic anemometers or fibre optic sensors (e.g., if electric fields could interfere) at representative locations inside atmospheric simulation chambers. For the selection of the appropriate sensor, the measurement range, temperature, precision, accuracy and time resolution of the sensors required for the different purposes need to be considered. In cases where the simulation chamber is exposed to intense light radiation, the sensor needs to be protected e.g., by a shading cover. If the sensors are exposed to condensable compounds, latent heat release should be considered, especially for fast sensors with low heat capacities. Problems associated with condensation of water can be reduced by coating the sensor with inert e.g., polyfluorinated greases. Furthermore, potential impacts of sensor aging should be avoided by periodic (e.g., annual) calibration. This can be achieved for example in a temperature-controlled liquid bath, in which

S. Pandis · A. R. Rickard
Foundation for Research and Technology Hellas, Heraklion, Greece
e-mail: spyros@chemeng.upatras.gr

A. R. Rickard
e-mail: andrew.rickard@york.ac.uk

H. Saathoff
Karlsruhe Institute of Technology, Karlsruhe, Germany
e-mail: harald.saathoff@kit.edu

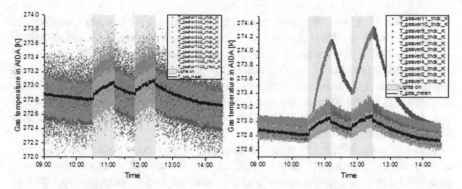

Fig. 2.1 Gas temperatures measured inside the AIDA aerosol and cloud chamber of KIT using chains of thermocouples (Ni–CrNi). Sensors are placed along the horizontal (left) and vertical (right) axis of the cylindrically shaped chamber. The initial temperature distribution is disturbed by switching on a LED light source on top of the chamber. Figure by Harald Saathoff ©, KIT

measurements are compared to those by certified reference sensors which are traceable to national standards. An example overview of potential temperature sensors is given e.g., by Lake Shore Incorporated. Calibrations should include the complete sensor chain including the same wiring as during chamber operation. An example of a temperature measurement inside the AIDA simulation chamber of KIT is shown in Fig. 2.1.

The type of sensor (thermocouple Ni–Cr–Ni) used in the AIDA chamber has a high precision and accuracy at a time resolution of seconds can be achieved. In addition to the overall temperature increase due to the illumination, several temperature sensors show impact of direct radiative heating by an average 0.05 K

The slightly higher temperatures measured by the two sensors (No. 10 and 11) that are placed at the horizontal (left) and a vertical (right) positions of the cylindrically shaped chamber indicate that warm air is trapped at the top of the vessel.

Measuring absolute or differential pressure for the atmospheric pressure range can be done with various types of sensors, which will not be reviewed here. An overview of potential pressure sensors is given for example by Avnet Inc. Some of the most robust and stable sensors are based on measuring changing capacitance (e.g., MKS Baratron). This type of sensor is insensitive to the specific gas mixture, can be temperature-stabilized for high precision measurements, and add typically just a heated stainless steel surface to the simulation chamber. The nature of the chamber environment means that the pressure should be uniform throughout its volume and therefore multiple pressure measurements are not required for most applications.

Measuring absolute or relative humidity in an atmospheric simulation chamber may require different approaches. If temperature and water concentrations are measured, the relative humidity can be calculated using the water vapour pressure over liquid water or ice. For this calculation, the corresponding vapour pressure formulations by Murphy and Koop (2005) are recommended. For temperatures below, 200 K the results by Nachbar et al. (2018a, b) should be used.

For atmospheric measurements often thin-film capacitive humidity sensors are applied. In chambers that work with atmospheric concentrations of reactive species such as the SAPHIR chamber of Forschungszentrum Jülich they offer precise and accurate measurements of relative humidity. However, this type of sensor is not recommended, if high concentrations of oxidizing reactants can get in contact with the sensors as chemical reactions may destroy the thin-film polymer sensors. In this case, metal oxide sensors can be used but they can also suffer from interaction with reactive or condensable compounds.

Absolute water mixing ratios can be measured by dew point mirror sensors which offer an inert e.g., rhodium or gold surface to the chamber contents. However, a successful dew point measurement requires that the major condensing species in the chamber is water and that the mirror is not contaminated e.g., with hygroscopic coatings or particles. Another advantage of the dew point mirror sensors is that they do not need a calibration as long as their temperature measurement is accurate. Among the various dew point mirror instruments several offer traceability to national standards e.g., via transfer standards at the manufacturer (e.g., MBW Calibration Ltd.).

If the sensor cannot be placed inside the chamber and a sampling tube must be used. The sampling tube may require heating to avoid water condensation e.g., if the temperature between the chamber and the instrument is varying. For measuring low water concentrations stainless steel tubing can be used but Teflon tubing should be avoided as it shows memory effects.

If the humidity inside a simulation chamber with condensed water (e.g., cloud droplets or water containing aerosol particles) is to be measured, spectroscopic methods such as FTIR or tuneable diode laser spectroscopy (TDLS) can be useful tools to obtain the condensed water (liquid and ice) and water vapour content. However, for each of these methods the background water concentration e.g., in the spectrometer or transfer optics needs to be treated carefully. An overview of several atmospheric hygrometers, their performance and potential connection to a simulation chamber, is given by Fahey et al. (2014). Tuneable diode laser spectroscopy offers fast and direct humidity measurement with good accuracy, if optical paths of sufficient lengths are available and even allows determination of water isotopes. Another very sensitive absorption method is the cavity ring-down spectroscopy (CRDS). Commercial instruments detecting water vapour in the infrared by CRDS are for example available from Picarro Inc. These instruments do not require calibration. Interferences can occur, if water vapor absorption lines overlap with absorption lines of trace gases that are present in a specific experiment.

It is obvious that the accuracy or precision needed for a certain variable depends on the application. If, for example, the relative humidity is required with an accuracy of 2% at 293 K the temperature needs to be measured with an accuracy of 0.16 K and the water vapour pressure has to be measured with an accuracy of 1%.

2.2 Determination of the Mixing Time and Dilution Rates

Considering the rate of Brownian diffusion of gases, mixing is often wrongly considered as a non-critical characteristic of chamber installation. On the contrary, because of the size of simulation chambers, reaching sufficient chemical homogeneity of the reactive mixture often takes a long time with respect to the rates of many chemical reactions occurring in the atmosphere.

The mixing time of air is a key parameter of a simulation chamber installation that will strongly impact the data analysis, because it is not reasonable to interpret data at a time resolution shorter than the mixing time. For gas phase chemistry, as most of chemical kinetic rate constants are directly proportional to the reactant concentrations, inhomogeneous concentrations can lead to false experimental data and/or strongly complicate the evaluation of experiments (Ibrahim et al. 1987). For particle phase studies, as condensation of semi-volatiles is highly non-linear with concentration, insufficient mixing can lead an incorrect estimation of secondary organic aerosol yields caused by local supersaturation (Schütze and Stratmann 2008).

Typical mixing times in atmospheric simulation chambers fall in the range of minutes, for example 1 min in the CESAM chamber with 4.2 m^3 (Wang et al. 2011) and 2 min in the SAPHIR chamber with 270 m^3 (Rohrer et al. 2005). Mixing is often achieved by fans made of inert material operated inside the chamber.

Schütze and Stratmann (2008) analysed the effect of operating one or two fans on the homogeneity of particle concentrations in a cylindrically shaped chamber (12.4 m^3 volume) using computational fluid dynamics. They found that inhomogeneities can also be induced by fans in the area, where the air is accelerated. Therefore, it is crucial to carefully choose locations of sampling points for instruments to not be affected by local inhomogeneities.

Apart from the impact on the bulk simulated atmosphere homogeneity, the mixing of air in a simulation chamber also impacts the exchange of energy (Voigtländer et al. 2012) and interactions of matter with the walls. Strong mixing not only increases the level of turbulence in the chamber potentially leading to non-linear effects (Ibrahim et al. 1987), but can also increase the wall loss rate of semi-volatile compounds by increasing the rate of collisions with the chamber wall. Furthermore, in Teflon film chambers, turbulent mixing may lead to movements of the chamber film that can favour the build-up of electrostatic charges and thereby increase the probability that particles are lost on the Teflon film (Wang et al. 2018a, b). Therefore, there is an optimum compromise between homogeneity and wall loss with respect to mixing that is specific for the shape and volume of each chamber.

To determine the mixing time in a simulation chamber, a non-reactive gaseous species that can be measured with a high time resolution can be injected at a single point in the chamber. The species needs to be detected at several positions in the chamber. It is not recommended to use spatially integrated measurements such as in situ spectrometric techniques as they often tend to underestimate the mixing time by spatially averaging the concentration. Moving the sampling point in repeated experiments, simultaneous detection at several points in one experiment and varying

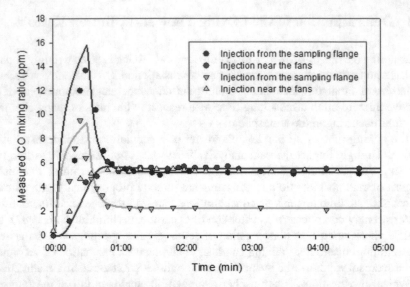

Fig. 2.2 Time series of CO concentrations sampling at one location after injection of CO at various injection points in the CESAM chamber. Lines are the results of modelling the mixing in the chamber. (Reused with permission from Wang et al. (2011) Open access under a CC BY 3.0 license, https://creativecommons.org/licenses/by/3.0/)

the point of injection further increases the precision, with which the mixing time can be determined. As an example, Fig. 2.2 shows the time series of CO concentration detected by an infrared gas filter correlation CO monitor at various sampling points after a point injection in the CESAM chamber (Wang et al. 2011).

Numerical modelling can further help understanding the mixing in the chamber. Wang et al. (2011) divided the volume of the CESAM chamber into 4137 cubic cells each of which has a volume of 1 L and set up a multi-box-model. Time series of trace concentrations were modelled for each box using a kinetic solver (Facsimile™ software package—Curtis 1979) with a non-zero initial concentration in the box in which the injection was located. Assuming isotropic mixing, the exchange rate was adjusted to match the measured concentration time series (Fig. 2.2). For experiments, in which the fan was operated at full speed, these calculations gave a first-order exchange rate of (3 ± 0.5) s^{-1}, which corresponds to the average speed of the gases of 0.3 m s^{-1}. This set-up of model can also be used for the analysis of experiments with complex chemistry as chemical reactions can be added. However, this type of model is not suitable to describe microphysics of the chamber atmosphere. Another approach to gain knowledge of the fluid dynamics in the chamber, is to perform computational fluid dynamics (CFD) calculations. Schütze and Stratmann (2008) used such simulations (FLUENT model, ANSYS Inc., Canonsburg, PA, USA) for a chamber with a volume 12.3 m^3 that was divided into 7714 cells. By combining the simulation with the Fine Particle Model (FPM, Particle Dynamics GmbH, Leipzig, Germany, Wilck et al. 2002), they performed simulations of the growth of ammonium-sulphate particles in humid air at room temperature for experiments, in which clouds were generated by gas expansion. Similar calculations with the FLUENT model were performed by Voigtländer et al. (2012) for the CLOUD chamber at CERN that has a volume of 26.1 m^3. The authors found that two fans and sufficiently high fan speeds were necessary for a homogeneous mixing of particles and gaseous species.

Trace gases and particles in experiments in atmospheric simulation chamber that have a fixed volume or are kept at a constant pressure are typically diluted over the course of an experiment due to the need to replenish the air that is lost by the consumption of sampling instruments are leakages. Exceptions are chambers where the volume can reduce over the course of an experiment (Carter et al. 2005).

The rate of dilution typically scales with the volume of the chamber. The large EUPHORE (volume 200 m^3) and SAPHIR (volume 270 m^3) outdoor chambers consist of Teflon film that is kept slightly over-pressurized compared to ambient pressure. The replenishment flow to maintain the pressure leads to a dilution of trace gas and particle concentrations at a low percentage range per hour (Becker 1996; Karl et al. 2004). In the smaller steel CLOUD chamber (volume 26.1 m^3) the dilution is typically higher with 6–10% per hour (Hoyle et al. 2016).

Precise and accurate knowledge of the dilution rate is key in the data analysis and modelling of experiments. For this purpose, two strategies can be employed that can be simultaneously applied. In many chambers the flow rates of the replenishment flow are monitored by a mass flow controller from which the dilution rate can that directly calculated, if the volume of the chamber is known (Hoyle et al. 2016; Karl et al. 2004; Wang et al. 2011). An alternative approach is to monitor the concentration of a chemically inert gas that is injected at the start of the experiment. The dilution rate can be calculated from the continuous measurement of its concentration as it decreases over the course of the experiment solely due to dilution. For a chamber equipped with an FTIR spectrometer, SF6 is often used because of its strong infrared absorption lines which gives a clear spectral fingerprint. SF6 can also be monitored with gas chromatography equipped with electron capture detector (GC-ECD) (Fry et al. 2011). For chambers equipped with a Proton-Transfer-Reaction-Mass-Spectrometry (PTR-MS) instrument, hexafluorobenzene (HFB) is a suitable dilution tracer (Hunter et al. 2014). Small alkanes such as ethane or cyclohexane are less inert but measurable with gas chromatography with a flame ionization detector (GC-FID) and have been also used in some studies (Metcalf et al. 2013). However, care has to be taken that the chemistry of these tracers does not disturb the experiment. CO_2 which can be precisely measured by cavity ring-down spectroscopy (CRDS) has been used in experiments in the SAPHIR chamber. This is only applicable, if the replenishment flow is free of CO_2 and chemical production in the experiment is negligible. This is typically the case for experiments in the SAPHIR chamber, because air is produced from liquid nitrogen of oxygen and trace gas concentrations are within the range of ambient concentrations.

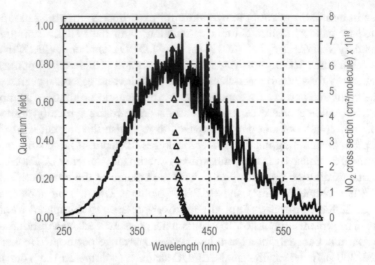

Fig. 2.3 Quantum yield (triangle) and absorption spectrum of NO$_2$ (line); spectral resolution 0.2–0.4 nm (Data from Burrows et al. 1998)

2.3 Determination of Photolysis Frequencies

Photolysis frequencies are important parameters for a quantitative understanding and modelling of photochemical processes in the atmosphere as well as in simulation chambers. For atmospheric measurements of photolysis frequencies, a range of suitable radiometric instruments have been developed and deployed (e.g., Hofzumahaus 2006; Hofzumahaus et al. 2002; Shetter and Müller 1999; Shetter et al. 2003). The most versatile method is spectroradiometry which can monitor spectral actinic flux densities with high time resolution as well as high spectral resolution in the relevant solar spectral range. Thoroughly calibrated instruments provide accurate photolysis frequencies for any photolysis process if the relevant molecular parameters of the molecule X—absorption cross sections σ and quantum yields ϕ—are known.

$$j(X) = \int \sigma(X)\phi F_\lambda(\lambda)d\lambda \qquad (2.3.1)$$

Recommendations of these parameters can be found in the literature (Fig. 2.3) (Atkinson et al. 2004; Burkholder et al. 2020; Keller-Rudek et al. 2013).

Considering possible radiation inhomogeneity, spectroradiometric measurements are often not able to provide a satisfying absolute light intensity estimation for the whole chamber, especially in indoor chambers. Chemical actinometry is therefore often used to determine the mean light intensity. Chemical actinometry is an independent method to determine photolysis frequencies by monitoring the change of the chemical composition induced by radiation. For atmospheric measurements of photolysis frequencies, chemical actinometry has rarely been applied because the experimental setup is comparatively extensive and process specific. Moreover, chemical actinometry has mostly been confined to the determination of photolysis frequencies $j(NO_2)$ and $j(O^1D)$. Nevertheless, chemical actinometry plays an important role in the validation of radiometric techniques (e.g., Hofzumahaus et al. 2004; Kraus et al. 2000; Shetter et al. 2003) and it is an integrated measure of the UV light intensity in simulation chambers (Bohn et al. 2005).

Most simulation chambers are commonly equipped with instruments for the detection of nitric oxide (NO), nitrogen dioxide (NO_2) and ozone (O_3) that are suitable to perform $j(NO_2)$ and $j(O^1D)$ actinometry experiments. *In-situ* spectrometric techniques such as Differential Optical Absorption Spectroscopy (DOAS), Tunable Diode Laser Spectroscopy (TDLAS) or Fourier-Transfer Infrared (FTIR) spectroscopy are highly recommended as they provide unambiguous and direct quantification of these species and can give integrated values over a large fraction of the chamber volume. On-line gas analysers for ozone (absorption technique) and NO_x (chemiluminescence technique) can also be used with confidence provided that the chamber is well mixed and that care is taken to ensure that sampling is performed at a point that is representative of the whole chamber. For NO_2 detection by chemiluminescence a photolytic conversion of NO_2 to NO is recommended, because molybdenum-converters can be affected by other species such as HONO, HNO_3 and organic nitrates (Dunlea et al. 2007). However, photolytic conversion is potentially affected by a negative interference at high VOC levels due to the efficient NO/NO_2-conversion through peroxy radicals formed in the photolysis of photolabile VOCs (Villena et al. 2012). The choice of instrumentation therefore depends on the type of experiment. As an alternative to chemiluminescence instruments cavity-based absorption methods can be used for the direct detection of NO_2.

Dedicated experiments under suitable conditions are required to obtain useful results. Moreover, a determination of $j(NO_2)$ or $j(O^1D)$ by chemical actinometry alone is not sufficient to characterize the photolytic properties of a simulation chamber. Rather a combination of techniques is required: spectroradiometry can provide actinic flux density spectra of the light source (artificial or the sun) which can then be scaled up or down to match the photolysis frequencies determined by chemical actinometry. Therefore, actinometry can be used to track the changing chamber radiometric conditions over time. Here, we focus on $j(NO_2)$ actinometry.

For sunlit chambers the radiation field inside can become inhomogeneous by shadows cast by structural elements of the chamber or instrumental set-ups, the influence of chamber walls through reflection, scattering and absorption, as well as by internal reflections. A radiometric point measurement inside the chamber, even with an ideal 4π sr field of view, may therefore not be representative for the entire chamber volume, an effect which is irrelevant for most other atmospheric measurements. Moreover, the chamber effects will depend on atmospheric conditions, most importantly on solar zenith and azimuth angles and the presence or absence of clouds. Therefore, sunlit chambers require both a continuous monitoring by radiometric devices, ideally a spectroradiometer (inside or outside the chamber) and suitable corrections accounting for specific *in-situ* chamber effects. These potentially time-dependent corrections can be determined relatively easily by chemical actinometry.

For chambers using artificial light sources the situation is simpler at a first glance. Unless lamps are dimmed or switched on and off, the radiation field can be considered independent of time (except for lamp aging effects on longer timescales). On the other hand, spatial inhomogeneity can be more pronounced compared to sunlit chambers dependent on the illumination technique, e.g., the use of single lamps, a collimated beam in tube-shaped chambers or all-around systems of tubular fluorescent lamps. In the best case, it is sufficient to occasionally record lamp spectra at a selected point within the chamber and perform chemical actinometry to derive adequate scaling factors. In contrast to sunlit chambers, spectral irradiance measurements are suitable as well. However, if different lamp types are combined, it is necessary to determine the corresponding photolysis frequencies separately.

Gradients in the radiation fields can result in gradients in short-lived species concentrations, which need to be considered for any chamber. Active mixing is a means to reduce such concentration gradients. In the following, we assume that trace gas concentrations are homogeneous, so that the chemical composition probed at any location is representative for the entire chamber. In this case, results represent a chamber-mean of actinic radiation.

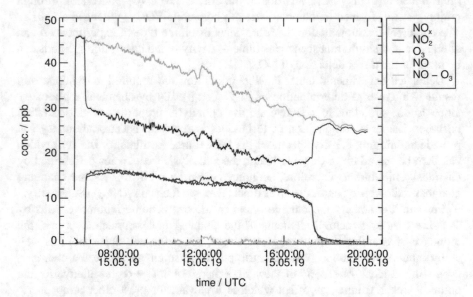

Fig. 2.4 Concentrations of NO_x, NO_2, NO, O_3 and the difference $NO-O_3$ during an actinometric experiment in SAPHIR at Forschungszentrum Jülich on a mostly clear-sky day. The chamber roof was opened around 06:30 and closed shortly after 17:30

2.3.1 Chemical Actinometry in Air

The photolysis of NO_2 is mostly UV-A driven and leads to the formation of nitric oxide (NO) and ground state oxygen atoms ($O(^3P)$):

$$NO_2 + h\nu \rightarrow NO + O(^3P) \quad (\lambda < 420\,nm) \tag{R2.3.1}$$

Under tropospheric conditions, the photolysis is followed by a fast and quantitative formation of ozone (O_3) in the reaction with an oxygen molecule (O_2):

$$O(^3P) + O_2 + M \rightarrow O_3 + M \tag{R2.3.2}$$

M is a third-body reaction partner. A chemical actinometer for atmospheric measurements of $j(NO_2)$ typically consists of a quartz flow-tube where a known concentration of NO_2 (mixing ratio in the ppm range) in synthetic air or O_2 is exposed to sunlight for a short period of time (ca. 1 s) (Shetter et al. 2003). The NO concentration produced during the exposure time is then a direct measure for $j(NO_2)$:

$$j(NO_2) = 1/[NO_2]\,d[NO]/dt \approx 1/[NO_2]\Delta[NO]/\Delta t \tag{2.3.2}$$

The presence of O_2 avoids interferences from the fast reaction $O(^3P) + NO_2 \rightarrow NO + O_2$. Moreover, at short exposure times the influence of the comparatively slow $NO + O_3$ back-reaction is negligible:

$$NO + O_3 \rightarrow NO_2 + O_2 \tag{R2.3.3}$$

For simulation chambers this concept of actinometry is not applicable because short exposure times are not feasible. However, under typical simulation chamber conditions, Reactions R2.3.1–R2.3.3 lead rapidly to a photochemical equilibrium or photostationary state (PSS). The relaxation time constant of this equilibrium is on the order of minutes depending on the values of $j(NO_2)$ and concentrations of trace gases. In the atmosphere, this equilibrium can be strongly affected by the presence of peroxyl radicals (HO_2 and RO_2), which also convert NO into NO_2 without consuming O_3. The so-called Leighton ratio $\varphi = j(NO_2)[NO_2]/(k_3[NO][O_3])$ (Leighton 1961) is a common measure for the deviation from the purely NO_x/O_3 determined equilibrium owing to peroxyl radical perturbations. On the other hand, in the absence of interfering reactions, the Leighton ratio is unity under steady-state conditions. Accordingly, $j(NO_2)$ can be calculated from the equilibrium concentrations of O_3, NO_2 and NO, and the (temperature dependent) rate constant of the $NO + O_3$ (2.07×10^{-12} exp($-1400/T$), IUPAC) Reaction R2.3.3:

$$j(NO_2) = k_{2.3.3}[NO][O_3]/[NO_2] \tag{2.3.3}$$

Figure 2.4 shows an example of an actinometric $j(NO_2)$ experiment in the sunlit simulation chamber SAPHIR at Forschungszentrum Jülich. Around 45 ppbv of NO_2 was injected into the dark chamber shortly before 6:00. In the illuminated chamber, a fast decrease of NO_2, and a corresponding rapid increase of NO and O_3 concentrations were observed. The NO_x (=NO + NO_2) concentration remained nearly constant during this quick initial adjustment of the photochemical equilibrium, as expected. The chamber roof remained open for approximately 12 h. During this time, the NO_x concentration slowly decreased mainly caused by dilution. However, this decrease is slow compared to the relaxation time constant of the photochemical equilibrium. Accordingly, at any time NO_2, NO and O_3 concentrations are in a steady-state equilibrium and $j(NO_2)$ can be calculated according to Eq. (2.3.3), taking into account the measured gas-phase temperature.

In Fig. 2.5 the resulting time-dependent photo-stationary state $j(NO_2)$ (PSS) is compared to predictions resulting from a radiometric measurements (SR) that are used to calculate $j(NO_2)$ inside the chamber with a radiation transfer model. The model is fed by measurements of spectral actinic flux densities of direct and diffuse radiation outside the SAPHIR chamber (Bohn and Zilken 2005). The SR approach correctly predicts the typical shape of the diurnal variation of $j(NO_2)$ inside SAPHIR on this clear-sky day. However, absolute values of $j(NO_2)$ predictions need to be significantly scaled down to match the actinometric data. This scaling factor also increases with time. This can be explained by an increasing degree of staining and mechanical degradation of the chamber walls caused by many years of outdoor

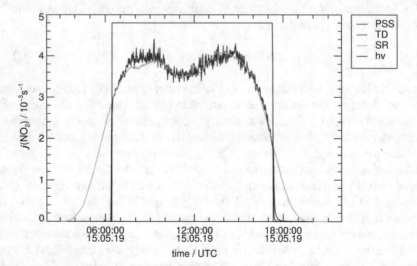

Fig. 2.5 $j(NO_2)$ determined by actinometry (photo-stationary approach PSS, time dependent model TD) and predicted by a combination of outdoor measurements of spectral actinic flux densities and a radiation transfer model (SR) scaled to match the actinometric data. The rectangular box (h) indicates the illumination period of the SAPHIR chamber

residence and operation. Therefore, the scaling factor needs to be regularly determined in experiments of this type, in order to correctly scale the SR data during other photochemical experiments. The radiometry/model approach for SAPHIR (Bohn and Zilken 2005) is quite specific and not directly transferable to other sunlit chambers for which a simple spectroradiometer measurement inside or outside the chamber may be sufficient depending on the geometry of the setup.

Experiments such as the one shown in Fig. 2.4 can be employed in any chamber where the mean residence time is sufficiently long compared to the relaxation time constant of the photochemical equilibrium. For chambers with artificial lights the concentrations may also change with time because of dilution but the photolysis frequency should remain constant—an additional test for the validity of the approach. If no time dependence of photolysis frequencies is expected, experiments can be kept much shorter. Even under non-equilibrium conditions during periods where $j(NO_2)$ changes more rapidly than in the example above, the actinometric approach works reliably, if concentration changes are analysed numerically. This time-dependent (TD) approach is described in more detail elsewhere (Bohn et al. 2005). Moreover, the dark periods after the experiments can be analysed by testing if the decays of NO and O_3 are consistent with the rate constant of Reaction 2.3.3. Analytical solutions for the analysis of the decays are described in Bohn et al. (2005) including conditions for which NO and O_3 concentrations are not only determined by the initial NO_2 concentration and dilution of trace gases are taken into account.

It should be noted that the experiment shown in Fig. 2.4 starts with a small excess of 2 ppbv O_3 from a previous experiment and ends with an about 1 ppbv excess of NO that can be explained by the production of NO from photolysis of nitrous acid (HONO) that is formed inside SAPHIR (Rohrer et al. 2005). In the experiment shown in Fig. 2.4, approximately 3 ppb of HONO was generated and photolyzed over the course of the experiment. OH that is also formed in the photolysis of HONO predominantly reacted with NO_2 which merely led to a small increase of the total loss of NO_x, but this had no significant influence on the steady-state concentrations of the NO_x/O_3 equilibrium that is established much faster.

Effects on the steady-state equilibrium concentrations can be minimized by:

- Reducing the HONO source in the chamber (e.g., low humidity in chambers made of Teflon film), because NO produced from HONO photolysis adds to the total concentration of nitrogen oxides in the chamber. Only, if the NO produced from HONO photolysis is small compared to the initial NO_2 concentration, it can be neglected in the calculations (Eq. 2.3.3).
- The absence of additional OH reactants that could produce HO_2 or RO_2, which reacts with NO, so that the equilibrium between NO_2 and NO is shifted to NO_2 and ozone is produced.
- Using high NO_2 concentrations that ensure that any OH is scavenged in the reaction with NO_2 to prevent the production of HO_2 or RO_2.

Optimum experimental conditions need to be carefully chosen for a specific chamber to avoid chemical interferences.

2.3.2   Chemical Actinometry in Nitrogen

When minimizing the radical sources that may affect the steady state equilibrium, the photolysis of NO_2 in pure nitrogen as a bath gas is often recommended to avoid the complexity arising from secondary chemistry in the determination of the NO_2 photolysis frequency in a simulation chamber. This method has long been known in atmospheric research for the calibration of UV sources and is well described by Holmes et al. (1973) and Tuesday (1961).

A fully oxygen-free atmosphere can be difficult to achieve in chambers, because air may enter the chamber by small leakages or permeation. Leakage of air into the chamber can be minimized by operating the chamber at a pressure slightly above atmospheric pressure or by a second wall around the chamber, so that the gap can be flushed with nitrogen. The latter also minimizes permeation.

Similar to the actinometry experiment in air described in the previous section, NO_2 is injected into the chamber, but the chamber is filled with pure nitrogen. In the absence of oxygen, no ozone is produced from the photolysis of NO_2, so that only NO and NO_2 concentrations need to be precisely measured. A high time resolution of instruments is needed, because the time resolved consumption of NO_2 and production of NO is observed. An initial NO_2 mixing ratio within the range of 0.1–1 ppmv is recommended, as the photolysis of NO_2 can be rather fast. A high initial concentration ensures that the time period that can be used for the evaluation is sufficiently long.

The decay of NO_2 and production of NO is mostly determined by the NO_2 photolysis reaction (Reaction R2.3.1), but oxygen atoms formed from in the photolysis significantly accelerate the loss NO_2:

$$O(^3P) + NO_2 \rightarrow NO + O_2 \qquad (R2.3.4)$$

$$O(^3P) + NO_2 + M \rightarrow NO_3 + M \qquad (R2.3.5)$$

The highest impact on the NO_2 loss is due to the formation of NO (Reaction R2.3.4) because the reaction rate constant of Reaction R2.3.4 is 5 times higher than that of Reaction R2.3.5 at room temperature and ambient pressure. Both reactions are only relevant in experiments in nitrogen because molecular oxygen is missing as a reaction partner for the oxygen atom (Reaction R2.3.2). For the same reason, also the reaction of oxygen atoms with NO impacts the temporal behaviour of NO and NO_2 in this type of experiment:

$$O(^3P) + NO + M \rightarrow NO_2 + M \qquad (R2.3.6)$$

This reaction together with the NO_2 photolysis reaction are responsible that eventually a photo-stationary state is established between NO and NO_2 concentrations. Rather small effects are expected from the formation of oxygen in Reaction R2.3.4, which would allow to form ozone (Reaction R2.3.2), but the reaction rate constant is too small for producing significant ozone concentrations at oxygen concentrations formed in the system.

The production of nitrate radicals in Reaction R2.3.5 further complicates the chemistry, because a small fraction of the nitrogen oxides is converted to NO_3 and N_2O_5 right after the start of the NO_2 photolysis, but converted back to NO_2 and NO at later times of the experiment with increasing NO concentrations due to the following reactions:

$$NO_2 + NO_3 + M \rightleftarrows N_2O_5 + M \qquad (R2.3.7)$$

$$NO_3 + NO \rightarrow NO_2 + NO_2 \qquad (R2.3.8)$$

Depending on the radiation in the chamber, NO_3 may be additionally photolyzed and NO_3 chamber wall loss could be significant.

For these reasons, the photolysis of NO_2 in nitrogen is not a first-order loss process. In Holmes et al. (1973) the following equation is derived to describe the decay rate of NO_2:

$$\frac{-2 j_{NO_2}}{\frac{dln[NO_2]}{dt}} = 1 + \frac{k_{2.3.5}[M]}{k_{2.34}} + \frac{k_{2.3.6}}{k_{2.3.4}} \frac{[M][NO]}{[NO_2]} + \frac{k_{2.3.2}}{k_{2.3.4}} \frac{[M][O_2]}{[NO_2]} \qquad (2.3.4)$$

The meaning of this equation is that as first approximation each photolyzed NO_2 molecule produces 2 NO molecules due to the direct formation of NO (Reaction R2.3.1) and the subsequent reaction of the oxygen atom with NO_2 (Reaction R2.3.4). The other terms are corrections that are needed due to other competing reactions of the oxygen atom (Reaction R2.3.2, R2.3.5 and R2.3.6).

Equation 2.3.4 can be used to calculate $j(NO_2)$ from the measured $[NO_2]$ decay. Figure 2.6 shows measured [NO], $[NO_2]$ and NO_x ($[NO_2] + [NO]$) concentrations in an experiment for the determination of $j(NO_2)$ in the CESAM atmospheric simulation chamber using 460 ppbv of gaseous NO_2 diluted in nitrogen. The total NO_x concentration is nearly unchanged during the experiment, because NO_2 to NO reactions (Reaction R2.3.1 and R2.3.4) dominate the reaction system. The total NO_x concentration decreases approximately by a 5% in this experiment. This loss can be attributed to the dilution of the reaction mixture due to sampling by the monitors. It is recommended to monitor the dilution rate, so that measured concentrations can be corrected for dilution.

Figure 2.5 shows the evaluation of the actinometry experiment describing the logarithm of the measured NO_2 concentration by integrating Eq. 2.3.4. The decreasing loss rate is due to the competition of the reaction of NO with the oxygen atom (Reaction R2.3.5), which gains in importance due to the increasing NO concentration while NO_2 is being photolyzed. Deviations of the calculations are likely due to neglecting further NO_3 chemistry (Reaction R2.3.7 and R2.38.). A first approximation of the photolysis frequencies can be calculated by taking only the first data points after the light is switched on. Assuming that the impact of the reaction of NO with oxygen

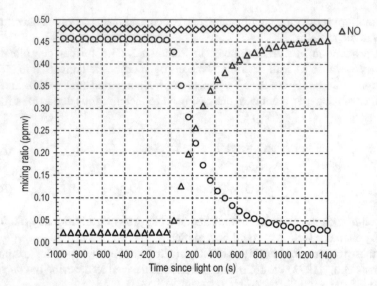

Fig. 2.6 NO$_2$, NO and NO$_x$ concentrations versus reaction time during NO$_2$ photolysis in nitrogen in the CESAM chamber (© Jean-François Doussin, personal communication)

atoms is negligible, because NO has not yet formed from the photolysis of NO$_2$, NO$_2$ decays as first-order loss process, so that a linear behaviour of the logarithm of the NO$_2$ concentration is expected Fig. 2.6). The photolysis rate can be calculated from the slope using Eq. 2.3.4 with [NO] = [O$_2$] = 0 (Fig. 2.7).

Because Eq. 2.3.4 does not consider the impact of NO$_3$ chemistry and the photostationary state between [NO$_2$] and [NO] concentrations that is established at later times of the experiment, it is recommend applying this calculation only for the time right after NO$_2$ photolysis has started. As an alternative, box-model calculations may be applied to determine the photolysis frequency by adjusting its value such that the NO$_2$ decay is best described. However, uncertainties may occur, if the oxygen concentration due to leakages is not known or NO$_3$ wall loss leads to a significant loss of nitrogen oxides during the experiment.

Both data analysis procedures exhibit a significant sensitivity to oxygen concentration which can be difficult to control in a large reactor. They are also very sensitive to the precision of the NO$_x$ measurement and to the potential interferences of NO$_y$ species. This is why, when routinely applied, it is recommended that data analysis is performed only on the first few data points. This avoids giving undue weighting to data acquired at the end of the experiment when NO$_2$ concentrations are close to the detection limit, NO$_y$ species arising from secondary chemistry may have accumulated and when O$_2$ concentration may have increased due to leaks. Results which show a deviation from those expected from the analysis described below may indicate that the O$_2$ concentration is not low enough.

j_{NO2} can be evaluated by considering short time steps and neglecting oxygen equation in Eq. 2.3.4 which becomes

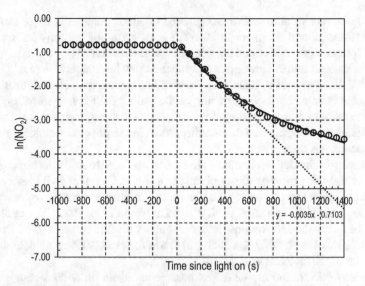

Fig. 2.7 NO_2 concentration time series in a typical actinometry experiment in nitrogen in the CESAM chamber. The dashed line is the result of a linear regression using the first four data points after the light had been switched on. The solid line is the result of the integration of Eq. 2.3.5 assuming a negligible initial oxygen concentration (© Jean-François Doussin, personal communication). In the present example, the initial slope method provides a $j(NO_2)$ value of $(2.3 \pm 0.2) \times 10^{-3}$ s^{-1}, while the use of the Eq. 2.3.5 yields $j(NO_2) = (2.8 \pm 0.1) \times 10^{-3}$ s^{-1}

$$j_{NO_2} = -\frac{1}{2\Delta t}\left\{1 + \frac{k_{2.3.5}[M]}{k_{2,34}} + \frac{k_{2.3.6}}{k_{2,3,4}}\frac{[M][NO]}{[NO_2]}\right\} \cdot \ln\left(\frac{[NO_2]_0}{[NO_2]}\right) \qquad (2.3.5)$$

If only NO_2 measurements are available, Eq. 2.3.5 may be transformed under the assumption that NO and NO_2 contain almost all the NO_x at any time ($[NO] = [NO]_0 + [NO_2]_0 - [NO_2]$).

2.4 Gas-Phase Wall Losses of Species

Significant wall loss can be observed for gaseous inorganic as well as organic compounds. The wall loss rate can be highly variable and is specific for a compound and the chamber. Therefore, wall loss needs also to be considered in the evaluation of experiments as part of the chamber auxiliary mechanism (Sect. 2.6).

For inorganic compounds, the wall uptake can be observed to be irreversible, reversible and/or even reactive, and is commonly measured for ozone and nitrogen containing compounds specifically for NO and NO_2 (Grosjean et al. 1985; Wang et al. 2014). Grosjean et al. (1985) reported little or no significant wall loss for most tested organic compounds. If they observed wall loss the loss appeared to be irreversible. More recent work has shown that this loss can be significant and also reversible for low volatile and/or oxygenated organic compounds (e.g., Loza et al. 2010; Matsunaga and Ziemann 2010). The wall losses of gaseous inorganic and organic compounds can occur on the same timescales as their gas-phase oxidation and gas/particles mass transfer processes and can therefore be competitive (e.g., Grosjean et al. 1985; Krechmer et al. 2016). These wall losses are expected to depend on: (1) the chamber characteristics (e.g., nature of the walls, geometry, age/history, surface to volume ratio, mixing time and procedure), (2) the environmental conditions (e.g., temperature, irradiation, relative humidity) and (3) the physicochemical properties of the compounds themselves.

Atmospheric simulation chambers are extensively used to study the homogeneous and/or multiphasic evolution of gas-phase compounds. Chamber wall losses can thus affect experimental results on (1) the kinetic and mechanistic studies of compounds in the gas phase (e.g., Bertrand et al. 2018; Biermann et al. 1985; Yeh et al. 2014) and (2) the formation and composition of secondary organic aerosols (e.g., Krechmer et al. 2016; Kroll et al. 2007; La et al. 2016; McVay et al. 2014; Pathak et al. 2008; Shiraiwa et al. 2013; Zhan et al. 2014).

In order to determine the wall loss parameters of an individual compound, its concentration in the gas phase is usually measured over time in a clean, dark chamber under constant environmental conditions. The preparation of the clean chamber is done as described in Chap. 3. and the wall loss characterization experiment is similar to the blank experiment described in Sect. 2.6 In the simplest case, the decay can be fitted to a function that describes an irreversible first order loss process, if a decrease of the gaseous concentration is observed after correcting for dilution. A parameterization can also include physical conditions or chemical properties such as equilibrium concentrations or saturation vapour pressure of the specific compound.

Prior to each characterization experiment, the chamber is prepared using usual cleaning, conditioning and filling protocols (Chap. 3). If the chamber allows, experiments should be performed at a fixed temperature and relative humidity but may need to be varied in a series of experiments. The wall losses of gaseous compounds are either studied for compounds that are directly injected into the chamber (e.g., Huang et al. 2018; Loza et al. 2010; Matsunaga and Ziemann 2010; Shiraiwa et al. 2013; Yeh and Ziemann 2014a, b, 2015; Zhang et al. 2014) or produced in the chamber from the oxidation of parent compounds (e.g., Huang et al. 2018; Krechmer et al. 2016; Zhang et al. 2015).

The injection of single compounds or mixtures of different compounds is typically done into the dark chamber using common procedures (Chap. 4). To avoid competition of gas/wall loss with gas/particle partitioning, the injected quantity of the compound should be below its saturation vapour pressure once inside the chamber, thereby preventing particle nucleation occurring. Ideally, a known quantity of the compound is injected into the chamber and is homogeneously mixed instantaneously, so that mixing of the compound and wall loss are separated in time. However, depending on the volatility of the compound, the injection duration can vary from minutes to hours, so that both processes may need to be taken into account for the determination of the wall loss rate.

If the wall loss of oxidation products that are not available as pure compounds need to be characterized, the precursors are first injected into the chamber. Concentrations should again be low enough that aerosol formation does not play a role for the species loss. Depending on the specific chamber, oxidation needs to be initiated for example by injection of ozone or hydroxyl-radical precursors. If photo-oxidation is required to produce oxygenated products, the chamber air needs to be exposed to light (lamps or sunlight). In the ideal case, oxidation is stopped after a few seconds for example by switching off lights, when a sufficiently high concentration of products is formed, and the decay of the target species can be used to determine the wall loss. If oxidation cannot be stopped or if there is a reversible loss of the compound, all processes need to be considered such as gas-phase production and equilibrium between the gas-phase and deposition on the wall.

Depending on the wall loss rate, concentrations need to be monitored over a few hours (2–15 h). It is worth noting that interactions of the compound with the walls can also occur in the inlet line of instruments (Deming et al. 2019; Krechmer et al. 2016; Liu et al. 2019; Pagonis et al. 2017). Delays and underestimations of concentrations can therefore be observed. Therefore, passivation of the inlet lines is recommended.

If the loss of compound A (gas phase concentration $[A_g]$) is irreversible, the process is described by a first order loss rate constant k_{gw}:

$$\frac{d[A_g]}{dt} = -k_{gw}[A_g] \tag{2.4.1}$$

If there is no production or injection of the compound during the time of observation and if wall loss is the only relevant process, the observed decay can be fitted to a single-exponential function that directly gives the first order loss rate constant.

In the case of a reversible loss process, the transfers of a gaseous species A to the wall (wall reservoir concentration $[A_w]$), and back to the gas phase, can be described by the gas phase first order loss rates k_{wg} and k_{gw}, respectively:

$$\frac{d[A_g]}{dt} = k_{wg}[A_w] - k_{gw}[A_g] \tag{2.4.2}$$

In the case that there are no other relevant production or destruction processes concurrently happening, Eq. (2.4.2) can be solved. Boundary conditions are that the sum of concentrations in the gas-phase and the wall reservoir equals the initial concentration $[A]_0$ that has been injected or produced by oxidation and $[A_g]_{eq}$ is the equilibrium gas-phase concentration that is eventually obtained:

$$[A_g](t) = \left([A]_0 - [A_g]_{eq}\right)\exp(-(k_{gw} + k_{wg})t) + [A_g]_{eq} \tag{2.4.3}$$

The fit of the observed concentration time series results in the effective wall loss rate of $k_{Weff} = (k_{gw} + k_{wg})$ and the equilibrium concentration $[A_g]_{eq}$. The initial concentration may be fixed to measured values or can also be obtained by the fitting procedure.

As an alternative to fitting to an analytic solution, the observed time profile of the measured concentration can be described by optimization of parameters in a numerical box model, which can allow for taking additional loss and production processes into account. The wall loss rate constant can be optimized using standard optimization procedures to minimize the difference between measured and modelled concentration time series.

Figure 2.8 shows the wall deposition velocities (i.e., the first order loss rate corrected by the surface (S) to volume (V) ratio of the simulation chamber calculated as $v_{gw} = k_{gw} V/S$) measured in various EUROCHAMP simulation chambers for several organic compounds as a function of their saturation vapour pressure (P^{sat}). The deposition velocity appears to depend on (i) the chamber characteristics which cover a large diversity of wall materials, surface to volume ratios and mixing times, (ii) the organic species properties, such as the saturation vapour pressure, and (iii) the environmental conditions, such as the relative humidity. For example, the walls of the chambers made of aluminium like the AIDA chamber at Karlsruhe Institute of Technology may induce a constant loss of gas-phase compounds, depending on molecular properties and wall temperature. In chambers made of Teflon, the wall loss rates often correlate with the vapour pressure. However, the effect of wall loss can be minimized, if the chamber has a high volume to surface ratio like the SAPHIR chamber made of Teflon film (270 m^3) at Forschungszentrum Jülich.

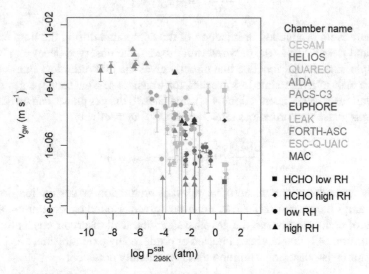

Fig. 2.8 Wall deposition velocities (v_{gw}) of gaseous organic compounds measured in EUROCHAMP chambers as a function of the saturation vapour pressure (P^{sat}). The organic species properties were estimated with the GECKO-A tool (e.g., Valorso et al. 2011) using the Nannoolal method to calculate the saturation vapour pressure at 298 K (Nannoolal et al. 2008). The loss rate also highly depends on the wall material and size of the individual chamber and the type of molecule, so that there is no unique value for the wall deposition velocity for a specific saturation vapour pressure value. (Figure from EUROCHAMP-2020 Deliverable 2.7, www.eurochamp.org)

2.5 Particle Wall Losses

Aerosol processes in the atmosphere can have typical timescales ranging from a few seconds up to several days. In order to investigate such processes under reasonably representative atmospheric conditions, simulation chambers used for the investigation of physical, chemical or biological transformations of aerosol particles should enable a sufficiently long particle lifetime. In addition to their interaction with each other and with the gases in suspension, the lifetime of aerosol particle in chambers may be substantially controlled by wall losses resulting from the combination of adsorption, deposition, diffusion and mixing processes, gravitational settling and electrostatic attraction, all depending on particle and wall properties. The physical wall loss of particles in closed vessels such as chambers will vary with the particle size and will depend on (i) the chamber shape, (ii) the mixing regime (especially for small particles), (iii) the density of the considered particles, (iv) the electrostatic state of the wall.

Clearly it is important to understand particle loss rates in any experiments aiming to characterize the dynamic evolution of the distribution of particles, which can range from characterization of formation and transformation processes through to the determination of optical properties and their dependence on particle mixing-state to investigation of gas-aerosol interactions such as in the formation of secondary organic aerosol (SOA). By definition, simulation chambers are volumes enclosed by walls and interactions of particles with their surfaces cannot be assumed negligible. The advantages of knowing the size-dependent wall losses of particles are substantial and various. The losses will determine the reduction in lifetime for contribution of the particles to properties or processes of interest in the experiment. For example, reduction in the particle lifetime will reduce the condensation sink it presents for gas-aerosol interaction (such as SOA formation). Similarly, optical extinction by a diminishing particle population will be commensurately reduced. Consequently, knowledge of the losses will enable more confidence to be ascribed to measured properties and inferred processes. For example, model-measurement comparison of the dynamical evolution of the particle population will enable model processes such as condensational growth, and properties such as vapour pressure, to be constrained so long as particle wall losses (as well as vapour–wall interactions) are known. Conversely, the consequences of neglecting to characterize particle wall losses is a substantially compromised ability to interpret any experiments aiming to capture properties or processes that depend on particle size distributions.

Provided that the aerosol in the chamber is well mixed, in the absence of any other process, the rate coefficient for the wall loss, β_i, can be represented as a simple first order loss:

$$dN_i/dt = -\beta_i N_i, \tag{2.5.1}$$

where N_i is the number concentration of particles of size class in the chamber (Crump et al. 1983). The coefficient β_i for each size class can be obtained by integrating Eq. (2.4.1) to give (Fig. 2.9):

$$-\ln(N_i/N_i, 0) = \beta_i t \qquad (2.5.2)$$

The requirement to quantify and account for the wall losses of particles in simulation chambers used for aerosol experiments is universal, though the extent of the required characterization is, to some extent, application dependent. This has led to a pragmatic variety of approaches to wall loss determination that broadly fit into two classifications:

- wall loss quantification using deliberate characterization experiments (denoted "seed injection" methodologies)
- wall loss quantification using chamber experiments (denoted as "In-experiment" methodologies).

Moreover, the methodology is dependent on both the chamber geometry and materials used in its construction. Specifically, rigid fixed-geometry structures of conductive materials (e.g., aluminium or steel), rigid fixed-geometry structures of insulating materials (e.g., glass) and flexible variable-geometry structures of insulating materials (e.g., Teflon) will each require and be best suited to particular approaches.

Whilst the methodology may vary, a common requirement for each approach is the availability of well-characterized and calibrated particle sizing and counting instrumentation and an appropriate source of particles.

The instrumentation normally comprises:

- a mobility sizing instrument coupled to a particle counter(s) (either a differential or scanning particle mobility sizer, SMPS or DMPS, coupled to a condensation particle counter with the appropriate size cut-off), and/or

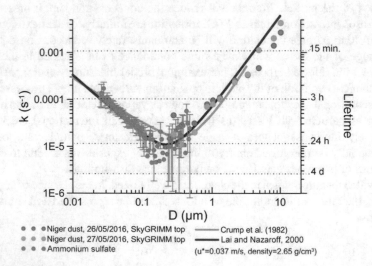

Fig. 2.9 Size dependent wall losses directly obtained from measurements of the first order decay of polydisperse particles nebulized or mechanically generated and injected into the CESAM chamber (Lamkaddam thesis 2017)

- an optical particle spectrometer with the appropriate configuration (normally backscatter of coherent or broadband white light, but forward scatter may be more appropriate for cloud droplets).

The particle source is usually a nebulizer of some sort, capable of generating salt solution aerosol from a quantified stock. Selection of the source and stock will depend on the desired particle size, breadth of distribution and number, and of course on composition (e.g., brush generator or fluidised bed may be more appropriate for soot or dust characterisations respectively.

Maintenance of chamber facilities and transport of smaller chambers for field campaigns can induce electrostatic charges on chamber walls and increase particle wall-losses. This generates a "disturbed" chamber that can introduce significant uncertainty in the particle wall-loss rates. The recovery time can even be months after the disturbance if natural charge dissipation is the only action that reduces the charges on the walls (Wang et al. 2018a, b).

An electrostatic eliminator device (fan or air gun) can reduce the induced charges on the chamber walls significantly faster that the natural charge dissipation process. Such an electrostatic fan was used in the chambers of FORTH laboratories to reduce the induced charges after maintenance and handling of several different chambers.

In order to determine size-dependent wall loss rates, two approaches for particle generation were for example employed in the CESAM chamber. Polydisperse $(NH_4)_2SO_4$ particles were nebulized from an aqueous saline solution to provide sub-micron particles and test dusts were mechanically generated for super-micron particles. Total number concentration was held below 10^4 cm^{-3} to minimize the collision probability and so as not to require a correction for coagulation. The number size distribution was measured as a function of time (with a SMPS for sub-micron particles and OPC for super-micron particles). A first order decay fit following Eq. 2.7.1 was fitted to the time evolution of each size-bin.

Wang et al. (2011) reported that particle lifetime in the stainless steel CESAM chamber ranges from 10 h to 4 days depending on particle size distribution, enabling the chamber to provide satisfactorily high-quality data on aerosol aging processes and their effects. More recently Lamkaddam (2017) has studied the physical wall loss rate as a function of particle size. Submicron ammonium sulphate particles were generated in small number to minimize coagulation and mineral dust were used for supermicron particles. The vertical air velocity was experimentally measured in the chamber and its value was used as the u* parameter in the Lai and Nazarof parameterisation (Lai and Nazaroff 2000). Plotting the particles wall loss frequency as a function of size in a log–log plot will yield a typical V-shape curve when electrostatic charges are not significant (Lai and Nazaroff 2000). Owing to its stainless steel construction, this is expected and is found to be the case for CESAM and, as shown in Fig. 2.5, the size-dependent wall loss compares satisfactorily with previous literature (Crump et al. 1983; Lai and Nazaroff 2000). Above all, even if developed for parallelepiped volumes, the Lai and Nazarrof parameterization has shown excellent agreement by just introducing the correct CESAM chamber dimensions, the measured u* and the correct density for particle material without any further adjustment or fitting of the model to the data points. The same approach was successfully adopted at the ChAMBRe facility (Massabò et al. 2018).

A range of approaches have been used to conduct an extensive characterization of the wall losses in the participating chambers. Pragmatic approaches to the characterization of each infrastructure has led to a variety of techniques according to the chamber and experimental type, instrumental availability and application-specific requirements. This diversity across the infrastructures has led to considerable expertise across the scientific applications and continuous contribution to the state-of-the-science in characterization of wall losses in simulation chambers. Best practice and model code have been shared, though design and adoption of a standardized protocol is still challenging. The recommendation is that particle wall losses are characterized as far as possible, and standardization is adopted as soon as the state-of-the-science allows. A more straightforward approach appears possible for rigid chambers constructed of the conducting material. Whilst novel mitigation approaches show promise, electrostatically enhanced particle loss in flexible plastic film chambers requires further investigation.

2.6   Characterization of the Chamber State by Gas-Phase Reference Experiments

Chamber-specific properties need to be characterized, in order to take into account the chamber background reactivity in any experimental evaluation procedure. This allows the separation of the chamber-specific chemical processes from the underlying chemistry that is being studied in experiments. They can be put into auxiliary mechanisms that complement chemical mechanisms to perform chemical modelling of chamber experiments. These auxiliary mechanisms are essential to make results from experiments carried out in different chambers comparable and transferable to the atmosphere.

Chamber auxiliary mechanisms contain a number of specific features to account for chamber properties that often arise from effects of the chamber walls. Primarily, these features consider (Fig. 2.10):

- Adsorption/desorption of nitrogen oxide species (NO_y, including HONO, N_2O_5 and HNO_3) and reactive organic species to/from the chamber walls.
- Deposition of aerosol to the chamber walls.
- Dilution of chamber trace constituents through leaks and gas removal by instruments.

Many of the chamber-specific processes can change over time due to memory effects from previous experiments carried out in the chamber. Hence experiments to characterize the processes should be performed regularly, for example at the

Fig. 2.10 Illustration of interaction of trace gases with the wall of simulation chambers

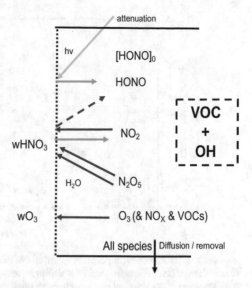

beginning/end of intensive experimental campaigns. Characterization of wall loss is discussed in detail in Sects. 2.4 and 2.5.

2.6.1   Chamber Blank Experiments

Chamber blank experiments are used to assess impurities in the background air matrix as well as degassing of species from the chamber walls. The chamber is prepared as is typically done in most experiments, starting by cleaning the chamber and filling with pure air (Chap. 3). Concentrations of trace gasses are observed throughout the experiments, so that their release from the chamber walls can be parameterized. Wall sources of compounds in a chamber are often photolytic and can also be affected by the amount of water vapour present. Hence experiments are performed under light and dark conditions, and at the upper and lower limits of the typical operating range for relative humidity in the chamber (e.g., Rohrer et al. 2005, Zador et al. 2006). Figure 2.11 shows examples of reference blank experiments carried out to determine the wall sources of formaldehyde (HCHO) in the EUPHORE chamber. Chamber blank experiments can also be used for the determination of wall loss processes (see Sects. 2.6 and 2.7).

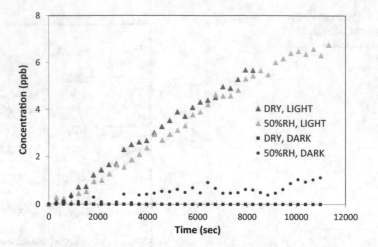

Fig. 2.11 Chamber blank experiments carried out in the EUPHORE chamber in order to determine formaldehyde (HCHO) wall sources at different relative humidity values (RH) and under light and dark conditions. This behaviour is generally interpreted as a photochemical O_3 production from background reactivity (unmeasured species that are formed in the sunlit chamber and could be released from the chamber wall). © EUPHORE

2.6.2   Reference Experiments Using Well Known Chemical Systems

Reference experiments with well-known chemical systems can be used to regularly evaluate if the chemistry of the system under investigation can be separated from chamber effects. The measured time series of trace gases and radicals can be compared and contrasted to chamber simulations performed by chemical box modelling. The model must include the chemistry of trace gases present in the reference experiment and the auxiliary mechanism that describe chamber-specific processes. Several types of experiments are described below.

2.6.3   Experiments with Mixtures of NO_x in Air

Because of the importance of NO_x for atmospheric chemistry, the behaviour of nitrogen oxides in the chamber is often characterized in the blank experiment. Nitrogen oxide species (NO_y) are known to be emitted into the gas phase from photolytic production on chamber walls, including NO, NO_2, HONO, HNO_3 (Rohrer et al. 2005; Zador et al. 2006). These species can also be inter-converted between each other, both in the gas phase and on the walls. For example, nitrogen dioxide may convert to nitrous acid and nitric acid ($NO_2 \rightarrow$ aHONO + bHNO_3). This heterogeneous chemistry can be affected by both light and relative humidity. Experiments in which NO_x is added to a clean chamber can be used to explore the rates of interconversion of NO_y species driven by the walls.

Chamber wall materials are typically chosen to be chemically inert. Effects from the wall material are often related to the chemical nature of adsorbed compounds arising from previous experiments. Because most of the chemical systems studied in simulation chambers lead to the formation of oxidized species, chambers walls generally exhibit an oxidative potential (Bloss et al. 2005; Hynes et al. 2005; Metzger et al. 2008).

Some the studies (Bloss et al. 2005; Metzger et al. 2008) show that the consumption of NO_2 is coupled with the formation of small quantities of HONO. In Metzger et al. (2008) the wall loss rate of NO_2 was $(1.05 \pm 0.35) \times 10^{-6}$ s^{-1}. In some cases, however, the chamber walls can be reductive as shown for metal chambers (Wang et al. 2011).

It is recommended to follow a protocol for chamber blank experiments similar to that described in Wang et al. (2011):

- Injection of 50–200 ppbv of NO_2.
- Monitor in the dark for 1 h.
- Irradiate the chamber air for 1 h.
- Monitor in the dark for 30 min.
- Monitor NO, NO_2, O_3, HCHO, HONO and radicals (if available).
- Systematic studies with changing RH are recommended.

An example of this type of reference experiment from the CESAM chamber is given in Fig. 2.12 After NO_2 had been injected into the chamber, a continuous loss of NO_2 was observed in the dark that was accompanied by a slow production of NO. When the lights were turned, NO_2 is photolyzed reaching a photo-stationary state that is established between NO_2, NO and O_3 concentrations within approximately 5 min. During the phase, when the chamber air was irradiated chamber wall effects lead to slow production of NO that is interpreted as NO_2 conversion on the wall. Ozone concentrations decreased due to the increase of NO, but also wall loss played a role. Consequently, when the lights were turned off the concentration of NO remained high, because the available ozone concentration was not sufficient to convert all NO back to NO_2. In addition, the sum of NO and NO_2 was lower compared to the initially injected NO_2 concentration due to chamber wall loss.

2.6.4 Photochemical Oxidation of CO/Methane

Radical concentrations in chambers are often impacted by chamber processes. For example, a major source for hydroxyl radicals (OH) in Teflon chambers is often the photolysis of nitrous acid (HONO) that is emitted from the chamber walls. In addition, radicals may be lost on the chamber walls or react with organic species that are released from the chamber wall but may not be quantified. Reference experiments are useful to test if radical sources and sinks are understood.

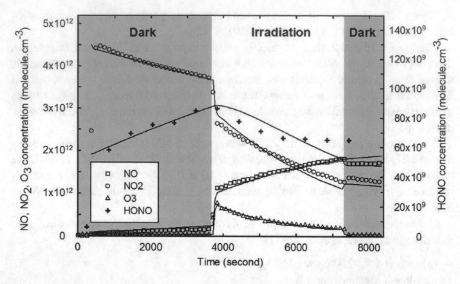

Fig. 2.12 Examples of a NO_x-air experiment carried out in the CESAM chamber, with initial injection of ca. 200 ppbv NO_2. (Reused with permission from Wang et al. (2011) Open access under a CC BY 3.0 license, https://creativecommons.org/licenses/by/3.0/)

The experimental procedure is similar to that of a blank experiment. The chamber is exposed to light, in order to trigger photolytic processes. The radical source can be the photolysis of nitrous acid released from the chamber walls. Ozone can also be injected to produce radicals from its photolysis. The chamber air would typically be humidified in the experiment, because water vapour is often needed to produce radicals. Due to the presence of sources of unknown OH reactants in the chamber, it is recommended to add an OH reactant. CO converts OH radical to hydroperoxyl radicals (HO_2) and methane converts OH to methylperoxyl radicals (CH_3O_2). In the presence of nitric oxide (NO) a radical reaction chain is initiated in which ozone is produced in the chamber. Steady-state equilibrium concentrations of radicals are rapidly established owing to the short chemical lifetime of radicals. OH reactant concentrations are chosen such that OH radical equilibrium concentrations are above the limit of detection of instruments detecting radicals. Results from chemical box models can be compared to observed radical concentrations to test if chamber processes are appropriately taken into account. In addition, the ozone concentration increase can also be compared, because ozone is chemically produced in the reaction of peroxyl radicals with NO.

Figure 2.13 shows an example for a reference experiment with CO and CH_4 injections in the SAPHIR chamber with the specific aim to test if OH radical concentrations can be described and understood. Results from a chemical box model gives excellent agreement between measured and modelled radical concentrations, if chamber-specific processes such as sources for nitrous acid (HONO) and formaldehyde (HCHO) are included and adequately described.

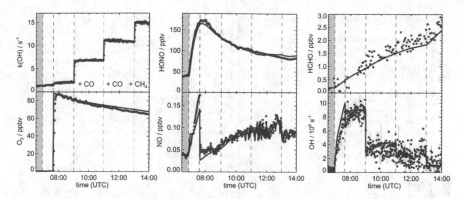

Fig. 2.13 Reference experiment in SAPHIR with injections of CO and CH$_4$ indicated by the increase in measured OH reactivity (kOH). Measurements (blue dots) are compared to results of a chamber chemical box model (red lines) that include chamber-specific properties such as chamber sources of nitrous acid (HONO) and formaldehyde (HCHO). Grey areas indicate times when the chamber was kept in the dark and vertical lines give times of injections

2.6.5   Photo-Oxidation of Propene in the Presence of NO$_x$

Experiments using more complex organic compounds with a well understood chemical oxidation mechanism, such as ethene (C$_2$H$_4$) or propene (C$_3$H$_6$), can be used to test the efficacy of the chamber auxiliary mechanism and can also be used to optimize/tune them (Bloss et al. 2005; Wang et al. 2011). Disagreements between the measured and modelled mixing ratios of precursor and product compounds are therefore assumed to be caused by chemistry driven by the chamber walls. A particularly clear chamber effect is the timing of the onset of removal of the VOC following the initial addition of the VOC and NO$_x$. The experiment begins with no addition of a radical source, with much of the initial reactivity in a chamber being driven by HONO coming off the chamber walls and being photolyzed to produce radicals which can react with the VOC. Hence this timing is a good indication of the rate of HONO production.

It is recommended to follow a similar protocol to that used in the work of Hynes et al. (2005) and Wang et al. (2011, 2014):

- Experiments carried out over a range of VOC–NO$_x$ concentration ratio of 0.6–17, e.g., injection of 500 ppbv C$_2$H$_4$ and 50–300 ppbv NO.
- Observation of trace gases concentrations including propene, NO, NO$_2$, ozone, HONO, HCHO, CH$_3$CHO, HCOOH, PAN, radicals, if available, during the photo-oxidation of C$_2$H$_4$ for 5 h.
- Observation of trace gas concentrations for 1 h in the dark.
- Studies with systematic changes of the relative humidity are recommended.

An example of a propene-NO$_x$ experiment carried out in the CESAM chamber is shown in Fig. 2.14.

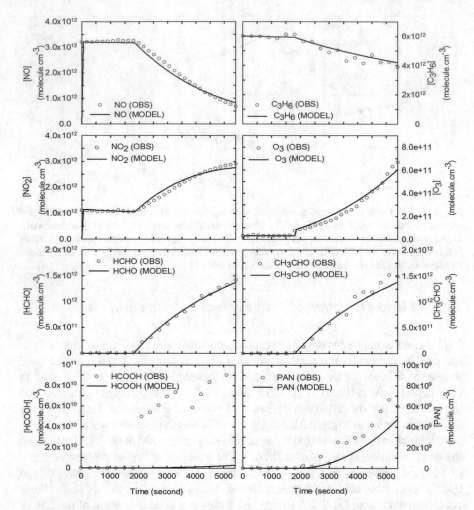

Fig. 2.14 The overall examples of a propene-NO_x experiment carried out in the CESAM chamber, with an initial injection of ~250 ppbv propene, 50 ppbv NO_2, 120 ppbv NO 200 ppbv NO_2, Wang et al. (2011). Comparison of simulated (solid lines) and experimental concentrations (symbols) for NO, NO_2, ozone, propene, formaldehyde, acetaldehyde, formic acid and peroxyacetyl nitrate (PAN) in a propene-NO_x-Air system. Solid lines are the results of modelling with an initial concentration of nitrous acid $[HONO]0 = 8$ ppbv. (Reused with permission from Wang et al. (2011) Open access under a CC BY 3.0 license, https://creativecommons.org/licenses/by/3.0/)

2.7 Characterization of the Chamber State by Aerosol-Phase Reference Experiments

Reference experiments with well-known SOA precursors can be used to regularly check the overall status and conditions of a reaction chamber. Characteristic SOA formation behaviour can be established for one or more precursors and checked at regular intervals to confirm the reproducibility of the SOA production behaviour, understand changes to chamber conditions or identify potential problems and address them. In this context, the primary purpose of aerosol-phase reference experiments is to provide a method to monitor SOA production behaviour in a specific reaction chamber over time and to understand how it is influenced by any changes to the chamber infrastructure conditions (e.g., wall material, lights) or operating procedures. Reference aerosol-phase experiments could potentially also be used to compare SOA formation behaviour in different reaction chambers assuming that fundamental differences in chemical and physical factors can be accounted for.

It is not possible for one specific aerosol-phase reference experiment to fulfil this purpose for all existing chambers. This is because chambers vary in many ways including their size, temperature and relative humidity range, mode of operation (e.g., batch, continuous-stirred tank or flow reactors), light source (e.g., natural, artificial, dark), oxidant environment (e.g., OH, O_3, NO_3) and suitability for the use of seed particles. These and other potential factors need to be considered when deciding the required number of aerosol-phase reference experiments for each reaction chamber. The remaining part of this section will provide a brief overview of the main types of SOA formation experiments and recommend aerosol-phase reference experiments for chamber operators to select from as deemed suitable for their needs.

SOA formation occurs when one or more VOCs are oxidized to produce products of adequately low volatility to condense into the particulate phase. This can be a result of progressive oxidation steps in multiple reactions leading to multi-generation products of sufficiently low volatility, or fast auto-oxidation products such as HOMs that condense in the particulate phase quickly. This route is typically dominated by gas-to-particle conversion processes and is widely investigated in the chamber studies. Alternatively, SOA formation may occur in the condensed phase, when water soluble VOCs dissolve in droplets or particles and are subjected to aqueous phase oxidation leading to SOA products that remain dissolved. The latter route is typically investigated in bulk studies, but recent developments in analytical capabilities have enabled studies on a single particle scale with moist aerosol.

SOA formation experiments are conducted either in the presence or absence of pre-existing particles. These are types of experiments are referred to as nucleation or seeded experiments, respectively. Nucleation experiments often require a sufficiently high initial concentration of the SOA precursor(s) so that adequate amounts of low volatility oxidation products accumulate up to the threshold of homogeneous nucleation to be reached. This threshold is dependent on the parent VOC and the volatility distribution of its oxidation products. This type of experiment is useful for studies investigating properties of pure SOA particles, in addition to those focusing on nucleation rates.

Seeded experiments, on the other hand, are frequently used to avoid limitations of or lack of ability to measure aerosol size distribution or mass associated with particles of small sizes. In this type of experiments, particles of know composition are introduced into chambers with controlled amount and know size distribution. Ammonium sulphate, ammonium bisulphate and sodium chloride particles are some of the examples commonly used in this type of experiments, where their aqueous solutions are nebulized into chambers with or without the use of a classifier (e.g., DMA) producing mono- or poly-dispersed particles with known size or distribution often around 80–100 nm. This ensures that the produced SOA materials are condensed onto particles within the measurement capability of most deployed aerosol instrumentation. Seeded experiments are also often used with VOC precursors of relatively low reactivity or those which require multiple oxidation steps or aqueous phase reactions to generate SOA. The use of seed particles reduces the loss of condensable vapours to the chamber walls by offering a competing condensation sink and facilitates SOA production. The ability to clearly distinguish between and quantify the mass of seed particles and SOA material is possible using online mass spectrometry techniques. This can also offer a direct method of mass wall loss decay rate of seed particles.

The choice for which reference experiment should be conducted in a chamber at regular interval should take into consideration the factors discussed so far in addition to the specific nature of the chamber and the types of experiments it is used for. The oxidant environment of a chamber plays fundamental role in its SOA formation characteristics as it influences both of its gas as well as the particle-phase chemistry. Therefore, both dark and photo-oxidation experiments should be considered when deciding on which reference experiment to conduct. These experiments could be designed to investigate SOA formation from a specific oxidant (e.g., OH, O_3) or to mimic atmospheric oxidation conditions such as day- or night-time chemistry, which involves more than one oxidant at a time (e.g., OH/O_3 or O_3/NO_3 etc.). Metrics such as SOA mass production and VOC decay should be regularly checked in dark experiments. Additional metrics should be included in the case of photo-oxidation experiments such as ozone formation behaviour. Establishing and tracking the behaviour of such metrics on a regular basis would provide useful reference knowledge to understand the overall chamber behaviour in terms of particle-phase formation.

The choice of oxidant is an important part of determining the SOA formation conditions and it is determined by the objectives of the undertaken research. Oxidation by one or a combination of hydroxyl radical, ozone and nitrate radical account for the majority of SOA formation studies in most of the existing chambers.

2.7.1 Reference Photo-Oxidation Experiments

Chamber experiments aim to mimic the degree of atmospheric oxidant exposure, which is the integral of the oxidant concentration and the experiment duration time. The latter is often limited by the residence time of the reactor. The choice of light type and characteristics are key components of each chamber's ability to achieve its oxidant concentration target. Most atmospheric chambers operate at OH concentration levels in the range of 10^6-10^7 molecules cm^{-3}, which are representative of daytime concentrations in most ambient environments. Depending on residence time and vapour wall loss rates, chambers typically simulate SOA formation corresponding to an oxidant exposure from a few hours to about a day or two.

In the atmosphere, OH radicals are produced from the reaction of H_2O with singlet oxygen atoms ($O(^1D)$) generated from O_3 photolysis at wavelengths <320 nm. In atmospheric chambers, the viability of this source is clearly dependent on the spectrum and wavelength-dependent intensities of the light source. This method of OH generation may be suitable in chambers equipped with light sources such as xenon-arc lamps. In this case, ozone is typically produced as a secondary product of the VOC and NO$_x$ chemistry.

$$O_3 + hv \rightarrow O_2 + O(^1D)$$

$$H_2O + O(^1D) \rightarrow 2OH$$

Other methods of OH production are needed in the case of chambers using blacklights with a spectrum peak at round 350 nm due to the lack of sufficient photon intensity required for a sustainable ozone photolysis. These methods include the use of sources such as HONO as, for example, formed from the chamber wall (Sect. 2.6) or H_2O_2.

The photolysis of HONO generates one OH radical and one NO radical. This means that the use of HONO as an OH source cannot be considered for experiments where NO$_x$-free or very low NO$_x$ conditions are needed. This is because NO is produced even if no additional NO$_x$ is added to the system.

$$HONO + hv \rightarrow OH + NO$$

A continuous source or multiple injections of HONO are typically required in this type of experiments in order to produce sufficiently high levels of OH radicals. This is due to the substantial photolysis rate of HONO.

Alternatively, H_2O_2 may be as a source of OH radicals. The photolysis of H_2O_2 produces two OH radicals, which subsequently react with H_2O_2, producing HO_2 radicals:

$$H_2O_2 + hv \rightarrow 2OH$$

$$OH + H_2O_2 \rightarrow HO_2 + H_2O$$

In this system, it is possible to maintain a steady OH concentration over a long period of time due to a combination of relatively slow H_2O_2 photolysis rate and suppression of OH propagation by reaction with H_2O_2 itself. Unlike HONO, H_2O_2 provides a NO$_x$-free source of OH.

Several VOCs could be used to conduct a reference photo-oxidation experiment using any of the oxidant sources mentioned so far. These include toluene, 1,3,5-trimethylbenzene (TMB) or α-pinene. However, in the case of chambers relying on the photolysis of O_3 for the production of OH, α-pinene experiments characterise SOA formed by a combination of ozone and OH oxidation. It is recommended that ammonium sulphate seed should be used in these explements for reasons discussed earlier in this section.

2.7.2 Reference Ozonolysis Experiments

For dark chambers without natural or artificial light sources, a reference VOC ozonolysis experiment should be conducted on a regular basis to check the overall conditions of the chamber. This serves to either confirm the reproducibility of SOA formation characteristics or provide insights into any chamber changes affecting its SOA production behaviour. Ozone reacts with VOCs containing unsaturated carbon bonds to form SOA. The gas-phase reaction of ozone with VOCs proceeds by the addition of ozone cross the C=C double bond to form an energy-rich primary ozonide, followed by decomposition of the primary ozonide to produce an energized carbonyl oxide species, known as the Criegee intermediate, and an aldehyde or ketone product. This is followed by unimolecular decay of the Criegee intermediate producing OH radicals. The OH yield from ozonolysis varies depending on the VOC molecular structure. Chamber experiments investigating the role of ozone chemistry in SOA formation often use compounds to react with the resulting OH radicals. These are known as OH scavengers and include CO, cyclohexane and 2-butanol. The use of such compounds is not needed for the purpose the reference ozonolysis experiment discussed here. This is because OH yield from ozonolysis of VOC is part of simulating night-time chemistry in the reaction chambers.

The choice of VOC precursors for carrying out a reference ozonolysis experiment is broad and includes atmospherically relevant compounds containing carbon double bonds and known to form SOA. Of these, α-pinene is commonly used in most chambers and is an obvious candidate. It is recommended that a dark ozonolysis of α-pinene is carried out in the absence of an OH scavenger. The use of ammonium sulphate seed is also recommended as discussed earlier in this section.

References

Atkinson, R., Baulch, D.L., Cox, R.A., Crowley, J.N., Hampson, R.F., Hynes, R.G., Jenkin, M.E., Rossi, M.J., Troe, J.: Evaluated kinetic and photochemical data for atmospheric chemistry: Volume I—gas phase reactions of O_x, HO_x, NO_x and SO_x species. Atmos. Chem. Phys. **6**, 1461–1738 (2004). https://doi.org/10.5194/acp-4-1461-2004

Avnet Inc.: https://www.avnet.com/wps/portal/abacus/solutions/technologies/sensors/pressure-sensors/core-technologies/. Accessed 01 June 2020

Baratron capacitance manometers: https://www.mksinst.com/n/baratron-capacitance-manometers. Accessed 01 June 2020

Becker, K.H.: The European photoreactor "EUPHORE". European Community, Final report (1996)

Bloss, C., Wagner, V., Bonzanini, A., Jenkin, M.E., Wirtz, K., Martin-Reviejo, M., Pilling, M.J.: Evaluation of detailed aromatic mechanisms (MCMv3 and MCMv3.1) against environmental chamber data. Atmos. Chem. Phys. **5**, 623–639 (2005). https://doi.org/10.5194/acp-5-623-2005

Bohn, B., Zilken, H.: Model-aided radiometric determination of photolysis frequencies in a sunlit atmosphere simulation chamber. Atmos. Chem. Phys. **5**, 191–206 (2005). https://doi.org/10.5194/acp-5-191-2005

Bohn, B., Rohrer, F., Brauers, T., Wahner, A.: Actinometric measurements of NO_2 photolysis frequencies in the atmosphere simulation chamber SAPHIR. Atmos. Chem. Phys. **5**, 493–503 (2005). https://doi.org/10.5194/acp-5-493-2005

Burkholder, J.B., Sander, S.P, Abbatt, J.P.D., Barker, J.R., Cappa, C., Crounse, J.D., Dibble, T.S., Hule, R.E., Kolb, C.E., Kurylo, M.J., Orkin, V.L., Percival, C.J., Wilmouth, D.M., Wine, P.H.: Chemical kinetics and photochemical data for use in atmospheric studies. Evaluation Numbber 19, Jet Proopulsion Laboratory, California Institute of Technology (2020)

Burrows, J.P., Dehn, A., Deters, B., Himmelmann, S., Richter, A., Voigt, S., Orphal, J.: Atmospheric remote-sensing reference data from GOME: Part 1. Temperature-dependent absorption cross-sections of NO_2 in the 231–794 nm range. J. Quant. Spectrosc. Radiat. Transfer **60**, 1025–1031 (1998). https://doi.org/10.1016/S0022-4073(97)00197-0

Carter, W.P.L., Cocker, D.R., III., Fitz, D.R., Malkina, I.L., Bumiller, K., Sauer, C.G., Pisano, J.T., Bufalino, C., Song, C.: A new environmental chamber for evaluation of gas-phase chemical mechanisms and secondary aerosol formation. Atmos. Environ. **39**, 7768–7788 (2005)

Crump, J.G., Flagan, R.C., Seinfeld, J.H.: Particle wall loss rates in vessels. Aerosol Sci. Techn. **2**, 303–309 (1983)

Curtis, A.R.: The FACSIMILE numerical integrator for stiff initial value problems. Computer science and systems division AERE Harwell, Harwell, Oxfordshire, Unclassified AERE—R 9352 (1979)

Deming, B., Pagonis, D., Liu, X., Day, D., Talukdar, R., Krechmer, J., de Gouw, J.A., Jimenez, J.L., Ziemann, P.J.: Measurements of delays of gas-phase compounds in a wide variety of tubing materials due to gas–wall interaction. Atmos. Meas. Tech. **12**, 3453–3461 (2019). https://doi.org/10.5194/amt-12-3453-2019

Dunlea, E.J., Herndon, S.C., Nelson, D.D., Volkamer, R.M., San Martini, F., Sheehy, P.M., Zahniser, M.S., Shorter, J.H., Wormhoudt, J.C., Lamb, B.K., Allwine, E.J., Gaffney, J.S., Marley, N.A., Grutter, M., Marquez, C., Blanco, S., Cardenas, B., Retama, A., Ramos Villegas, C.R., Kolb, C.E., Molina, L.T., Molina, M.J.: Evaluation of nitrogen dioxide chemiluminescence monitors in a polluted urban environment. Atmos. Chem. Phys. ACP **7**, 2691–2704 (2007)

Fahey, D.W., Gao, R.S., Möhler, O., Saathoff, H., Schiller, C., Ebert, V., Krämer, M., Peter, T., Amarouche, N., Avallone, L.M., Bauer, R., Bozóki, Z., Christensen, L.E., Davis, S.M., Durry, G., Dyroff, C., Herman, R.L., Hunsmann, S., Khaykin, S.M., Mackrodt, P., Meyer, J., Smith, J.B., Spelten, N., Troy, R.F., Vömel, H., Wagner, S., Wienhold, F.G.: The AquaVIT-1 intercomparison of atmospheric water vapor measurement techniques. Atmos. Meas. Tech. **7**, 3177–3213 (2014). https://doi.org/10.5194/amt-7-3177-2014

Fry, J.L., Kiendler-Scharr, A., Rollins, A.W., Brauers, T., Brown, S.S., Dorn, H.P., Dubé, W.P., Fuchs, H., Mensah, A., Rohrer, F., Tillmann, R., Wahner, A., Wooldridge, P.J., Cohen, R.C.: SOA from limonene: role of NO_3 in its generation and degradation. Atmos. Chem. Phys. **11**, 3879–3894 (2011). https://doi.org/10.5194/acp-11-3879-2011

Grosjean, D.: Wall loss of gaseous pollutants in outdoor simulation chambers. Environ. Sci. Tech. **19**, 1059–1065 (1985)

Hofzumahaus, A.: Measurement of photolysis frequencies in the atmosphere. In: Heard, D.E. (ed.) Analytical Techniques for Atmospheric Measurement. Blackwell Publishing, pp. 406–500 (2006)

Hofzumahaus, A., Kraus, A., Kylling, A., Zerefos, C.S.: Solar actinic radiation (280–420 nm) in the cloud-free troposphere between ground and 12 km altitude: measurements and model results. J. Geophys. Res. **107**, PAU 6-1–PAU 6-11 (2002). https://doi.org/10.1029/2001JD900142

Hofzumahaus, A., Lefer, B.L., Monks, P.S., Hall, S.R., Kylling, A., Mayer, B., Shetter, R.E., Junkermann, W., Bais, A., Calvert, J.G., Cantrell, C.A., Madronich, S., Edwards, G.D., Kraus, A., Müller, M., Bohn, B., Schmitt, R., Johnston, P., McKenzie, R., Frost, G.J., Griffioen, E., Krol, M., Martin, T., Pfister, G., Röth, E.P., Ruggaber, A., Swartz, W.H., Lloyd, S.A., Van Weele, M.: Photolysis frequency of O_3 to $O(1D)$: Measurements and modeling during the International Photolysis Frequency Measurement and Modeling Intercomparison (IPMMI), J. Geophys. Res. **109** (2004). https://doi.org/10.1029/2003JD004333

Holmes, J.R., O'Brien, R.J., Crabtree, J.H., Hecht, T.A., Seinfeld, J.H.: Measurement of ultraviolet radiation intensity in photochemical smog studies. Environ. Sci. Tech. **7**, 519–523 (1973). https://doi.org/10.1021/es60078a002

Hoyle, C.R., Fuchs, C., Järvinen, E., Saathoff, H., Dias, A., El Haddad, I., Gysel, M., Coburn, S.C., Tröstl, J., Bernhammer, A.K., Bianchi, F., Breitenlechner, M., Corbin, J.C., Craven, J., Donahue, N.M., Duplissy, J., Ehrhart, S., Frege, C., Gordon, H., Höppel, N., Heinritzi, M., Kristensen, T.B., Molteni, U., Nichman, L., Pinterich, T., Prévôt, A.S.H., Simon, M., Slowik, J.G., Steiner, G., Tomé, A., Vogel, A.L., Volkamer, R., Wagner, A.C., Wagner, R., Wexler, A.S., Williamson, C., Winkler, P.M., Yan, C., Amorim, A., Dommen, J., Curtius, J., Gallagher, M.W., Flagan, R.C., Hansel, A., Kirkby, J., Kulmala, M., Möhler, O., Stratmann, F., Worsnop, D.R., Baltensperger, U.: Aqueous phase oxidation of sulphur dioxide by ozone in cloud droplets. Atmos. Chem. Phys. **16**, 1693–1712 (2016). https://doi.org/10.5194/acp-16-1693-2016

https://www.nist.gov/system/files/documents/2020/03/24/gmp-11-calibration-intervals-20190506. pdf, Accessed 16 Dec 2020

Huang, Y., Zhao, R., Charan, S.M., Kenseth, C.M., Zhang, X., Seinfeld, J.H.: Unified theory of vapor–wall mass transport in Teflon-walled environmental chambers. Environ. Sci. Tech. **52**, 2134–2142 (2018). https://doi.org/10.1021/acs.est.7b05575

Hunter, J.F., Carrasquillo, A.J., Daumit, K.E., Kroll, J.H.: Secondary organic aerosol formation from acyclic, monocyclic, and polycyclic alkanes. Environ. Sci. Techol. **48**, 10227–10234 (2014). https://doi.org/10.1021/es502674s

Hynes, R.G., Angove, D., Saunders, S., Haverd, V., Azzi, M.: Evaluation of two MCM v3.1 alkene mechanisms using indoor environmental chamber data. Atmos. Enviro. **39**, 7251–7262 (2005). https://doi.org/10.1016/j.atmosenv.2005.09.005

Ibrahim, S.S., Bilger, R.W., Mudford, N.R.: Turbulence effects on chemical reactions in smog chamber flows. Atmos. Environ. **21**, 2609–2621 (1987). https://doi.org/10.1016/0004-698 1(87)90192-2

Karl, M., Brauers, T., Dorn, H.-P., Holland, F., Komenda, M., Poppe, D., Rohrer, F., Rupp, L., Schaub, A., Wahner, A.: Kinetic Study of the OH-isoprene and O_3-isoprene reaction in the atmosphere simulation chamber, SAPHIR. Geophys. Res. Lett. **31**, L05117 (2004)

Keller-Rudek, H., Moortgart, G.K., Sander, R., Sörensen, R.: The MPI-Mainz UV/VIS spectral atlas of gaseous molecules of atmospheric interest. Earth Syst. Sci. Data **5**, 365–373 (2013). https://doi.org/10.5194/essd-5-365-2013

Kraus, A., Rohrer, F., Hofzumahaus, A.: Intercomparison of NO_2 photolysis frequency measurements by actinic flux spectroradiometry and chemical actinometry during JCOM97. Geophys. Res. Lett. **27**, 1115–1118 (2000). https://doi.org/10.1029/1999GL011163

Krechmer, J., Pagonis, D., Ziemann, P.J., Jimnez, J.L.: Quantification of gas-wall partitioning in Teflon environmental chambers using rapid bursts of low-volatility oxidized species generated in situ. Environ. Sci. Tech. **50**, 5757–5765 (2016). https://doi.org/10.1021/acs.est.6b00606

Lai, A., Nazaroff, W.: Modeling indoor particle deposition from turbulent flow onto smooth surfaces. J. Aerosol Sci. **31**, 463–476 (2000). https://doi.org/10.1016/s0021-8502(99)00536-4

Lake Shore Cryotronics, Inc.: https://www.lakeshore.com/resources/sensors (2019). Accessed 16 Dec 2020

Lamkaddam, H.: Study under simulated condition of the secondary organic aerosol from the photooxydation of n-dodecane: Impact of the physical-chemical processes, Université Paris-Est (2017)

Leighton, P.A.: Photochemistry of air pollution. Academic Press, New York (1961)

Leskinen, A., Yli-Pirilä, P., Kuuspalo, K., Sippula, O., Jalava, P., Hirvonen, M.R., Jokiniemi, J., Virtanen, A., Komppula, M., Lehtinen, K.E.J.: Characterization and testing of a new environmental chamber. Atmos. Meas. Tech. **8**, 2267–2278 (2015). https://doi.org/10.5194/amt-8-2267-2015

Liu, X., Deming, B., Pagonis, D., Day, D.A., Palm, B.B., Talukdar, R., Roberts, J.M., Veres, P.R., Krechmer, J.E., Thornton, J.A., de Gouw, J.A., Ziemann, P.J., Jimenez, J.L.: Effects of gas–wall interactions on measurements of semivolatile compounds and small polar molecules. Atmos. Meas. Tech. **12**, 3137–3149 (2019). https://doi.org/10.5194/amt-12-3137-2019

Loza, C.L., Chan, A.W.H., Galloway, M.M., Keutsch, F.N., Flagan, R.C., Seinfeld, J.H.: Characterization of vapor wall loss in laboratory chambers. Environ. Sci. Technol. **44**, 5074–5078 (2010). https://doi.org/10.1021/es100727v

Massabò, D., Danelli, S.G., Brotto, P., Comite, A., Costa, C., Di Cesare, A., Doussin, J.F., Ferraro, F., Formenti, P., Gatta, E., Negretti, L., Oliva, M., Parodi, F., Vezzulli, L., Prati, P.: ChAMBRe: a new atmospheric simulation chamber for aerosol modelling and bio-aerosol research. Atmos. Meas. Tech. **11**, 5885–5900 (2018). https://doi.org/10.5194/amt-11-5885-2018

Matsunaga, A., Ziemann, P.J.: Gas-Wall partitioning of organic compounds in a Teflon film chamber and potential effects on reaction product and aerosol yield measurements. Aerosol Sci. Technol. **44**, 881–892 (2010). https://doi.org/10.1080/02786826.2010.501044

MBW Calibration Ltd.: www.mbw.ch (2020). Accessed 16 Dec 2020

Metcalf, A.R., Loza, C.L., Coggon, M.M., Craven, J.S., Jonsson, H.H., Flagan, R.C., Seinfeld, J.H.: Secondary organic aerosol coating formation and evaporation: chamber studies using black carbon seed aerosol and the single-particle soot photometer. Aerosol Sci. Technol. **47**, 326–347 (2013). https://doi.org/10.1080/02786826.2012.750712

METEOMET: https://www.meteomet.org/ (2020). Accessed 16 Dec 2020

Metzger, A., Dommen, J., Gaeggeler, K., Duplissy, J., Prevot, A. S.H., Kleffmann, J., Elshorbany, Y., Wisthaler, A., Baltensperger, U.: Evaluation of 1,3,5 trimethylbenzene degradation in the detailed tropospheric chemistry mechanism, MCMv3.1, using environmental chamber data. Atmos. Chem. Phys. **8**, 6453–6468 (2008). https://doi.org/10.5194/acp-8-6453-2008

Murphy, D.M., Koop, T.: Review of the vapour pressures of ice and supercooled water for atmospheric applications. Quart. J. r. Meteorol. Soc. **131**, 1539–1565 (2005). https://doi.org/10.1256/qj.04.94

Nachbar, M., Duft, D., Leisner, T.: Volatility of amorphous solid water. J. Phys. Chem. B **122**, 10044–10050 (2018a). https://doi.org/10.1021/acs.jpcb.8b06387

Nachbar, M., Duft, D., Leisner, T.: The vapor pressure over nano-crystalline ice. Atmos. Chem. Phys. **18**, 3419–3431 (2018b). https://doi.org/10.5194/acp-18-3419-2018

Nannoolal, Y., Rarey, J., Ramjugernath, D.: Estimation of pure component properties: Part 3. Estimation of the vapor pressure of non-electrolyte organic compounds via group contributions and group interactions. Fluid Phase Equilib. **269**, 117–133 (2008). https://doi.org/10.1016/j.fluid.2008.04.020

NIST 2019: See Good Measurement Practices GMP 11 & 13: https://www.nist.gov/system/files/documents/2019/06/21/gmp-13-ensuring-traceability-20190621.pdf

Pagonis, D., Krechmer, J.E., de Gouw, J., Jimenez, J.L., Ziemann, P.J.: Effects of gas–wall partitioning in Teflon tubing and instrumentation on time-resolved measurements of gas-phase organic compounds. Atmos. Meas. Tech. **10**, 4687–4696 (2017). https://doi.org/10.5194/amt-10-4687-2017

Rohrer, F., Bohn, B., Brauers, T., Brüning, D., Johnen, F.J., Wahner, A., Kleffmann, J.: Characterisation of the photolytic HONO-source in the atmosphere simulation chamber SAPHIR. Atmos. Chem. Phys. **5**, 2189–2201 (2005). https://doi.org/10.5194/acp-5-2189-2005

Saathoff, H., Möhler, O., Schurath, U., Kamm, S., Dippel, B., Mihelcic, D.: The AIDA soot aerosol characterisation campaign 1999. J. Aerosol Sci. **34**, 1277–1296 (2003). https://doi.org/10.1016/s0021-8502(03)00363-x

Schütze, M., Stratmann, F.: Numerical simulation of cloud droplet formation in a tank. Comput. Geosci. **34**, 1034–1043 (2008). https://doi.org/10.1016/j.cageo.2007.06.013

Shetter, R.E., Junkermann, W., Swartz, W.H., Frost, G.J., Crawford, J.H., Lefer, B.L., Barrick, J.D., Hall, S.R., Hofzumahaus, A., Bais, A., Calvert, J.G., Cantrell, C.A., Madronich, S., Müller, M., Kraus, A., Monks, P.S., Edwards, G.D., McKenzie, R., Johnston, P., Schmitt, R., Griffioen, E., Krol, M., Kylling, A., Dickerson, R.R., Lloyd, S.A., Martin, T., Gardiner, B., Mayer, B., Pfister, G., Röth, E.P., Koepke, P., Ruggaber, A., Schwander, H., van Weele, M.: Photolysis frequency of NO_2: measurement and modeling during the International Photolysis Frequency Measurement and Modeling Intercomparison (IPMMI). J. Geophys. Res. **108** (2003). https://doi.org/10.1029/2002JD002932

Shetter, R.E., Müller, M.: Photolysis frequency measurements using actinic flux spectroradiometry during the PEM-Tropics mission: instrumentation description and some results. J. Geophys. Res. **104**, 5647–5661 (1999). https://doi.org/10.1029/98JD01381

Tuesday, C. S.: The atmospheric photooxidation of trans-2-butene and nitric oxide. In: Chemical Reactions in the Lower and Upper Atmosphere, pp. 1–49. Interscience, New York, S.Y. (1961)

Valorso, R., Aumont, B., Camredon, M., Raventos-Duran, T., Mouchel-Vallon, C., Ng, N.L., Seinfeld, J.H., Lee-Taylor, J., Madronich, S.: Explicit modelling of SOA formation from α-pinene photooxidation: sensitivity to vapour pressure estimation. Atmos. Chem. Phys. **11**, 6895–6910 (2011). https://doi.org/10.5194/acp-11-6895-2011

Villena, G., Bejan, I., Kurtenbach, R., Wiesen, P., Kleffmann, J.: Interferences of commercial NO_2 instruments in the urban atmosphere and in a smog chamber. Atmos. Chem. Tech. **5**, 149–159 (2012). https://doi.org/10.5194/amt-5-149-2012

Voigtländer, J., Duplissy, J., Rondo, L., Kürten, A., Stratmann, F.: Numerical simulations of mixing conditions and aerosol dynamics in the CERN CLOUD chamber. Atmos. Chem. Phys. **12**, 2205–2214 (2012). https://doi.org/10.5194/acp-12-2205-2012

Wang, J., Doussin, J.F., Perrier, S., Perraudin, E., Katrib, Y., Pangui, E., Picquet-Varrault, B.: Design of a new multi-phase experimental simulation chamber for atmospheric photosmog, aerosol and cloud chemistry research. Atmos. Meas. Tech. **4**, 2465–2494 (2011). https://doi.org/10.5194/amt-4-2465-2011

Wang, X., Liu, T., Bernard, F., Ding, X., Wen, S., Zhang, Y., Zhang, Z., He, Q., Lü, S., Chen, J., Saunders, S., Yu, J.: Design and characterization of a smog chamber for studying gas-phase chemical mechanisms and aerosol formation. Atmos. Meas. Tech. **7**, 301–313 (2014). https://doi.org/10.5194/amt-7-301-2014

Wang, N., Jorga, S.D., Pierce, J.R., Donahue, N.M., Pandis, S.N.: Particle wall-loss correction methods in smog chamber experiments. Atmos. Meas. Tech. **11**, 6577–6588 (2018a). https://doi.org/10.5194/amt-11-6577-2018

Wang, N., Kostenidou, E., Donahue, N.M., Pandis, S.N.: Multi-generation chemical aging of α-pinene ozonolysis products by reactions with OH. Atmos. Chem. Phys. **18**, 3589–3601 (2018b). https://doi.org/10.5194/acp-18-3589-2018

Wilck, M., Stratmann, F., Whitby, E.: A fine particle model for fluent: description and application. In: Sixth International Aerosol Conference, Taipei, Taiwan, pp. 1269–1270 (2002)

WMO: World meteorological organization, guide to instruments and methods of observation, WMO-No. 8, Geneva (2018). ISBN: 978-92-63-10008-5

Yeh, G.K., Ziemann, P.J.: Identification and yields of 1,4-hydroxynitrates formed from the reactions of C8–C16 n-alkanes with OH radicals in the presence of NO_x. J. Phys. Chem. A **118**, 8797–8806 (2014a). https://doi.org/10.1021/jp505870d

Yeh, G.K., Ziemann, P.J.: Alkyl nitrate formation from the reactions of C8–C14 n-alkanes with OH radicals in the presence of NO_x: measured yields with essential corrections for gas–wall partitioning. J. Phys. Chem. A **118**, 8147–8157 (2014b). https://doi.org/10.1021/jp500631v

Yeh, G.K., Ziemann, P.J.: Gas-wall partitioning of oxygenated organic compounds: measurements, structure–activity relationships, and correlation with gas chromatographic retention factor. Aerosol. Sci. Technol. **49**, 727–738 (2015). https://doi.org/10.1080/02786826.2015.1068427

Zador, J., Turányi, T., Wirtz, K., Pilling, M.J.: Measurement and investigation of chamber radical sources in the European Photoreactor (EUPHORE). J. Atmos. Chem. **55**, 147–166 (2006). https://doi.org/10.1007/s10874-006-9033-y

Zhang, X., Cappa, C.D., Jathar, S.H., McVay, R.C., Ensberg, J.J., Kleemann, M.J., Seinfeld, J.H.: Influence of vapor wall loss in laboratory chambers on yields on secondary organic yields. Proc. Nat. Acad. Sci. **11**, 5802–5807 (2014). https://doi.org/10.1073/pnas.1404727111

Zhang, X., Schwantes, R.H., McVay, R.C., Lignell, H., Coggon, M.M., Flagan, R.C., Seinfeld, J.H.: Vapor wall deposition in Teflon chambers. Atmos. Chem. Phys. **15**, 4197–4214 (2015). https://doi.org/10.5194/acp-15-4197-2015

Chapter 3
Preparation of Simulation Chambers for Experiments

David Bell, Jean-François Doussin, and Thorsten Hohaus

Abstract When setting up a simulation chamber experiment it is essential, in order to ensure meaningful results, to start with a well-controlled chemical system. Coming after the chapter dealing with the requested careful characterization of the simulation chamber, the present chapter describes the preparation of the chamber before running an experiment. It includes various chamber cleaning protocols, the preparation of a clean chamber atmosphere (the reacting mixture) and a series of protocols for blank experiments. Indeed, having a clean atmosphere in a simulation chamber, as free as possible from both particulate and gaseous impurities, is essential to ensure high quality experimental results. As it may not be possible to have a perfectly clean chamber, blank experiments are crucial to both assess chamber cleanliness, account for impurities and establish uncertainties of the observed phenomena. In the present chapter, various cleaning protocols which involve the oxidation of the impurities, dilution, temperature degradation/evaporation, but the evacuation or manual cleaning are described as well. The various techniques to generate clean gas mixture mostly clean O_2, N_2 or water vapor, are discussed. Finally, complementarily to the reference experiments proposed in Chap. 2, blank experiments to characterize walls chemical inertia, chamber-dependent radical sources or the presence of water-soluble species are also described.

Having a clean atmosphere in a simulation chamber, as free from both particulate and gaseous impurities as possible is essential to ensure high quality experimental results that have a meaningful impact. It may not be practical or possible to

D. Bell
Paul Scherrer Institute, Würenlingen, Switzerland
e-mail: david.bell@psi.ch

J.-F. Doussin (✉)
Centre National de la Recherche Scientifique, Paris, France
e-mail: jean-francois.doussin@lisa.ipsl.fr

T. Hohaus
Forschungszentrum Jülich, Jülich, Germany
e-mail: t.hohaus@fz-juelich.de

© The Author(s) 2023   113
J.-F. Doussin et al. (eds.), *A Practical Guide to Atmospheric Simulation Chambers*,
https://doi.org/10.1007/978-3-031-22277-1_3

have a perfectly clean chamber and therefore blank experiments are crucial to both assess chamber cleanliness and account for impurities. Hence, cleaning processes and regular blank experiments are required to have a chamber that can generate meaningful and reproducible results.

The goal of chamber cleaning is to recreate the state of the chamber at the beginning of the previous experiment and to remove all unwanted species that were formed or injected over the course of a previous experiment. When operating in a batch mode, the chamber needs to be cleaned prior to each experiment in order to start in a state where there are minimal particles or gas phase species present. The goal is obviously to eliminate species from previous experiments down to levels that will not affect the consecution of the planned experiment. However, when operating in a continuous flow mode (where all reactants are continually added), then chamber cleaning should occur on a systematic cycle to verify that no build-up of unwanted contaminants is occurring.

The goal of blank experiments is to assess if a chamber is sufficiently clean or not and, if during a campaign, chamber cleanliness is an issue hampering the interpretation of results. The simplest blank experiment is certainly exposing a chamber only filled with clean air to the source of light and carefully monitor compounds build-up coming out directly from walls release or indirectly from the decompositon of sticky compounds trapped on the walls. A complementary typical blank experiment proceeds by conducting an experiment, but only with the addition of oxidant i.e., without the addition of a volatile organic compound (VOC). This makes possible to evaluate if a chamber artefact/contaminant is reacting with the oxidant rather than the VOC as desired. If this is the case, and there is significant production of low-volatility material then organic aerosol can be formed and observed.

3.1 Chamber Cleaning Protocols

Each chamber has its own chamber cleaning techniques that has been specially developed for its unique setup and may vary depending on the specific requirements of the coming experiments. Despite significant differences between the setup of each chamber, their desired studies, and their volume $(1–270 \, \text{m}^3)$, the principles of chamber cleaning are similar across all facilities. The steps for cleaning a chamber typically include:

- Creating an oxidative environment via O_3 and/or OH radicals.
- Dilution of gases and particles with a source of clean air.
- Creating a humid environment, as it has been shown that competitive adsorption of water allow the release of stuck species when chambers are exposed to high RH clean air.
- Proceed until all relevant VOCs, oxidants, inorganic compounds, and particle number concentration are below specifically set thresholds (e.g., 1 ppb, limit of detection, 1–10 #/cc).

- If unsuccessful, physically clean the chamber or replace the chamber (last resort).

As a result, typical measurements required during the course of chamber cleaning include: relevant gas monitors, a VOC detector such as a proton transfer reaction mass spectrometer (PTR-MS) or GC-FID, scanning mobility particle sizers (SMPS), and/or a condensation particle counter (CPC).

3.2 Chamber Cleaning Concepts

General non-invasive chamber cleaning methods rely upon using temperature, dilution, and oxidation to remove unwanted species from the chamber environment. The three aspects of the chamber that need to be cleaned are the air inside of the chamber, the walls of the chambers, and any leftover material on the sampling lines.

3.2.1 Oxidation

Oxidation of unwanted species present in the chamber is primarily used to turn heavy molecule sticked on the walls into smaller molecules (including CO and CO_2) through molecular fragmentation that can be easily eliminated thanks to their higher volatility. If the contamination is not that severe then one or two cleaning cycles should be sufficient to remove the contamination from the system. Of course, the evaluation of the level of cleanliness of the chamber requires well defined test or blank experiments to decide whether to continue or stop the cleaning process, which will be discussed below.

Oxidation typically proceeds by addition of O_3 or H_2O_2 in large quantities, which is photolyzed to produce OH radicals that rapidly react with most volatile organic compounds (VOCs), and produce molecules with lower volatility. Lower volatility species will condense on pre-existing particles, form new particles, or stick to the walls of the chamber. If the molecules have sufficiently low-volatility then they will be effectively removed from the chamber assuming they remain relatively unreactive on the walls of the chamber, or are removed from the chamber via dilution.

Ozonolysis of the remaining chamber contaminants is probably the most common procedure. It is nevertheless not without any drawbacks. First, whilst ozone will react with most unsaturated species, compounds such as aromatics and saturated hydrocarbons will remain unaffected. Second, ozonolysis may also lead to lower volatility products (Atkinson 2000) which may contribute to the organic contaminants burden of the chamber.

In order to favor the fragmentation of the contaminant carbon skeleton, and so to lead to lighter products that will be eliminated from chamber atmosphere, the exposure to oxidizing gases such as ozone can be further completed in the case of indoor chamber by switching on the irradiation system or exposing to the sun for outdoor chambers. In particular, the photolysis of ozone in large concentration in the presence of water produce a series of oxidizing radicals such as OH, $O^{(3P)}$, $O^{(1D)}$ that are extremely reactive toward most contaminants in the chamber (Atkinson 2000). Direct photolysis of contaminants carrying chromophores (e.g., aldehydes and ketones) also leads to fragmentation and the production of lighter products, which will contribute to the chamber cleaning.

3.2.2   Dilution

Dilution is the other process used to clean the chamber. For an inflatable chamber, this process is achieved by continually flushing the chamber with purified air (see Sect. 3.4) in order to remove contaminants/particles. Removal of particles is directly related to their wall loss lifetime and the lifetime of chamber dilution. Depending on the volume of the chamber, the chamber dilution lifetime may or may not be shorter than the wall loss lifetime of the chamber. Generally, smaller chambers can be cleaned more quickly via dilution than larger chambers because the flows required to clean large chambers are substantial. For instance, a 8 m^3 collapsible chamber at the Paul Scherrer Institute (PSI) can be cleaned from particles after flushing for 6 h, and this can observed in Fig. 3.1 which shows how the particle number concentration changes as a function of time. In this demonstration, the chamber volume had been reduced to ~4 m^3. The red line corresponds to the smoothed particle concentration over a 20 min. window. Over the duration of cleaning, shown in Fig. 3.2 the particle number concentration reached near background levels (5 #/cc) at ~24:00.

Regarding the gas phase, a typical non-linear rate at EUPHORE is a removal of half of the concentration in 30 min. Some gases can be particularly tricky to clean from the chamber due to their "stickiness". These include some small molecules, such

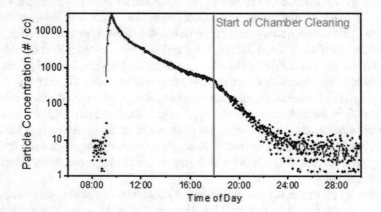

Fig. 3.1 Particle number concentration from an experiment with NH$_4$SO$_4$ seeds during photo-oxidation of Toluene +OH in the 8 m^3 chamber at the Paul Scherrer Institute

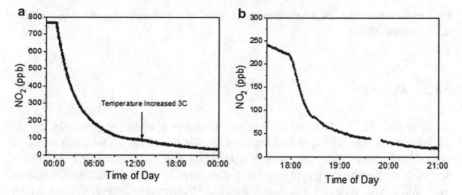

Fig. 3.2 **a** An experiment where 770 ppb of NO_2 had been added to the chamber, and chamber dilution at 18–21 °C over the course of 24 h. **b** Chamber dilution of NO_2 during a temperature ramp from 15 to 33 °C

as: NO_2, HCl, formic acid, ammonia, amines, etc. When different types of "sticky" VOCs are used (or formed) in the chamber it will be necessary to spend time to troubleshoot how to best clean the chamber after such experiments. For instance, in preliminary experiments performed in the PSI 8m3 chamber, with a crystalline N_2O_5 source, large amounts of NO_2 and HNO_3 were formed. This results from the reactions: $N_2O_5 \rightarrow NO_2 + NO_3$ and $N_2O_5 + 2\,H_2O \rightarrow 2\,HNO_3$. In some initial experiments, large quantities of N_2O_5 were added to the chamber resulting in significant concentrations of NO_2. Figure 3.2a shows the NO_2 concentrations during the cleaning cycle. In the example described here, the temperature of the chamber enclosure was initially 18 °C and was increased to 21 °C. Over 24 h of chamber flushing the NO_2 concentration never decreased below 50 ppb, and continued cleaning was required as a result. In subsequent experiments, the temperature was increased to 30 °C overnight to drive NO_y off the chamber walls and into the gas phase. Figure 3.2b shows cleaning of NO_2 with dilution taking place alongside a temperature ramp from 15 to 33 °C. Here NO_2 is efficiently cleaned out overnight and reaches reasonable concentrations after 3 h of dilution. The initial loss of NO_2 is due to expanding the chamber from ~3 to ~6 m^3.

For evacuable chambers, dilution cleaning is often achieved by pumping the chamber down to high vacuum by mean of oil-free pumping systems involving combination of special rotary pumps, roots pump and/or turbomolecular pumps (Barnes et al. 1994; Doussin et al. 1997; Wang et al. 2011). Sometime the systems are completed with series of sorption pumps (De Haan et al. 1999). In comparison to dilution, evacuating the chamber does not only allow the possibility to quickly replace potentially contaminated air with clean air, but also by reducing the total pressure in the chamber, it helps evaporating the low volatility species adsorbed on the walls. This advantage is nevertheless somewhat limited by the fact that the saturating vapor pressure of the species involved in SOA formation are often orders of magnitude lower that the best vacuum achieved in simulation chambers (Schervish and Donahue 2020).

3.2.3 Baking

As seen with in the "oxidation" section, cleaning a chamber generally means combining dilution with a process aiming at pushing the gas-wall partitioning of contaminants towards the gas phase. Raising the chamber temperature can increase that partitioning into the gas phase. Some chambers have the possibility to increase their wall temperature, for example the CESAM chamber (Créteil, France) where the walls are raised to 60 °C for several hours during the cleaning/pumping procedure. The NIES chamber (Tsukuba, Japan) has the capability to reach 200 °C and so to efficiently evaporate semi-volatile species on the surface of the walls. The benefit of raising chamber wall temperature has been more thoroughly described by Schnitzhofer et al. (2014) during the first runs of the CLOUD chamber (CERN). The CLOUD chamber was filled with synthetic air and was heated to 100 °C for 2 days at atmospheric pressure. No significant difference appears between 100 and 5 °C as total VOCs concentration measurements remain in the 1ppbv range. These authors recognize, nevertheless, that the chamber cleanliness benefited from a heating cycle, when a specific VOC had been added to the chamber for experimental reasons. From this last protocol, it can be recommended that a baking procedure when possible should be coupled with a low pressure evacuation of the chamber for a greater efficiency of low volatility species adsorbed on the wall.

3.2.4 Manual Cleaning

The above-described protocols have shown their efficiency in many instances. Nevertheless, in some cases such as the first use of a rigid chamber or after particularly dirty experiments (e.g., soot or mineral dust use, high concentration experiment, bio-aerosol study…) or before a particularly sensitive experiments, it is often needed to physically enter the chamber to manually clean it before applying a more common cleaning procedure. Indeed, some contaminants (dust, soots, bio-aerosol) may exhibit such a low vapor pressure that evacuation or dilution or baking would have a very limited cleaning efficiency. The presence of machining grease on the new material or heavy adsorbed chemicals arising from oxidation experiments may lead to similar issues.

Generally, the manual cleaning of chambers involves the use of significant quantities of both organic solvents and ultrapure water in order to remove both organic and ionic contaminants. Due to the fact that staff will be exposed to the chemical used, the toxicity of the chosen cleaning agent has to be minimized—often absolute ethanol is used. Safety should be considered carefully in manual cleaning processes,

Fig. 3.3 Manual cleaning of
an indoor evacuable stainless
steel chamber. ©
LISA-CNRS

for example check that chamber and laboratory are well ventilated, do not allow lone
working, ensure appropriate personal protection equipment (PPE) (see Fig. 3.3).
Appropriate PPE will also prevent contamination of the chamber by human material
(hair, cells…), the proper personal protection equipment has to be used. Ultra clean
lint free tissues must be employed. Do not be tempted to take short-cuts; always
remove or protect internal fittings such as White cell mirrors!

The benefit of this procedure being mostly to remove low- to non-volatile species,
after a manual cleaning, the chamber must be considered as heavily contaminated
by the solvent used. Generally, the solvent will be a volatile chemical and will be
eliminated through flushing or pumping with or without baking the chamber. Traces
of the cleaning solvent must then be systematically sought for, in blank experiments
following manual cleaning.

Another possibility is to clean the chamber by applying hot pressurized water
steam directly on the walls with a vaporizer to facilitate the removal of sticky contaminants. On a second step, milli-Q grade water is sprayed to help drag dirty water from
the walls.

3.3 Preparation of a Clean Chamber Atmosphere

Producing a well-controlled environment not only implies working in a clean
chamber but also to be able to fill this reactor with well-controlled matrix i.e., to
fill the chamber with clean air. Again, various technological set-up are currently in
operation. They mostly depend on the required clean air flow and so of volume of
the chamber or the type of experiment (i.e., batch flow operation requires more clean
air than static chamber operation).

In one of the biggest chambers in the world, the EUPHORE chamber, a high-
volume clean air set-up has successfully been in operation for several decades. The
EUPHORE facility comprises of two 200 m^3 chambers. Each chamber can be filled
with air from a separate air purification system. For pressurizing the filter system

to ca. 6 bar, a screw compressor (Mannesmann, Type Ralley 110 AS) is used. This type of compressor is suitable for continuous operation. After the compressor the air is passed through a condensate trap to separate oil and water from the air. The emulsion is separated in an oil/water separator. Two pressure tanks with a volume of $1m^3$ each are used as a buffer reservoir to reduce the switch frequency of the compressor. The air is dried in adsorption driers (Zander, Type HEA 1400) with an air throughput of ca. 500 m^3/h. These driers are filled with a molecular sieve type 4A (ECO 30%, MOL 70%). With this, a pressure dew point of -70 °C is reached and the CO_2 content is reduced significantly. With the help of a charcoal adsorber, NO is eliminated and oil vapor as well as non-methane hydrocarbons are reduced, e.g., benzene and toluene are below the detection limits of the instruments, 70 ppt and 40 ppt (3 standard deviations), respectively. The air inlets are located in the center of the chambers. After passing a pressure reduction valve the clean air is blown into the chamber via silencers of bespoke design. Due to its dimension and the noise when in operation, this device is located in a dedicated room located next to the smog chamber laboratories.

At the SAPHIR chamber, another large (>270 m^3) outdoor chamber, synthetic air is produced by mixing evaporated liquid nitrogen and liquid oxygen (Linde, purity $> 99.9999\%$). Mixing is done with flow-controllers to ensure specific mixing ratios of oxygen and nitrogen. A metal tank of several cubic meter serves as reservoir for the consumption of the chamber, but also instruments. It is filled with the mixed synthetic air with a two-point pressure control loop. A similar route toward synthetic air production is used to produce the clean chamber environment found at the Cosmics Leaving OUtdoor Droplets (CLOUD) chamber at CERN (Duplissy et al. 2010).

At the CESAM chamber (Wang et al. 2011), synthetic air produced from the mixture of high purity O_2 originating from commercial cylinder (Air Liquide®, Alphagaz® class 1) and nitrogen produced from the evaporation of a pressurized liquid nitrogen tank. Similar to the SAPHIR process, this nitrogen source is cost-effective and free from trace gas such as VOCs or NO_x, but it exhibits a contamination of ca. 200 ppb of carbon monoxide (± 100 ppb depending on pressure service or delivery lot). However, due to its low reactivity compared to atmospheric processes and its very high vapor pressure this was not considered as a major inconvenience either for ozone production studies or for aerosol chemistry studies. Further, blank reactivity experiments (see below) account for the consequences of such a contamination.

The chambers at the Paul Scherrer Institute (8–27 m^3) utilize an AADCO 250 series (AADCO Instruments, Inc. USA) air purifiers coupled to high pressure air lines. This air purification system provides zero air with background of trace gases including: $O_3 < 1$ ppb, CO < 6 ppb, $NO_x < 100$ ppt, organic contamination ~4 ppb (Paulsen et al. 2005). These contamination levels are sufficient to study SOA formation, but not clean enough for new particle formation studies such as that mentioned above in CLOUD.

In some specific cases, specials care to the cleanliness of the background air must be taken. It is especially the case for experiments involving nucleation event studies where results may be extremely sensitive to H_2SO_4, NH_3 and Extremely Low

Volatility OC (ELVOC) background concentrations. When relevant these species must be targeted by the analytical techniques involved in the background characterization. These techniques must be extremely sensitive as their target must be monitored in the sub-ppt range to avoid any impact on the results. This is particularly the case for NH_3 background blanks. Bianchi et al. (2012) and Brégonzio-Rozier et al. (2016) have shown that ammonia contamination in the ppt range were not uncommon in simulation chamber. Being ubiquitous at these low concentrations, prone to permeation because of its small size and possibly formed by reduction at the wall of the stainless chamber, ammonia elimination from chamber atmosphere is a particular challenge. Ammonia has been discussed as having a dramatic impact on the nucleation rate (Ball et al. 1999; Benson et al. 2011; Korhonen et al. 1999). It can also play a role on condensation growth if there is an attempt to use low or modest amounts of acidic seeds. Similarly, previous wall HNO_3 contamination can rapidly lead to nitrate buildup in particles.

When humidifying the chamber, significant quantities of water vapor have to be injected to adjust the relative humidity of the simulated atmosphere. As an example, saturating with water a 20 °C atmosphere requires more than 17 g-per-cubic-meter of water which make water the most abundant gas right after nitrogen (N_2) and oxygen (O_2) in most of the atmospheric simulation experiment. In consequence, using the highest purity water is often desirable as soluble species can often be introduced during water evaporation.

3.4 Control and Blank Experiments

Similar to the chamber cleaning protocols, the protocols for blank experiments are specifically developed for each chamber. However, certain procedures are observed by most of the chamber protocols which can be used as a general guideline on how to check for the cleanliness and status of a chamber. Overall, to ensure a basic understanding of the status of a chamber prior to an experiment, most chambers are monitoring the following conditions while oxidants are introduced and/or produced in the chamber:

- Concentration of oxidants (typically O_3 and/or OH)
- Concentration of inorganic compounds (typically NO, NO_2, SO_2)
- Concentration of volatile organic compounds (VOCs)
- Aerosol number concentrations and size distribution.

In general, the chamber is regarded as clean when the concentration of most of the compounds measured falls below the detection limit of the monitoring instrument. Aside from the direct measurement of contaminant concentration after a cleaning procedure, more dynamic protocols take advantage from the atmospheric processes themselves to characterize invisible (or unmeasured) contamination. Indeed, even if an initial characterization of the chamber state through measurements remains a clear prerequisite for any simulation run, major contamination affecting the results of the planned experiments are often not detectable.

As an example, in spite of very high cleanliness levels, it was observed at the CLOUD chamber from CERN (Duplissy et al. 2010) that a small rise of wall temperature over a short time interval almost always gave rise to a spontaneous burst of freshly nucleated particles. This effect most probably due to trace vapors (sulfur dioxide, sulfuric acid and/or organic compounds) previously below the instruments detection limits and who, when released from the walls of the chamber, contributed to nucleation.

"Blank" or "control" experiments are hence critical part of the experimental strategy to such an extent that they need to be carefully analyzed and stored together with the experiments themselves. Similarly, to the chamber cleaning protocols, the protocols for blank experiments may be specific to a chamber but they are always tightly related to the objectives of the experiments. There is certainly a significant diversity. However, certain procedures are observed by most of the chamber protocols for a common type of experiment and can be used as a general guideline on how to check for the state of a chamber.

3.4.1 Walls Chemical Inertia

The walls of a chamber are a vital aspect of any experiment taking place. The walls represent a large surface area which facilitates interfacial reactions and consequently can be a reactive sink or source of any gas phase species. Therefore, characterizing the walls' oxidative or reductive potential represents a fundamental task in any chamber blank experiment. Adding chemically sensitive species (e.g., O_3, NO, NO_2, etc....) to the chamber and following their time series can be a useful blank experiment to determine the role the chamber walls are playing. The lifetime of sensitive species such as ozone or NO, in a 'clean' chamber filled with air is often considered as indicators of the chemical inertia of the chamber walls. (Leskinen et al. 2015; Wang et al. 2011). The clean-air and NO_x system has been studied in Teflon chambers in a number of studies (Bloss et al. 2005; Metzger et al. 2008; Wang et al. 2014). It has been proven as a very sensitive system to detect the release of NO_y species from the wall or unknown radical sources. As these processes are often related to wall cleanliness, these blank experiments provide useful insight on chamber walls physico-chemical behavior. Moreover, as the NO_x/air/light chemical system lies in the heart of tropospheric chemistry oxidation scheme, it is now promoted, in addition, as reference experiments that need to be carried out regularly not only to check for chamber contamination but also to set the chamber auxiliary mechanism parameters. Protocols are hence recommended in Chap. 2.

3.4.2 Chamber Dependent Radical Sources

The history of the chamber is also important when considering the chemical inertia of the walls. Indeed, walls are not only sinks for reactive species or products, they are also well known sources for species that can more or less directly affect the radical balance of a simulation experiments. In particular, HONO and HCHO are among the most common wall emitted species that will give rise to OH radical through photolysis. For low NO_x experiment, irradiating a mixture of a reactive VOC (such as propene) and air while checking that ozone formation remain negligible is a good diagnostic. Another option could be being able to model the production of O_3 in high NO_x experiments with a reactive VOC (e.g., propene) (refer to protocol provided in Chap. 2).

To characterize an invisible organic reactivity of the chamber background, one can photolyse a "clean air" atmosphere (Hynes et al. 2005) and control for the formation of any relevant species (see Sect. 2.4, Chap. 2). This procedure must be carried out under typical relative humidity conditions as humidity is known to affect release of some adsorbed contaminants. Ozone is certainly a good target for such a blank experiment due to the amplification of the ozone production through radical cycle. HCHO or formic acid as termination products of oxidation processes are also common species arising under such conditions.

Aside from gas phase processes, an undetected organic reactivity of the chamber background can also affect aerosol formation. These blank experiments are especially important to carry out when focusing on weak secondary aerosol producers such as isoprene. As experiments are performed semi-volatile and low volatility oxygenated organics can build up on the chamber walls changing their effect on experiments. These species coming off the walls can act as a source of reactivity with O_3 or OH thereby resulting in the formation of SOA. Therefore, it is necessary to conduct blank experiments on a regular basis. Proper blank experiments are performed in the presence of an oxidation source (OH or O_3) and seed aerosol, which provides a surface for low-volatility species to condense. For example, after a series of experiments in the PSI chamber where a polymer-mix was injected into the chamber, it was necessary to check the cleanliness of the chamber.

Sources of VOC contaminations can also result from the presence of undesired components of the chamber itself. In experiments at the CLOUD chamber in CERN, plastic material used in both sampling lines and the O_3 generator itself were responsible for the production VOCs that correlated with the presence of O_3 (CLOUD3 and 4 in Fig. 3.4). Likely from the reaction of O_3 with the material itself. Once the plastic parts were removed from the chamber and a new O_3 generator was built out of quartz and stainless steel then the production of VOCs was minimized and there was no longer a strong correlation with the presence of O_3.

Similarly, when studying SOA formation from VOC oxidation, it is often recommended to set-up the oxidation process (e.g., ozonolysis or OH oxidation) in the presence of seeds aerosol, absence of any VOC and to monitor, as a background formation, the aerosol formation and growth. (Leskinen et al. 2015).

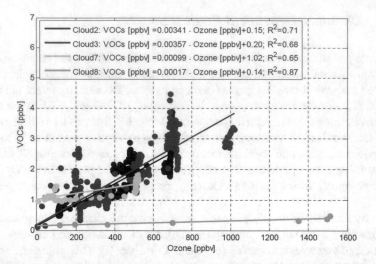

Fig. 3.4 Correlation between VOCs (C1-C3) measured by the PTR-MS. In CLOUD 2 and 3 there were plastic parts present on the ozone generator and other instruments around the chamber. After their replacement, the contamination was significantly diminished. (Reused with permission from Schnitzhofer et al. 2014 Open access under a CC BY 3.0 license, https://creativecommons.org/lic enses/by/3.0/)

3.4.3   Soluble Species Affecting Potential Aqueous SOA Formation

The scope of simulation chamber use has been extended to the investigation of cloud assisted aerosol formation (Ervens et al. 2011). Cloud assisted SOA formation is extremely difficult to control and is a very sensitive process to potential water-soluble contaminants. Brégonzio-Rozier et al. (2016) found out that it was necessary to perform a thorough manual cleaning involving the use ultrapure ethanol, followed by bathing the walls with large quantities of ultrapure water, and completed by baking the wall to 60 °C and overnight pumping at a secondary vacuum. They also implemented an experimental sequence including, before each experiment, a cleaning session followed by a "blank" experiment. Considering that the overall goal of their study was to quantify aqSOA formation trigger by a cloud event, these "blank" experiments consisted of triggering cloud formation events in the 'clean' chamber only filled up with ultrapure air at RH close to 100% (Brégonzio-Rozier et al. 2016). An example of these tests aiming at quantifying a potential background aqSOA formation is given in Fig. 3.5. It can be seen that even if the cleaning protocol was not able to totally suppress the formation of particles through cloud processing of impurities, it was able to reduce its extent by a factor of ca. 5 and to bring it close to the instrument detection limits. The authors organized the curation of their experiments, together with the related "blank" experiments, in order to take into account this artifact in their data analysis.

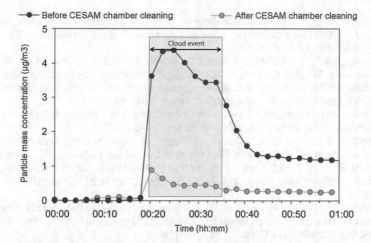

Fig. 3.5 Effect of cleaning a cloud chamber on background aqueous SOA formation during a cloud event (adapted from Brégonzio-Rozier thesis 2013). The mass concentration is deduced from SMPS measurement assuming spherical particle and a density of 1. The particles are dried to below 30% RH before injection in the differential mobility analyzer

Finally, protocols for preparing the chambers differ strongly with regard to chambers (material, size, specialization of each chamber) and even more with regard to the scientific objective of the planned experiments. Cleaning procedure must take this into account as well as the experiments that were previously carried out. Every time it is possible, the experiments themselves must be accompanied with control/blank experiments aiming at evaluating the cleanliness and status of the chambers as well as the chemical background reactivity of the reactive mixture (bath gas, purity of water, of precursors…).

Preparing a chamber for conducting robust atmospheric simulation experiments, is not only about applying carefully standard cleaning protocols and reactive mixture recipes, but also about implementing the full traceability of the experimental conditions, that implies the curation of blank experiments datasets, together with the precise preparation protocol applied.

References

Atkinson, R.: Atmospheric chemistry of VOCs and NO$_x$. Atmos. Environ. **34**, 2063–2101 (2000)

Ball, S.M., Hanson, D.R., Eisele, L., McMurry, P.H.: Laboratory studies of particle nucleation: initial results for H$_2$SO$_4$, H$_2$O and NH$_3$ vapors. J. Geophys. Res. **104**, 23709–23718 (1999)

Barnes, I., Becker, K.H., Mihalopoulos, N.: An FTIR product study of the photooxidation of dimethyl disulfide. J. Atmos. Chem. **18**, 267–289 (1994). https://doi.org/10.1007/bf00696783

Benson, D.R., Yu, J.H., Markovich, A., Lee, S.-H.: Ternary homogeneous nucleation of H$_2$SO$_4$, NH$_3$, and H$_2$O under conditions relevant to the lower troposphere. Atmos. Chem. Phys. **11**, 4755–4766 (2011). https://doi.org/10.5194/acp-11-4755-2011

Bianchi, F., Dommen, J., Mathot, S., Baltensperger, U.: On-line determination of ammonia at low pptv mixing ratios in the CLOUD chamber. Atmos. Meas. Tech. **5**, 1719–1725 (2012). https://doi.org/10.5194/amt-5-1719-2012

Bloss, C., Wagner, V., Jenkin, M.E., Volkamer, R., Bloss, W.J., Lee, J.D., Heard, D.E., Wirtz, K., Martin-Reviejo, M., Rea, G., Wenger, J.C., Pilling, M.J.: Development of a detailed chemical mechanism (MCMv3.1) for the atmospheric oxidation of aromatic hydrocarbons. Atmos. Chem. Phys. **5**, 641–664 (2005). https://doi.org/10.5194/acp-5-641-2005

Brégonzio-Rozier, L.: Formation d'Aérosols Organiques Secondaires au cours de la photooxydation multiphasique de l'isoprène, University of Paris Est Créteil (2013)

Brégonzio-Rozier, L., Giorio, C., Siekmann, F., Pangui, E., Morales, S.B., Temime-Roussel, B., Gratien, A., Michoud, V., Cazaunau, M., DeWitt, H.L., Andrea, T., Monod, A., Doussin, J.-F.: Secondary organic aerosol formation from isoprene photooxidation during cloud condensation–evaporation cycles. Atmos. Chem. Phys. **16**, 1747–1760 (2016). https://doi.org/10.5194/acp-16-1747-2016

De Haan, D.O., Brauers, T., Oum, K., Stutz, J., Nordmeyer, T., Finlayson-Pitts, B.J.: Heterogeneous chemistry in the troposphere: experimental approaches and applications to the chemistry of sea salt particles. Int. Rev. Phys. Chem. **18**, 343–385 (1999)

Doussin, J.F., Ritz, D., DurandJolibois, R., Monod, A., Carlier, P.: Design of an environmental chamber for the study of atmospheric chemistry: new developments in the analytical device. Analusis **25**, 236–242 (1997)

Duplissy, J., Enghoff, M.B., Aplin, K.L., Arnold, F., Aufmhoff, H., Avngaard, M., Baltensperger, U., Bondo, T., Bingham, R., Carslaw, K., Curtius, J., David, A., Fastrup, B., Gagné, S., Hahn, F., Harrison, R.G., Kellett, B., Kirkby, J., Kulmala, M., Laakso, L., Laaksonen, A., Lillestol, E., Lockwood, M., Mäkelä, J., Makhmutov, V., Marsh, N.D., Nieminen, T., Onnela, A., Pedersen, E., Pedersen, J.O.P., Polny, J., Reichl, U., Seinfeld, J.H., Sipilä, M., Stozhkov, Y., Stratmann, F., Svensmark, H., Svensmark, J., Veenhof, R., Verheggen, B., Viisanen, Y., Wagner, P.E., Wehrle, G., Weingartner, E., Wex, H., Wilhelmsson, M., Winkler, P.M.: Results from the CERN pilot CLOUD experiment. Atmos. Chem. Phys. **10**, 1635–1647 (2010). https://doi.org/10.5194/acp-10-1635-2010

Ervens, B., Turpin, B.J., Weber, R.J.: Secondary organic aerosol formation in cloud droplets and aqueous particles (aqSOA): a review of laboratory, field and model studies. Atmos. Chem. Phys. **11**, 11069–11102 (2011)

Hynes, R.G., Angove, D.E., Saunders, S.M., Haverd, V., Azzi, M.: Evaluation of two MCM v3.1 alkene mechanisms using indoor environmental chamber data. Atmos. Environ. **39**, 7251–7262 (2005)

Korhonen, P., Kulmala, M., Laaksonen, A., Viisanen, Y., McGraw, R., Seinfeld, J.H.: Ternary nucleation of H_2SO_4, NH_3, and H_2O in the atmosphere. J. Geophys. Res. **104**, 26349–26353 (1999)

Leskinen, A., Yli-Pirilä, P., Kuuspalo, K., Sippula, O., Jalava, P., Hirvonen, M.R., Jokiniemi, J., Virtanen, A., Komppula, M., Lehtinen, K.E.J.: Characterization and testing of a new environmental chamber. Atmos. Meas. Tech. **8**, 2267–2278 (2015). https://doi.org/10.5194/amt-8-2267-2015

Metzger, A., Dommen, J., Gaeggeler, K., Duplissy, J., Prevot, A.S.H., Kleffmann, J., Elshorbany, Y., Wisthaler, A., Baltensperger, U.: Evaluation of 1,3,5 trimethylbenzene degradation in the detailed tropospheric chemistry mechanism, MCMv3.1, using environmental chamber data. Atmos. Chem. Phys. **8**, 6453–6468 (2008). https://doi.org/10.5194/acp-8-6453-2008

Paulsen, D., Dommen, J., Kalberer, M., Prévôt, A.S.H., Richter, R., Sax, M., Steinbacher, M., Weingartner, E., Baltensperger, U.: Secondary organic aerosol formation by irradiation of 1,3,5-Trimethylbenzene:NO_x:H_2O in a new reaction chamber for atmospheric chemistry and physics. Environ. Sci. Technol. **39**, 2668–2678 (2005)

Schervish, M., Donahue, N.M.: Peroxy radical chemistry and the volatility basis set. Atmos. Chem. Phys. **20**, 1183–1199 (2020). https://doi.org/10.5194/acp-20-1183-2020

Schnitzhofer, R., Metzger, A., Breitenlechner, M., Jud, W., Heinritzi, M., De Menezes, L.P., Duplissy, J., Guida, R., Haider, S., Kirkby, J., Mathot, S., Minginette, P., Onnela, A., Walther, H., Wasem, A., Hansel, A., The Cloud Team.: Characterisation of organic contaminants in the CLOUD chamber at CERN. Atmos. Meas. Tech. **7**, 2159–2168 (2014). https://doi.org/10.5194/amt-7-2159-2014

Wang, J., Doussin, J.-F., Perrier, S., Perraudin, E., Katrib, Y., Pangui, E., Picquet-Varrault, B.: Design of a new multi-phase experimental simulation chamber for atmospheric photosmog, aerosol and cloud chemistry research. Atmos Measur Tech **4**, 2465–2494 (2011)

Wang, X., Liu, T., Bernard, F., Ding, X., Wen, S., Zhang, Y., Zhang, Z., He, Q., Lü, S., Chen, J., Saunders, S., Yu, J.: Design and characterization of a smog chamber for studying gas-phase chemical mechanisms and aerosol formation. Atmos. Meas. Tech. **7**, 301–313 (2014). https://doi.org/10.5194/amt-7-301-2014

Chapter 4
Preparation of Experiments: Addition and In Situ Production of Trace Gases and Oxidants in the Gas Phase

David M. Bell, Manuela Cirtog, Jean-François Doussin, Hendrik Fuchs, Jan Illmann, Amalia Muñoz, Iulia Patroescu-Klotz, Bénédicte Picquet-Varrault, Mila Ródenas, and Harald Saathoff

Abstract Preparation of the air mixture used in chamber experiments requires typically the injection of trace gases into a bath gas. In this chapter, recommendations and standard protocols are given to achieve quantitative injections of gaseous, liquid or solid species. Various methods to produce ozone, nitrate radicals and hydroxyl radicals are discussed. Short-lived oxidants need to be produced during the experiment inside the chamber from pre-cursor species. Because highly reactive oxidants like hydroxyl radicals are challenging to detect an alternative method for the quantification of radical concentrations using trace molecules is described.

4.1 Introduction

In order to obtain useful data from simulation chamber experiments, reliable and reproducible additions of reactants such as volatile organic compounds (VOCs) and inorganic compounds are required. The most suitable method depends on the physical

D. M. Bell
Paul Scherrer Institute, Villingen, Switzerland

M. Cirtog · B. Picquet-Varrault
University of Paris East Créteil, Créteil, France

J.-F. Doussin
Centre National de la Recherche Scientifique, Paris, France

H. Fuchs (✉)
Forschungszentrum Jülich, Jülich, Germany
e-mail: h.fuchs@fz-juelich.de

J. Illmann · I. Patroescu-Klotz
Bergische Universität Wuppertal, Wuppertal, Germany

A. Muñoz · M. Ródenas
Fundación Centro de Estudios Ambientales del Mediterráneo, Paterna, Spain

H. Saathoff
Karlsruhe Institute of Technology, Karlsruhe, Germany

© The Author(s) 2023
J.-F. Doussin et al. (eds.), *A Practical Guide to Atmospheric Simulation Chambers*,
https://doi.org/10.1007/978-3-031-22277-1_4

129

properties of the compound and on the specific chamber geometry and operational mode. The injection method should ensure homogeneous mixing, avoid contamination or memory effects of the reaction mixtures and avoid interferences of secondary processes. In addition, the goal of the specific experiment and properties of instruments used for the analysis (sensitivity and sampling frequency of instruments) need to be considered. Thus, the injection procedure needs to be carefully chosen and should be well characterized.

Typically, the injection of stable compounds is achieved by transporting small amounts of reagents into the chamber together with a high gas flow of nitrogen or synthetic air using a specialized inlet system. Although gaseous and liquid reagents can be often directly injected into the chamber (Fig. 4.1), homogeneous mixing can be challenging, if reagents are introduced at one point in the chamber. Most chambers are therefore equipped with fans to ensure rapid mixing. Because oxidants such as OH and NO_3 radicals are short-lived and highly reactive so that they cannot be stored, they need to be produced during the experiment from stable precursor compounds.

In this chapter, the addition of gaseous, liquid, and solid organic and inorganic compounds into a chamber and methods for the in situ production of oxidants are described.

4.2   Injection of Gaseous Compounds

Small amounts of pure gases (<10 cm^3) can be directly injected with a gas-tight syringe straight through a septum in the inlet system of the chamber (Fig. 4.1). The gas is then flushed into the chamber together with a high flow of the bath gas. Alternatively, the injection port with the septum can be directly mounted on the chamber wall. In order to ensure rapid mixing, it is useful to have the injection port located close to a fan (Fig. 4.1). For an accurate estimation of the injected volume, the dead volume of the needle of the gas syringe needs to be determined, because the gas in the needle is also injected into the chamber. The dead volume can be measured by weighing the syringes when only the needle is filled with pure water whose density was determined in advance (for the appropriate laboratory temperature).

The resulting mixing ratio ($c_{reactant}$) of the injected compound inside the chamber can be calculated using the ratio of the injected volume ($V_{reactant}$) and the volume of the reaction chamber ($V_{chamber}$):

$$c_{reactant} = \frac{V_{reactant}}{V_{chamber}} \tag{4.2.1}$$

The accurate preparation of mixtures at the ppbv level can be difficult if a small sample volume is required. The manual injection with gas-tight syringes is not always precise and can limit the reproducibility of experiments. As an alternative, six-way valve systems (not shown on Fig. 4.1) replacing the syringe injection port allow more accurate and precise injections of reagent volumes as low as a few tenths of microliters.

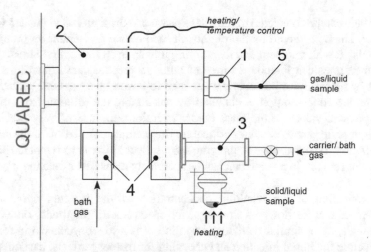

Fig. 4.1 Means of additions of stable molecules into the atmospheric chamber QUAREC (University Wuppertal): (1) injection port for syringes, (2) heated injection block, (3) alternative device for the addition of low volatile liquid or solid samples, (4) valves for bath gases and (5) syringe. Schematics by Iulia Patroescu-Klotz, BUW

High concentrations of reactive gases in the chamber, for example, needed to scavenge OH radicals in ozonolysis experiments, can be achieved by directly flowing pure gases or gas mixtures from gas cylinders equipped with pressure regulating valves into the chamber via an inlet system. The gas flow into the chamber can be controlled by flow metres or flow controllers and pressure gauges. The volume of the injected reagent can be calculated using the measured flow rate (F) and the injection time (t):

$$V_{reactant} = F \times t \tag{4.2.2}$$

In this case, the accuracy of the injected volume is limited by the accuracy of the measured flow rate and the injection time. If flow controllers based on thermal measurement of the flow rate are used, the thermal capacity of the specific gas may need to be considered to derive the actual flow rate.

Alternatively, the reagent gas samples can be prepared (purified and/or dried) in a vacuum line. The reagent gas can then be prepared in a glass bulb. The total concentration can be calculated from its volume (V_{bulb}) and the pressure in the bulb (p_{bulb}) and the chamber ($p_{chamber}$) measured with an appropriate manometre:

$$c_{reactant} = \frac{V_{bulb} \times p_{bulb}}{V_{chamber} \times p_{chamber}} \tag{4.2.3}$$

The bulb can be connected to the inlet system and the material is flushed into the chamber. The bulb should be thoroughly cleaned before use, either by pumping or by cleaning it with a solvent and then drying it in an oven for several hours.

Vacuum lines can include a series of bulbs to prepare small quantities of the reagent by successive dilution into a gas bath followed by a pressure reduction. In this case, the concentration is obtained by calculating the dilution from the ratio of the pressure values in the glass bulbs. Therefore, the use of very precise and reliable pressure gauges becomes critical for the accurate calculation of the resulting concentration. The advantage of applying successive dilution steps is that the pressure in the bulbs remains in a range that can be accurately measured by standard pressure gauges.

Determination of the resulting concentration of the injected compound in the chamber can be done in situ, e.g., by spectrometric methods, online (gas-chromatography or similar methods) or offline. It is also worth checking for blank values before the initial injection and checking for memory effects, particularly for 'sticky' substances. The use of short and heated Teflon tubes in the inlet system is recommended for the injection of these compounds. Preparing diluted mixtures can also help to achieve a quantitative injection. Injections via flow controllers should be avoided for these compounds because they are prone to memory effects and may become clogged or damaged.

4.3   Injection of Compounds from Liquid Sources

Reagents that are liquid at room temperature can be injected into the simulation chamber with similar systems like gaseous reagents. The optimum method depends on the vapour pressure and stability of the compound. It is worth noting that the partial pressure of the compound inside the chamber needs to be lower than the saturation vapour pressure of the reagent at the chamber temperature.

The simplest method is using a syringe, with which the compound is injected into a bath gas stream in a heated inlet system, where the liquid rapidly evaporates and is transported into the chamber. This is particularly applicable for moderate to low vapour pressure compounds. Caution is recommended for sticky compounds. The resulting concentration in the reaction chambers is

$$c_{reactant} = \frac{V_{inj.} \times \rho \times P \times N_A}{M \times V_{chamber}} \qquad (4.3.1)$$

with $V_{inj.}$ = injected volume of the liquid compound; ρ = density of the liquid compound; P = purity of the compound sample; N_A = Avogadro's number; M = molar mass of the compound; $V_{chamber}$ = volume of the chamber.

The precision and reproducibility of injections using syringes with volumes in the low micro-litre range are low. Part of the liquid can be lost due to evaporation from the glass syringe between preparation and injection and the dead volume of the needle affects the injected volume similar to the injection of gaseous compounds (Sect. 4.2). For calibration purposes, pure compounds can be dissolved in a suitable solvent, in order to obtain high-diluted solutions that ensure that the desired amount of organic compound is transferred without losses into the gas phase inside the chamber (Etzkorn et al. 1999). This can increase the accuracy, with which the injected volume is known, and minimize losses. Caution has to be taken that the solvent does not affect the experimental results nor the measurement of trace gases. There are also commercial solutions available for the evaporation of compounds solved in liquids into an air flow. An example is the Liquid Calibration Unit (LCU) provided by the company IONICON that was designed for the calibration of mass spectrometer instruments. Flows are accurately controlled by mass flow controllers to produce well-defined concentrations of the reactants in the air that is provided. However, using flow controllers is only feasible for some solvents and works best for water. This limits the application of the LCU to certain compounds.

The injection of liquid samples can also be achieved using permeation sources. Permeation tubes are commercially available or can be custom-built. A constant permeation rate is achieved if the tube is kept at a constant temperature. The compound is flushed at a well-defined rate from the permeation source into the chamber (Tumbiolo et al. 2005). The resulting concentration of the compound in the chamber can be accurately calculated from measured flow rates and the permeation rate that is determined from reference permeation sources or the measured weight loss of the compound in the permeation source.

One other method to introduce low-volatility compounds into the chamber is using a glass vial that is connected to the inlet system through a glass/stainless steel line (Fig. 4.2). The vapour above the liquid is flushed with the bath gas into the chamber. Evaporation can be enhanced by heating the vial. Lines are recommended to be cleaned before use, for example, by evacuating them or flushing them with clean dry air.

Similar to gaseous compounds, a vacuum line can be used to generate gaseous reagent samples in a glass bulb containing a well-defined concentration. The liquid reagent sample is placed in a glass vial that is attached to the line (Fig. 4.2). In order to remove volatile impurities from the sample, the reactant can be frozen with liquid nitrogen and the air is pumped away from the sample. The valve between the finger and the vacuum line is then closed and the Dewar flask is removed to evaporate impurities from the frozen sample. The sample is then again frozen with liquid nitrogen and the gaseous impurities are pumped away. Several cycles of this procedure may be required to achieve a high purity of the sample.

After the cleaning procedure, the reagent sample is thawed and the vapour above the liquid can be transferred into a glass bulb attached to the preparation line. The concentration can be determined from the pressure in the glass bulb as described for the preparation of gaseous mixtures (Eq. 4.2.3). Once the desired pressure is achieved both the finger and the bulb are disconnected from the prep-line and any residual reagent sample is then pumped away. The reagent in the bulb can be flushed into the chamber.

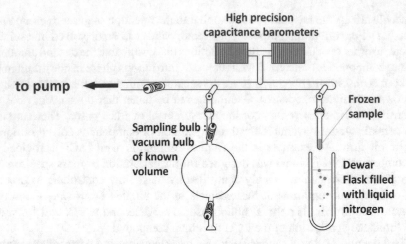

Fig. 4.2 Schematic representation of the vacuum line system used to remove impurities from liquid compounds and to prepare a glass bulb containing a well-defined concentration of the reactant

The effective concentration of low volatile compounds in the gas phase in the chamber can be lowered due to deposition on the surface of the inlet system and on the chamber wall near the inlet or due to condensation on particles. Re-evaporation can affect experiments at later times. The concentration can also be lower than expected if thermally unstable compounds decompose in the heated inlet or compounds form dimers if high concentrations are injected (for example, formic acid or acetic acid). These effects can be avoided or minimized by applying an appropriate temperature of the inlet system, by warming up the entire line system and by adjusting the flow with which the evaporated compound is transported into the chamber. Using a well-controlled heating system is recommended rather than using a heat gun. For example, the port for the injection of compounds in the QUAREC chamber at the University Wuppertal is housed in a metal casing which can be heated up to a temperature of 60 °C.

Water vapour is an essential part of the gaseous compounds in the atmosphere and its presence can affect various gas-phase reactions and specifically particles. Although it is a liquid at room temperature and therefore can be injected with similar methods as described above, the amount of water that needs to be introduced to reach atmospheric concentrations is much higher than that of other trace gases. Therefore, small impurities in the water used for the humidification can be an important source of contaminations in the experiment. Special precautions need to be taken not only regarding the generation of the pure water using, for example, a Milli-Q-water or Nanopure water device but also in the selection of all materials in the humidification system as well as of cleaning procedures and intervals. Particle formation can be an issue in the humidification system. They can be removed by heated stainless steel filters in the inlet system. In addition, atmospheric relative humidity is typically within the range of 20–80% so that a substantial fraction of air in the chamber needs to be exchanged to yield atmospheric water concentrations without condensation in the humidification system, even if the system is heated to moderate temperatures. Therefore, the humidity of air cannot easily be changed during an experiment without diluting all trace gases.

Methods to humidify the air are:

- Boiling pure water and transporting the steam with a high gas flow into the chamber.
- Using a commercial humidification system such as the Control–Evaporation–Mixing (CEM) system by the company Bronkhorst. This system sprays a controlled flow of pure liquid water into a heated volume, in which the evaporated water is mixed with a controlled flow of dry air.
- Evaporating pure water into an evacuated chamber and flushing the water vapour into the chamber as described above. The system can be either made of glass or stainless steel and is recommended to be heated.
- Using a temperature-controlled Nafion tube. In this system, water molecules are transported through a membrane, if pure liquid water is on one side of the membrane and a dry gas flow is on the other side of the membrane.

4.4 Injection of Compounds from Solid Sources

Solid compounds can be introduced by flushing the vapour phase above the solid sample into the chamber similar to the method applied for liquids (Fig. 4.1). Controlling the temperature allows to adjust the concentration of the fraction of the compound in the gas phase. Heating can be achieved with a heat gun but needs to be carefully applied. Care must be taken to avoid that small parts of the solid material are flushed into the chamber due to thermal turbulence in the vial specifically if the vial is heated.

The resulting concentration of the reactant in the chamber can be calculated from the weight loss of the solid sample in the vial:

$$c_{\text{reactant}} = \frac{m \times P \times N_A}{M \times V_{\text{chamber}}} \tag{4.4.1}$$

with m = weight loss of the compound in the vial; P = purity of the compound sample; N_A = Avogadro's number; M = molar mass of the compound; V_{chamber} = volume of the chamber.

Removing impurities and preparation of glass bulbs containing a well-defined concentration of the compound can be similarly done as described for liquids except that there is no need for freezing (Fig. 4.2).

The transfer of the reactant from solid sources is often not quantitative due to condensation and deposition on the walls of the inlet system so that the resulting concentration in the chamber is lower than calculated from the weight loss. Similar to the injection from liquid sources, this can be minimized by heating the inlet system, but specifically, compounds that are solid at room temperature may also be thermally unstable at higher temperatures.

In some very specific cases, the thermal decomposition of a solid polymer can be used to produce gaseous monomers. In these cases, the solid polymer is placed in the heated vial and the gaseous monomer above the solid sample can be flushed into the chamber. An example is the decomposition of solid paraformaldehyde $(CH_2O)_n$ to formaldehyde (HCHO).

The injection of dissolved solids in aerosol particles is described in Chap. 5.

4.5   Production of Hydroxyl Radicals (OH)

The hydroxyl radical (OH) radical is the main oxidant of trace gases in the atmosphere (Stone et al. 2012). In the troposphere, the primary source of OH is mainly the photolysis of ozone, formaldehyde (HCHO) and nitrous acid (HONO) and the ozonolysis of alkenes. The study of OH oxidation processes in simulation chambers can provide kinetic data (e.g. reaction rate coefficients, Sect. 7.3), product yields of individual reactions or can be used to test entire reaction schemes. Due to the very short lifetime and high reactivity of OH, radicals cannot be stored but must be generated during the experiment. Sources for OH radicals in chamber experiments are often photolytic reactions like those occurring in the atmosphere. Precursor compounds such as ozone are introduced in the gas mixture. OH radical production from precursors can be accompanied by the formation of other reactive species such as NO, for example, in the case of the photolysis of HONO that can affect the chemistry in the experiment.

It is crucial for the design of OH oxidation experiments to consider the type of precursor used for the primary radical production and the effect of radical termination and radical regeneration reactions, which are closely connected to the presence of nitrogen oxides (NO and NO_2) in the air mixture (Sect. 4.5.7).

4.5.1   OH Production from Ozone Photolysis

Ozone photolysis in the presence of water vapour is the main source of OH radicals in the atmosphere (Finlayson-Pitts and Pitts 1986):

$$O_3 + h\nu \rightarrow O(^1D) + O_2 \quad \lambda \leq 310\,nm \tag{R4.5.1}$$

$$O(^1D) + H_2O \rightarrow 2\,OH \tag{R4.5.2}$$

In order to make use of ozone photolysis as OH source in chamber experiments, ozone and water vapour needs to be present and UV radiation is required. Ozone can be provided by ozone generators (Sect. 4.8). Ozonolysis reactions of unsaturated compounds may affect the chemistry of the experiment and water vapour can also affect some chemical reactions. This needs to be considered in the evaluation of experiments. If experiments need to be done in dry air, hydrogen could be used instead of water vapour. This leads to the production of equal concentrations of OH and hydroperoxy radicals:

$$O(^1D) + H_2 \rightarrow OH + H \qquad\qquad (R4.5.3)$$

$$H + O_2 + M \rightarrow HO_2 + M \qquad\qquad (R4.5.4)$$

4.5.2   OH Production from Nitrous Acid Photolysis

Nitrous acid (HONO) is commonly used as an OH source in chambers. It photolyses in the near UV–visible spectral range:

$$HONO + h\nu \rightarrow OH + NO \quad \lambda \leq 400\,nm \qquad\qquad (R4.5.5)$$

This reaction produces concurrently OH and NO that affect the radical system (Sect. 4.5.7). HONO is reformed in the reaction of OH and NO, so that both are in a photo stationary state. The reaction of HONO with OH is a sink for OH that may compete with the reaction of OH with the target species for exceptionally high HONO concentrations:

$$HONO + OH \rightarrow H_2O + NO_2 \qaqquad\qquad (R4.5.6)$$

Due to the fast photolysis of HONO (lifetime in the range of several 10 min for atmospheric conditions), typically a continuous source of HONO is required to sustain the production of OH radicals during the experiment.

Nitrous acid is not commercially available. It can be synthesized by following the protocol adapted from Nash et al. (1968), Cox et al. (1974) and Burkholder et al. (1992). HONO is generated by the reaction of an aqueous solution of $NaNO_2$ with diluted sulfuric acid. For example, a diluted $NaNO_2$ solution (0.1–1%) is continuously added to a solution of sulfuric acid (30% by weight) with a flow rate of 0.24 ml/min. The reaction can be achieved in a closed 3-necked bulb that is continuously stirred by a magnetic stirrer (Fig. 4.3). The output needs to be directly flushed into the chamber (Sect. 4.2), because nitrous acid easily decomposes. Decomposition can be minimized, if the reaction is performed in the dark and at low temperature (e.g. in an ice bath). Traces of NO, NO_2 and H_2O formed by the self-reaction of HONO can be present in the resulting mixture (Chan et al. 1976). For example, the system

Fig. 4.3 HONO generation
system at EUPHORE

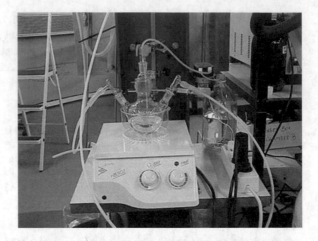

installed at the EUPHORE chamber delivers a constant flow of HONO that leads
to an increase of the HONO mixing ratio in the chamber (200 m³) of 1ppbv/min.
Similar amounts of NO and NO_2 are concurrently observed.

HONO is also known to be heterogeneously formed on Teflon surfaces. The exact
mechanism is not fully understood. The source requires radiation and is enhanced
with increasing temperature and relative humidity (Rohrer et al. 2005). Therefore,
HONO is typically continuously produced in Teflon chambers that are illuminated for
photochemistry experiments and can serve as OH source. Emission rates scale also
with the size of the chamber. Therefore, careful characterization of the HONO source
needs to be done in Teflon chambers that are used for photochemical experiments.

The HONO concentrations can be measured by LOPAP instruments (Kleffmann
et al. 2002) with high precision and accuracy. Recently, also cavity-based absorption
spectroscopy has been applied for the detection of HONO, but the sensitivity of
these instruments is less compared to that of LOPAP instruments (Min et al. 2016).
The HONO source can also be characterized in reference experiments (Chap. 2),
if the increase of the total reactive nitrogen oxide concentration (NO_x) is observed
assuming that HONO is the only relevant source for NO_x in the chamber (Rohrer
et al. 2005).

4.5.3   Production of OH from Alkyl Nitrite Photolysis

Photolysis of alkyl nitrites can be used as an OH source in chamber experiments.
The most widely used precursor is methyl nitrite. Its photolysis has been extensively
studied (Gray and Style 1952; Taylor et al. 1980; Niki et al. 1981; Atkinson et al.
1981):

$$CH_3ONO + h\nu \rightarrow CH_3O + NO \quad \lambda \leq 430\,nm \qquad (R4.5.7)$$

$$CH_3O + O_2 \rightarrow HCHO + HO_2 \qquad\qquad (R4.5.8)$$

$$HO_2 + NO \rightarrow OH + NO_2 \qquad\qquad (R4.5.9)$$

In contrast to ozone or H_2O_2 photolysis, low energy photons are required, thus reducing the probability that other organic compounds are concurrently photolysed. However, the by-products, formaldehyde and nitric oxide, can complicate the evaluation of experiments. For example, if the formaldehyde yield from the oxidation of organic compounds is to be determined, it is recommended to use a different alkyl nitrite such as isopropyl nitrite, for which the by-product of the photolysis is acetone. It is worth noting that OH radicals are indirectly generated from the photolysis of alkyl nitrite because HO_2 radicals can be converted to OH in the reaction with NO. This implies that methyl/alkyl nitrite photolysis should be considered as a HO_x source rather than as an OH source. In addition, the effect of the presence of NO_x on the chemical system needs to be considered (Sect. 4.5.7). Injection of additional NO is recommended to accelerate the conversion of HO_2 to OH (Reaction R4.5.9) and to suppress NO_3 and O_3 in their reactions with NO.

Alkyl nitrites, which are not commercially available, can be synthesized following an experimental protocol adapted from Gray and Style (1952) and described by Taylor et al. (1980): The alkyl nitrite is prepared by dropwise addition of sulfuric acid (50%) to a mixture of the corresponding alcohol and saturated aqueous natrium nitrite ($NaNO_2$). The reaction is performed at ice temperature (Fig. 4.4). A stream of oxygen-free N_2 carries the reaction products through a bubbler filled with tablets of potassium hydroxide (KOH) (or with a KOH solution) to remove acids. It is further flowed through anhydrous calcium chloride ($CaCl_2$) to remove traces of water. Finally, the alkyl nitrite is frozen in a trap that is kept at dry ice temperature (196 K). The purity of the alkyl nitrite can be increased by fractionally distillation in the vacuum. The crystals are colourless. If they melt, a lemon-yellow liquid is obtained. Alkyl nitrite can be kept for months in a freezer at a temperature of $-18\,°C$ but it is recommended to perform a distillation after having it stored for a long time to remove impurities which could have been formed by alkyl nitrite decomposition. The alkyl nitrite is then introduced into the chamber following the procedures described in Sects. 4.2 and 4.3. Because of its high volatility, injection as a liquid is not recommended, but to prepare a gaseous mixture in a glass bulb. An example of the evolution of trace gas concentrations during the photolysis of methyl nitrite in the CSA chamber at the CNRS-LISA is shown in Fig. 4.5 demonstrating that a number of different compounds are formed in the complex system.

4.5.4 Production of OH from Photolysis of Peroxides

The photolysis of hydrogen peroxide (H_2O_2) is one way to produce OH without the need for other reactants and without the chemical production of other reactive species (Calvert and Pitts 1966):

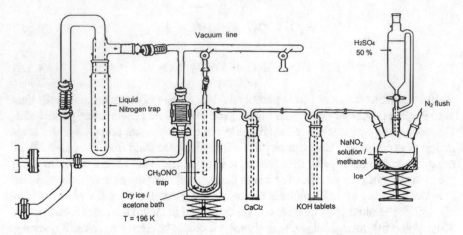

Fig. 4.4 Experimental setup for the synthesis of alkyl nitrites. (From Picquet 2000)

$$H_2O_2 + h \rightarrow 2OH \qquad \lambda < 300\,nm \qquad\qquad (R4.5.10)$$

However, the absorption cross section of this reaction is rather small. Even at shorter wavelengths around $\lambda = 250\,nm$ the value is only approximately 8×10^{-20} cm^2 molecule^{-1} and the value reduces to approximately 4×10^{-22} cm^2 molecule^{-1} at $\lambda = 350\,nm$ (IUPAC). Therefore, radiation at wavelengths $\lambda < 300$ is typically used so that photolysis of other compounds may need to be considered. In addition, high concentrations of H_2O_2 need to be injected so that the source needs to be very clean to avoid impurities affecting the experiment.

Commercial solutions of stabilized H_2O_2 with concentrations between 30 and 50% in water are often used. Higher purity solutions can be used but this requires additional purification which is dangerous as concentrated H_2O_2 solution is highly explosive. Therefore, also traces of water are concurrently injected. H_2O_2 loss at chamber wall can be significant so that the calculation of the OH production from the amount of injected H_2O_2 could lead to an overestimation of the radical source. OH concentrations can be obtained, for example, by using an OH tracer (Sect. 4.9).

The H_2O_2 solution can be introduced into the chamber by gently heating the solution, by bubbling dry nitrogen through the solution, or by direct injection with a syringe (Sect. 4.2). When bubbling a carrier gas through the H_2O_2 solution, water vapour is more efficiently taken up from the solution. As a consequence, the H_2O_2 concentrations in the solution is gradually increasing so that the H_2O_2 concentration that is transported into the chamber increases with time. Therefore, also the OH production rate in the chamber increases over time. A high concentration of H_2O_2 in the solution could become a safety problem so that caution with this method is recommended.

Radical production from the photolysis of organic peroxides can also be used as radical source in chamber experiments, but the subsequent chemistry is more complex compared to the photolysis of H_2O_2 due to the concurrent production of

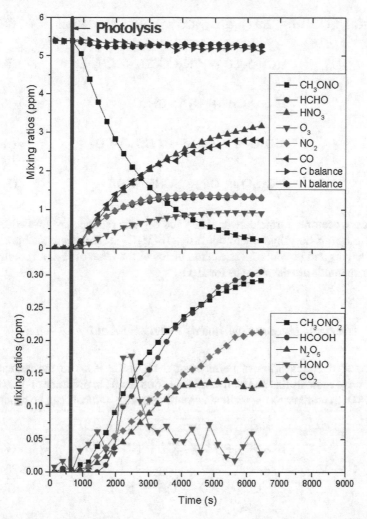

Fig. 4.5 Time profiles of reactant and products mixing ratios following irradiation of a mixture CH₃ONO in air in the CSA simulation chamber (Laboratoire Interuniversitaire des Systèmes Atmosphérique, LISA). NO is injected to enhance the conversion of HO_2 to OH (from Picquet 2000). Nitrate radicals and dinitrogen pentoxide (N_2O_5) can be formed in the experiment because NO_3 photolysis is small for the radiation of the lamps used in this experiment (wavelengths at 360 and 420 nm)

organic radicals and products. This is similar to the photolysis of alkyl nitrates, but OH is directly formed from the photolysis so that there is no need to convert peroxy radicals to OH in the reaction with NO. For example, photolysis of tert-butyl hydroperoxide proceeds by the following reactions (Calvert and Pitts 1966; Baasandorj et al. 2010):

$$(CH_3)_3COOH + h\nu \rightarrow (CH_3)_3CO + OH \quad \lambda \leq 320\,nm \quad (R4.5.11)$$

$$(CH_3)_3CO \rightarrow CH_3COCH_3 + CH_3 \quad (R4.5.12)$$

$$CH_3 + O_2 \rightarrow CH_3O_2 \quad (R4.5.13)$$

$$CH_3O_2 + CH_3O_2 \rightarrow 2\,CH_3O + O_2 \quad (R4.5.14)$$

$$CH_3O + O_2 \rightarrow HCHO + HO_2 \quad (R4.5.15)$$

In this case, acetone, formaldehyde, HO_2 and organic radicals are formed. Advantages of using tert-butyl hydroperoxide instead of H_2O_2 are that it is available at higher concentrations and its wall loss rate could be lower than that of H_2O_2. Handling and injection methods are the same as for H_2O_2.

4.5.5 Thermal Decomposition of Pernitric Acid

The thermal decomposition of pernitric acid HO_2NO_2 leads to the formation of hydroperoxyl radicals that can be further converted to OH in presence of NO (Barnes et al. 1982). In contrast to the methods described above, radicals can be produced in the dark:

$$HO_2NO_2 + M \rightleftharpoons HO_2 + NO_2 + M \quad (R4.5.16)$$

$$HO_2 + NO \rightarrow OH + NO_2 \quad (R4.5.17)$$

As shown by Graham et al. (1977, 1978), the unimolecular decay time of HO_2NO_2 at room temperature and atmospheric pressure is on the order of 10 s. However, because a thermal equilibrium is rapidly established, the effective lifetime can be on the order of hours. It is mainly determined by the loss rate of HO_x radicals in the chemical system because the depletion of the HO_2 concentration prevents the back-reaction of HO_2 with NO_2 to HO_2NO_2.

HO_2NO_2 can be synthesized following a protocol described by Bames et al. (1982). Pernitric acid is prepared by reacting concentrated H_2O_2 (85%) with NO under vacuum. Typically, a 1 L glass flask containing 50 ml H_2O_2 at 0 °C is evacuated to 1 hPa. The flask is then pressurized to 400 hPa with NO while the H_2O_2 is magnetically stirred. Due to the production of HO_2NO_2, the pressure rapidly decreases. After a few minutes, the pressure is approximately 70 hPa and the excess of NO can be pumped out. The procedure of filling the flask with NO is repeated several times.

HO_2NO_2 can be injected into the chamber or gas mixtures can be prepared as described above. Because HO_2NO_2 is unstable, impurities such as HNO_3, H_2O and NO_2 are difficult to avoid and may need to be considered in the interpretation of results.

4.5.6 OH Production from the Ozonolysis of Alkenes

Another method to produce OH radicals in the dark is the ozonolysis of alkenes that is also a significant radical source in the atmosphere (Johnson and Marston 2008). The complex reaction of ozone with alkenes leads to the formation of Criegee intermediates that can stabilize or rapidly decompose to radicals with the yield Π:

$$alkene + O_3 \rightarrow \Pi_{OH}OH + \Pi_{RO_2}RO_2 + products \qquad (R4.5.18)$$

In the case of 2,3-dimethyl-2-butene, the OH yield is high with a value of 0.93 ± 0.14 (Cox et al. 2020). Besides the high OH yield, its symmetric structure leads to the formation of a limited number of other products, which reduces the likelihood for interferences in the experiment:

$$2, 3\text{-dimethyl-2-butene} + O_3 \rightarrow OH + acetone + products \qquad (R4.5.19)$$

2,3-dimethyl-2-butene is a liquid at room temperature and can therefore be injected into the chamber as described above (Sect. 4.3). Ozone generation and injection is described in Sect. 4.8.

Due to the complexity of the ozonolysis reaction that is often only partly known in detail and the concurrent production of organic radicals and organic products that can be involved in the chemical system and particle formation the interpretation of experiments can be difficult. If ozone is the limiting reactant in the system, subsequent ozonolysis reactions could be partly avoided.

4.5.7 OH Production in the Presence of NO_x

Most of the methods to produce OH radicals require radiation and nitrogen oxides ($NO_x = NO_2 + NO$) in the presence or absence of ozone. The reaction of OH with organic compounds initiates a radical reaction chain, in which different radical reactions compete. Depending on the availability of reaction partners, different reaction pathways can dominate the chemical system. In addition, radiation used for the initial OH production could photolyse other compounds than the OH precursor. For example, nitrogen dioxide is often concurrently photolysed leading to the formation of ozone and nitric oxide (NO). Due to the fast back-reaction of NO with ozone, NO, O_3 and NO_2 concentrations form a photo-stationary state.

NO is also a reaction partner for peroxy radicals that are produced in the OH radical chain. The ozone that is produced from the subsequent photolysis of NO_2 leads a net increase of the ozone concentration, because no ozone has been consumed in oxidation of NO before. Therefore, also the ratio of NO_2/NO concentrations has the tendency to increase over the course of an experiment. Due to the strong coupling of radiation, radicals, nitrogen oxides and ozone, photochemistry experiments need to be carefully designed to ensure that the desired chemistry can be observed.

The OH concentration that is present in the experiment does not only depend on the production rate from radical precursors and the loss rate in its reaction with inorganic and organic reaction partners, but also on the rate of radical regeneration. The reaction of OH initiates a radical chain reaction. This radical reaction cycle includes reactions of nitric oxide (NO) with organic peroxy (RO_2) and hydroperoxy (HO_2) radicals and eventually reforms OH. Due to the short lifetime of radicals that range between a fraction of second and minutes, OH, HO_2 and RO_2 radical concentrations are quickly in a steady state. The presence of NO shifts the equilibrium toward OH and increases the number of OH regeneration cycles before competing reactions such as the reaction of OH + NO_2 and peroxy radical recombination reactions terminate the radical chain. Therefore, the addition of NO_x in chamber experiments is one method to enhance the oxidation rate of reactants. Due to the photo-chemical equilibrium between NO_2 and NO, this is most efficient, if the NO_2 photolysis rate is high. This can be achieved if the chamber is equipped with lamps providing the required radiation. Shifting the photo-chemical equilibrium to NO has also the advantage that the radical termination reaction of NO_2 with OH producing nitric acid is reduced. This reaction can otherwise limit the oxidation efficiency. Therefore, it is recommended to keep the NO_x (NO_2 + NO) concentration in a range that NO_2 concentrations are moderate.

Recently, the H-shift isomerization reactions of RO_2 have been recognized to be competitive with other radical reactions at atmospheric conditions. Subsequent reactions can also lead to the regeneration of radicals (Peeters et al. 2014) as well as to the production of highly oxidized molecules (HOMs) (Ehn et al. 2014).

Depending on the fate of organic peroxy radicals, different product distributions can be observed from the oxidation of organic compounds with OH. For example, if the dominant pathway is the reaction with NO, often aldehyde and carbonyl compounds are formed, whereas the reactions with HO_2 leads typically to the formation of hydroxyperoxides. In the atmosphere, recombination reactions of organic peroxy radicals are typically slower compared to its reactions with NO and HO_2. In chamber experiments, often high concentrations of OH and organic compounds are used to shorten reaction times and to accumulate measurable product concentrations, thereby increasing the production rate of RO_2 radicals. This could, however, drive the chemical regime towards RO_2 + RO_2 recombination reactions so that results may not be easily transferred to atmospheric conditions.

In experiments that aim for studying oxidation pathways that favour RO_2 + RO_2 or RO_2 + HO_2 reactions, the competing reaction with NO needs to be suppressed. This can be achieved by injecting ozone or by using an OH source that does not concurrently produce nitrogen oxides such as H_2O_2 photolysis (Sect. 4.5.4). However, the release of HONO in chambers that are made of Teflon film may limit the minimum concentration of nitrogen oxide species (Chap. 2). If the goal of the experiment is to investigate uni-molecular hydrogen shift reactions in RO_2 radicals all, bi-molecular RO_2 reactions need to be suppressed. In addition to minimizing the NO concentration, low reactant concentrations are required to reduce the loss due to RO_2 recombination reactions. This, however, may be limited by competing chamber effects interfering with the chemical system or by the sensitivity of instruments.

4.6 Production of Nitrate Radicals

The nitrate radical (NO_3) is photochemically unstable in the presence of visible light but can accumulate in the absence of sunlight. In consequence, NO_3 acts as a night-time oxidant in the atmosphere. Studying NO_3-initiated processes in simulation chambers allows the provision of both kinetic and mechanistic data of individual reactions or entire chemical reaction systems. They are typically performed in the dark to avoid photodissociation of NO_3. Like OH, NO_3 radicals need to be generated during the experiment and the reaction of NO_3 with organic compounds can initiate a reaction chain that needs to be considered in the evaluation of experiments. Several methods have been developed to produce NO_3 including reactions of halogens with nitric acid, photolysis of nitric acid or reaction of chlorine atoms with chlorin nitrate (Wayne et al 1991), but only few of them are suitable for chamber simulation experiments and are described in the following sections.

4.6.1 Production of NO_3 from the Gas-Phase Reaction of NO_2 and O_3

In the atmosphere, NO_3 radicals are produced from the gas-phase reaction of nitrogen dioxide and ozone. This process can be applied in chamber experiments by injecting NO_2 (Sect. 4.2) and O_3 (Sect. 4.8):

$$NO_2 + O_3 \rightarrow NO_3 + O_2 \qquad \text{(R4.6.1)}$$

Instead of injecting NO_2, also NO can be used to produce NO_2 from its reaction with ozone:

$$NO + O_3 \rightarrow NO_2 + O_2 \qquad \text{(R4.6.2)}$$

The advantage of using NO is that the purity of NO that is commercially available can be higher compared to that of NO_2.

The reaction rate constant at room temperature is relatively low with a value of $k_{4.6.1} = 3.2 \times 10^{-17}$ molecule^{-1} cm^3 s^{-1}. Due to the presence of NO$_2$, part of the produced NO$_3$ is further converted to dinitrogen pentoxide (N$_2$O$_5$) that is thermally labile:

$$NO_2 + NO_3 + M \rightleftarrows N_2O_5 + M \qquad (R4.6.3)$$

In most cases, NO$_3$ and N$_2$O$_5$ concentrations can be assumed to be in a thermal equilibrium.

The actual NO$_3$ concentration in the chamber is typically highly variable because NO$_2$ and O$_3$ required for the production are consumed and the N$_2$O$_5$ serves as a reservoir for NO$_3$ (Fig. 4.6). The formation of N$_2$O$_5$ can be minimized by using higher ozone than NO$_2$ concentrations (Mitchell et al. 1980). However, ozone itself can be an oxidant that could interfere becoming significant for the evaluation of the experiment if high ozone concentrations are present.

The loss rate of NO$_3$ due to oxidation of reactants in chamber experiments is typically much lower compared to that of the OH radical. Therefore, chamber wall loss reactions can compete and contribute significantly to the total loss of NO$_3$ in the experiments even in chambers with a large volume-to-surface ratio (Fig. 4.6, Dorn

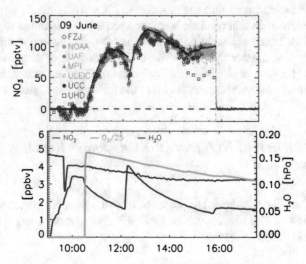

Fig. 4.6 NO$_3$ formation from the reaction of NO$_2$ and O$_3$ in the SAPHIR simulation chamber in Jülich, Germany, during the NO$_3$ radical intercomparison campaign. O$_3$ was injected once and NO$_2$ several times over the course of the experiment to enhance the production rate of NO$_3$. Despite its large volume of 270 m^3, NO$_3$ wall loss is significant so that the NO$_3$ concentration reaches maximum concentrations approximately 1 h after the last injection of reactant. Shortly before 16:00 the air mixture was exposed to sunlight leading to the rapid destruction of NO$_3$. (Reused with permission from Dorn et al. (2013) Open access under a CC BY 3.0 license, https://creativecommons.org/licenses/by/3.0/)

et al. 2013). In addition, N_2O_5 wall loss could impact the chemical system due to its thermal equilibrium with NO_3.

NO_3 radicals can also be formed in a reactor outside the chamber and the air mixture can be flushed into the chamber (Thuener et al. 2004). Concentrations can be chosen, such that one of the reactants, NO_2 or O_3, is in excess, whereas the other reactant is consumed in the reactor so that NO_3 production stops before the air mixture enters the chamber. The reaction time that would be required to consume both reactants would be much longer than the NO_3 lifetime in the reactor with respect to wall loss. If ozone is consumed in the reactor, this method allows for experiments without additional ozonolysis reactions. The reactor needs to be kept in the dark and it is recommended to use inert materials such as Teflon or SilcoTec® coated steel.

Because of the complex chemical system and the impact of wall loss, trace gas concentrations should be monitored. The detection of NO_2 and O_3 belongs typically to the standard repertoire of measurements with commercial instruments at chambers. The measurement of NO_2 and O_3 allows calculating the production rate of NO_3. The direct detection of NO_3 is typically done with custom-built instruments applying absorption spectroscopy (Dorn et al. 2013) but can be challenging due to is high reactivity and small concentrations. Detection of less reactive N_2O_5 by FTIR or cavity-based absorption spectroscopy can be used to calculate NO_3 concentrations from the thermal equilibrium between NO_3 and N_2O_5 (Reaction R4.6.3), if the NO_2 concentration and temperature are monitored in the experiment.

4.6.2 Production of NO_3 from the Thermal Decomposition of N_2O_5

N_2O_5 can be frozen as crystals at dry ice temperature. Therefore, NO_3 can be delivered to the chamber by first producing and storing frozen N_2O_5 and then injecting it into the chamber as described for solid compounds by flowing an air stream over the frozen crystal (Sect. 4.4). The evaporation rate can be controlled if the temperature of the cold trap containing the crystals can be varied.

While the air is heating up to the temperature in the chamber, NO_3 and NO_2 are produced from the thermal decomposition of N_2O_5 that has evaporated from the crystal (Reaction R4.6.3, Fig. 4.7). Therefore, this method provides ozone-free NO_3 (Atkinson et al. 1984a, b; Barnes et al. 1990; D'Anna et al. 2001; Spittler et al. 2006; Kerdouci et al. 2012; Slade et al. 2017). However, NO_2 is a by-product of the N_2O_5 decomposition. One challenge is to minimize impurities because the solid sample can contain other nitrogen oxide compounds present in the synthesis such as NO_2 and nitric acid.

N_2O_5 can be synthesized from the gas phase reaction of NO_2 and O_3 (Reactions R4.6.1 and R4.6.3) in a vacuum line and two glass traps (Fig. 4.7). N_2O_5 freezes out, if the gas mixture is flowed through a glass trap that is kept in a dry ice or a mixture of liquid nitrogen/ethanol bath. The temperature of the trap should be around 193 K

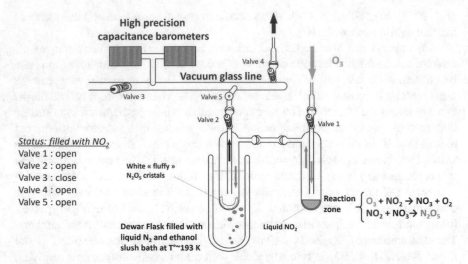

Fig. 4.7 Set-up for producing frozen N_2O_5 crystals using a vacuum preparation line

and must not be less than 163 K to avoid condensation of ozone. Instead of using liquid nitrogen/ethanol, also dry ice and an alcohol can be used to achieve a similar temperature.

In order to minimize impurities in the frozen sample, the following procedure is recommended:

- The entire vacuum line and traps should be cleaned to remove any traces of impurities and adsorbed water from the surfaces. As an anhydride, N_2O_5 reacts rapidly with water to form nitric acid. The traps are recommended to be dried in an oven for several hours. The vacuum with the two attached traps (Fig. 4.7) can be flushed with dry nitrogen and then pumped down to low pressure (10^{-3} hPa) for several hours.
- A Dewar flask containing a mixture of liquid nitrogen and ethanol at approximately 193 K is prepared and the first trap is cooled down. After closing the connection of the traps to the vacuum line the trap is filled with NO_2 (several hundred hPa) so that NO_2 condenses as white crystals. This procedure can be repeated several times to accumulate a sufficiently high amount of frozen NO_2.
- The Dewer flask is then moved to the other trap, NO slowly warms up and turns into a brown liquid and while ozone is flowed through the two traps. This is done at atmospheric pressure (Fig. 4.7). NO_2 that evaporates in the first trap reacts with ozone to form NO_3 and N_2O_5 that freezes out in the second trap. High ozone concentrations are required that can be produced with a silent discharge ozoniser fed with pure oxygen (Sect. 4.8). The total flow rate is recommended to be between 1.5 and 2 l/min so that the residential time in the trap is long enough to allow N_2O_5 to freeze as fluffy white crystals. All NO_2 must have been consumed before entering the second trap to prevent from NO_2 being frozen again

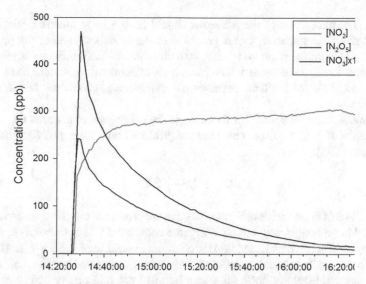

Fig. 4.8 Time series of N_2O_5, NO_3 and NO_2 after the injection of N_2O_5 from evaporating frozen N_2O_5 in the CSA simulation chamber at LISA-CNRS (©). NO_3 and NO_2 are produced from the thermal decomposition of N_2O_5. NO_3 and N_2O_5 concentrations are decreasing due to chamber wall loss (Fouqueau 2019)

in entering the second trap. This can be checked by the change of the gas colour from yellow/brown to colourless.

- The N_2O_5 cristals can be further purified by pumping on the bulb containing them, by removing the Dewar flask and connecting it to the vacuum pump. Under reduced pressure, as N_2O_5 has a much lower vapour pressure than NO_2, impurities are eliminated from the crystals that are formed. N_2O_5 crystals can be stored under vacuum at a temperature of –18 °C for several weeks.

The injection of NO_3 from frozen N_2O_5 method is, for example, applied in the CSA and CESAM atmospheric simulation chambers at Laboratoire Interuniversitaire des Systèmes Atmosphérique (LISA). As an example, Fig. 4.8 shows the time series of trace gases after injections of N_2O_5 at the start of the experiment. Significant wall leads to the consumption of N_2O_5 and NO_3.

4.7   Production of Cl Radicals

In the atmosphere, chlorine atoms are homogeneously formed from the photooxidation of chlorine compounds or from heterogeneous processes occurring, for example, on sea salt particles (Simpson et al. 2015). In chamber experiments, the production of chlorine atoms is based on the photolysis of various precursors, organic or inorganic halogenated species that are injected following the procedures described in Sects.

4.2–4.4. The photolysis of the precursor should have a large quantum yield at the wavelength of the radiation that is provided in the chamber. Neither the precursor nor products from the photolysis other than chlorine should interfere with the reaction system that is investigated. The photolysis of a precursor should have a large quantum yield for the available radiation and experiments should not be affected by by-products.

The most common source for Cl atoms is the photolysis of gaseous Cl_2 in air at wavelengths of $\lambda > 300$ nm. This reaction produces only Cl atoms (Atkinson and Aschmann 1985):

$$Cl_2 + h\nu \rightarrow 2\,Cl \tag{R4.7.1}$$

Photolysis of compounds other than Cl_2 that are present in the chamber experiment may need to be considered, but most organic compounds do not photolyse at these wavelengths. Cl_2 can be injected into the chamber as described in Sect. 4.2. Although a clean source, Cl_2 is prone to react directly with unsaturated species and sulphur compounds. Molecular chlorine is also a harmful gas and proper safety measures must be taken when handling it.

There are several other precursors that require radiation at lower wavelengths than Cl_2 to produce chlorine atoms from photolysis such as phosgene ($COCl_2$, carbonyl dichloride), 2,2,2-trichloroacetyl chloride (CCl_3COCl, Hass et al. 2020), chloroform and tetrachloro methane (Matheson et al., 1982). Among these, oxalyl chloride is most commonly used (Baklanov and Krasnoperoy 2001; Gosh et al. 2012):

$$(ClCO)_2 + h\nu \rightarrow CO + Cl + ClCO \tag{R4.7.2}$$

The photolysis of oxalyl chloride is a relatively clean source for chlorine atoms, because the only by-product is CO. It is commercially available as a liquid and does not require special safety measures. Oxalyl chloride can be injected into the reaction chamber following the procedures described in Sect. 4.3 by either direct injection with a syringe or by flowing dry air through a heated Pyrex glass bulb containing the liquid oxalyl chloride. If high energy-rich radiation is applied, the photolysis of organic compounds may need to be considered in the evaluation of experiments.

4.8 Production of Ozone

Ozone (O_3) plays an important role in tropospheric chemistry. Its photolysis is the most important source of OH radicals (Sect. 4.5.1) and it is an oxidant for organic compounds (Sect. 4.5.6). Ozone can be stored at low temperature as a solid, a liquid or can be adsorbed; however, storing ozone is not recommended due to the difficult handling and serious safety issues. Therefore, gaseous ozone is directly produced before injecting it into the chamber.

4.8.1 Photochemical Ozone Generation

Ozone can be generated by using an ultraviolet lamp emitting radiation at a wavelength of 185 nm (Fig. 4.9). Low-pressure discharge mercury lamps such as "pen ray Hg lamps" are often used. Their main emission line is at a wavelength of 254 nm, but they also emit 185 nm radiation to dissociate oxygen:

$$O_2 + h\nu \rightarrow 2\,O \tag{R4.8.1}$$

$$O + O_2 \rightarrow O_3 \tag{R4.8.2}$$

Pure oxygen or synthetic air is flowed through a glass bulb that is illuminated by the lamp. The lamp can be placed outside if the bulb is made of fused quartz glass. Shielding of all radiation including any stray light is required for safety reasons due to the high potential for damages of DNA, if the radiation hits the skin of humans. It is highly recommended to have all parts of the system made of glass or Teflon to avoid rapid destruction of ozone on surfaces.

The resulting ozone concentration highly depends on the residence time of the air in the photoreactive region, on the pressure, and on the lamp emission power and needs to be characterized for each design of a device. The maximum ozone production is mainly limited by the size of lamps and glass bulbs. For example, the device used at the EUPHORE chamber produces 23ppbv of ozone per min in a volume of $200m^3$.

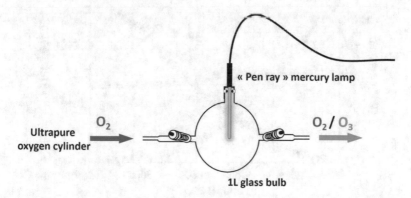

Fig. 4.9 Example of a photolytic ozone generation set-up

4.8.2   Ozone Generation by Electrical Discharge

Ozone can also be generated with ozone discharge generators that are commercially available. They provide a reproducible ozone production rate and have a high efficiency so that ozone mixing ratios in the percentage range in the output gas flow can be achieved. They are based on dissociating oxygen in an electric field. It is important to use high-purity oxygen to avoid artefacts from the concurrent dissociation or recombination reactions of other compounds or impurities. For example, using air containing nitrogen in addition to oxygen leads to the production of high concentrations of complex nitrogen oxides.

Some devices use a high-voltage electric arc between two electrodes, but they are not recommended because particles can be released from the surfaces of the electrode that are flushed into the chamber together with the ozone. Instead, "silent discharge" or "corona" ozone generators should be used. They produce a plasma between two dielectric electrodes. In a corona discharge ozone generator, the electrical discharge takes place in an air gap within the corona cell designed specifically to split the oxygen molecule for the ozone production. In this air gap, a dielectric is used to distribute the electron flow evenly across the gap (Gibalov and Pietsche 2006). These generators are not only very efficient, but they are also very robust which makes their use convenient for simulation chamber experiments.

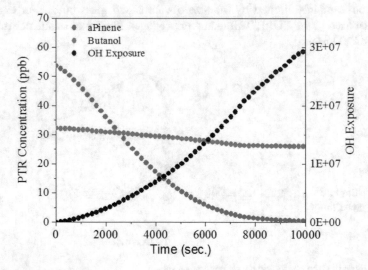

Fig. 4.10 OH exposure calculated from the measured time series of the butanol concentration in the 8 m^3 chamber at Paul Scherrer Institute during an experiment in which α-pinene is oxidized

4.9   Using OH Radical Tracers in Simulation Experiments

OH radical initiated chemistry is extensively used in atmospheric simulation chambers to study both gas-phase chemistry and SOA formation. However, because of its fast reaction rates, the OH radical is present only in low concentrations making its direct detection difficult. Typically, Laser Induced Fluorescence (LIF), Differential Optical Laser Absorption Spectroscopy (DOAS) or Chemical Ionization Mass Spectrometers (CIMS) instruments are used to measure OH radicals in the atmosphere or in an atmospheric simulation chamber (Schlosser et al. 2009). However, not all chambers have access to such an instrument to provide a direct measure of OH radicals. Therefore, having a robust method by which to infer OH radical concentration with typical chamber instrumentation is important to constrain processes occurring in the chamber and to be able to connect data obtained from atmospheric simulation chambers to the ambient atmosphere. Without having a measure of OH exposure, variations between experiments can easily be attributed to different processes when in reality only the formation or destruction of OH radicals have changed. This is especially necessary in cases of complex emissions where reproducibility of experiments can be difficult to achieve.

One way to track the OH concentrations during an experiment is to add an additional organic compound to the chamber that has an established OH reaction rate and to follow its decay throughout an experiment. This method was demonstrated by Atkinson et al., where they calculated the yields of formation of OH radicals from the ozonolysis of various terpenes (Atkinson et al. 1992). Yields of OH radicals were inferred by monitoring the products of an OH scavenger (for example, cyclohexanone and cyclohexanol from the reaction of OH with cyclohexane) as measured by a gas chromatography coupled with a flame ionization detection (Atkinson et al. 1992; Alam et al. 2011).

Another method that has been applied is the use of 1,3,5-trimethyl benzene (TMB) as a tracer molecule of OH radicals during chamber studies (Rickard et al. 1999). However, a downside to using TMB is the high potential for forming SOA, making it non-ideal for studies aiming at studying SOA since the tracer will be incorporated into the aerosol. This would suggest that smaller molecules would be better tracers for OH reactivity because they will not form SOA or be incorporated into it. Another consideration when choosing a tracer comes from possible overlap with the tracer and an oxidation product or fragment of the desired VOC. The final consideration is that the OH tracer should react sufficiently slow enough to remain in the chamber throughout the experiment and its reactivity with OH does not significantly compete with the reactivity of the target compounds.

To avoid these problems, d_9-butanol has been used as an OH tracer (Atkinson et al. 1992; Barmet et al. 2012; Stefenelli et al. 2019). This molecule does not overlap with other molecules or fragments, if organic compounds are detected, for example, with a proton-transfer mass spectrometer (PTR) because the deuteration shifts the parent ion to an even mass, as opposed to most molecules that have an odd mass when reacting with the reagent ion (H_3O^+ in case of the PTR instrument). Oxidation products of d$_9$-butanol are small molecules that have a high vapour pressure so that they do not partition into the aerosol phase and thereby do not interfere in experiments, in which particles are investigated.

If only one tracer is used, the OH exposure that is defined as OH concentration integrated over time can be determined from the measured tracer concentration:

$$OH\, Exposure(t) = \frac{[tracer]_0 - [tracer]_t}{k_{OH,tracer}} \qquad (4.9.1)$$

$k_{OH,tracer}$ is the reaction rate constant of the tracer with OH and $[tracer]_0$ is the initial tracer concentration (Fig. 4.10).

This method can be extended by using two tracers which have a significant different reaction rate constants in the reaction with OH. For instance, Stefenelli et al. (2019) chose naphthalene as a second tracer in addition to d$_9$-butanol. In this case, the OH exposure can be calculated from the ratio of the measured tracer concentrations:

$$OH\, Exposure(t) = \left(\frac{ln\left(\frac{d_9 butanol}{naphthalene}\right)_0 - ln\left(\frac{d_9 butanol}{naphthalene}\right)_t}{k_{OH,butanol} - k_{OH,naphthalene}} \right) \qquad (4.9.2)$$

If the tracers are lost by other processes than the reaction with OH during the experiment such as dilution, the estimated OH exposure would be too high, if these loss processes are not taken into account. In this case, loss rates (k_{loss}) of these processes need to be accurately known to correct measured concentration time series. An iterative correction procedure is, for example, described in Galloway et al. (2011):

$$[tracer]_t^{corr} = [tracer]_{t-1}^{corr} + [tracer]_t - [tracer]_{t-1} + [tracer]_{t-1} \Delta t k_{loss}$$

4.10 Using OH Scavengers in Simulation Experiments

In oxidation experiments, the coupling of different oxidants specifically in the production of OH radicals (Sect. 4.5) it is often challenging to disentangle the complexities of the chemical system in chamber experiments (Bianchi et al. 2016; Riva et al. 2019). For example, ozonolysis of alkenes produces OH radicals with yields between 0.13 and 1.15, (Rickard et al. 1999) so that the reaction of the organic compound with OH competes with the reaction with ozone in the experiment (Atkinson et al. 1992). The reaction with OH can be suppressed, if a OH scavenger is additionally injected as often done in many experiments investigating ozonolysis reactions (Docherty and Ziemann 2003; Donahue et al. 2005; Henry and Donahue 2011; Henry et al. 2012; Keywood et al. 2004). The concentration of the radical scavenger must be sufficiently high so that the majority of the OH radicals (e.g. more than 99%) react with the scavenger instead other reactants in the chemical system:

$$\frac{k_{OH+scavenger}[scavenger]}{k_{OH+scavenger}[scavenger] + k_{OH+reactant}[reactant]} > 0.99 \qquad (4.10.1)$$

Typically, it is sufficient to calculate the amount of scavenger that is required for the maximum concentration of the reactant, because the OH production rate, for example, from the ozonolyis of the reactant is expected to decrease while the reactant concentration is decreasing (Fig. 4.11). In some cases, also OH production from product species may need to be additionally considered.

The chemistry from the OH scavenger and from the products of its reaction with OH needs to be taken into account in the evaluation of experiments. It should not significantly interfere with the chemical system that is investigated in the experiment.

CO is often used as scavenger in ozonolysis studies because it is unreactive to ozone. The reaction with OH generates HO_2 and CO_2:

$$OH + CO + O_2 \rightarrow HO_2 + CO_2 \qquad (R4.10.2)$$

Typical other OH scavenger molecules are H_2, H_2O_2, and organic compounds such as alcohols, and alkanes.

There are no organic compounds produced from the reaction of the scavenger with OH, if CO, H_2 or H_2O_2 are used so that experiments are not affected by organic

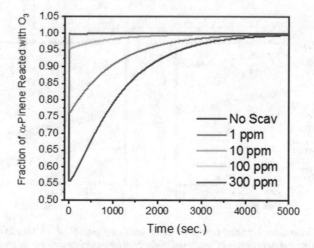

Fig. 4.11 Model calculations showing the effect of the injection of 1-butanol as OH scavenger in an experiment, in which the ozonolysis ($[O_3]_0 = 500$ ppbv) of α-pinene (100 ppvb) is investigated, on the temporal evolution of the fraction of α-pinene that reacts with ozone. Without scavenger, the fraction of OH reaction can be as high as 45%, whereas 300 ppmv 1-butanol is sufficient to scavenge all OH radicals so that the reaction of α-pinene with OH is suppressed

compounds produced from the scavenger. However, the reaction rate constants of their reactions with OH is relatively low so that large concentrations of the scavenger are required. This can cause safety hazards, because H_2 and CO are inflammable gases and CO is toxic.

An interference from H_2O_2 can occur in experiments investigating the aerosol phase, because H_2O_2 can be taken up on particles specifically at elevated relative humidity. Further impact from the scavenger molecules on the experiment can be caused by the radicals that are produced in their reaction with OH, because they take part in the radical reaction system in the experiment. Depending on the production rate of OH, significant concentrations of peroxy radicals can be produced.

The reaction of CO, H_2 and H_2O_2 with OH leads to the production of HO_2 (Reaction R4.10.2) so that HO_2 concentrations can be significantly higher compared to an experiment without scavenger (Fig. 4.12). This can shorten the lifetime of RO_2 radicals formed from the ozonolysis of organic compound due to their reaction with RO_2 and could potentially alter the product distribution (Keywood et al. 2004).

If an organic compound is used as OH scavenger, the impact of organic compounds needs to be considered. In addition, RO_2 formed from the scavenger could also affect the chemical system by increasing the rate of radical recombination reactions. Recent studies have also shown that the presence of RO_2 radicals formed from the scavenger + OH reaction pathway can result in the formation of dimers that include RO_2 derived from the scavenger. Therefore, the scavenger may also impact the formation of secondary organic aerosol (McFiggans et al. 2019; Zhao et al. 2018).

In general, it is recommended to perform sensitivity model calculations to estimate the impact of an OH scavenger on the results of the experiment.

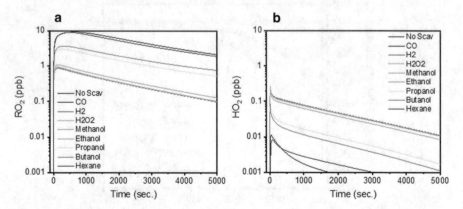

Fig. 4.12 Concentrations expected from model calculations using MCM 3.3.1 for RO_2 (**a**) and HO_2 (**b**) radicals during the ozonolysis of α-pinene in a chamber experiment for a mixture of α-pinene (100 ppbv), O_3 (500 ppbv). Model calculations are performed for different OH scavenger molecules (CO–30,000 ppmv, H2–2%, H_2O_2–500 ppmv, methanol–500 ppmv, ethanol–500 ppmv, propanol–500 ppmv, butanol–300 ppm, and hexane) demonstrating the effect on radical concentrations in the chamber experiment

References

Alam, M., Camredon, M., Rickard, A., Carr, T., Wyche, K., Hornsby, K., Monks, P., Bloss, W.: Total radical yields from tropospheric ethene ozonolysis. Phys. Chem. Chem. Phys. **13**, 11002–11015 (2011). https://doi.org/10.1039/c0cp02342f

Atkinson, R., Carter, W.P.L., Winer, A.M., Pitts, J.N.: An experimental protocol for the determination of OH radical rate constants with organics using methyl nitrite photolysis as an OH radical source. J. Air Pollut. Control Assoc. **31**, 1090–1092 (1981). https://doi.org/10.1080/00022470.1981.10465331

Atkinson, R., Aschmann, S.M., Winer, A.M., Pitts, J.N.: Kinetics of the gas-phase reactions of nitrate radicals with a series of dialkenes, cycloalkenes, and monoterpenes at 295 .+-. 1 K. Environ. Sci. Technol. **18**, 370–375 (1984a), https://doi.org/10.1021/es00123a016

Atkinson, R., Carter, W.P.L., Plum, C.N., Winer, A.M., Pitts, J.N., Jr.: Kinetics of the gas-phase reactions of NO_3 radicals with a series of aromatics at 296 ± 2 K. Int. J. Chem. Kinet. **16**, 887–898 (1984b). https://doi.org/10.1002/kin.550160709

Atkinson, R., Aschmann, S.M.: Kinetics of the gas phase reaction of Cl atoms with a series of organics at 296 ± 2 K and atmospheric pressure. Int. J. Chem. Kinet. **17**, 33–41 (1985). https://doi.org/10.1002/kin.550170105

Atkinson, R., Aschmann, S.M., Arey, J., Shorees, B.: Formation of OH radicals in the gas phase reactions of O_3 with a series of terpenes. J. Geophys. Res. Atmos. **97**, 6065–6073 (1992). https://doi.org/10.1029/92JD00062

Baasandorj, M., Papanastasiou, D.K., Talukdar, R.K., Hasson, A.S., Burkholder, J.B.: (CH3)3COOH (tert-butyl hydroperoxide): OH reaction rate coefficients between 206 and 375 K and the OH photolysis quantum yield at 248 nm. Phys. Chem. Chem. Phys. **12**, 12101–12111 (2010). https://doi.org/10.1039/c0cp00463d

Baklanov, A.V., Krasnoperov, L.N.: Oxalyl chloride a clean source of chlorine atoms for kinetic studies. J. Phys. Chem. A **105**, 97–103 (2001). https://doi.org/10.1021/jp0019456

Barmet, P., Dommen, J., DeCarlo, P.F., Tritscher, T., Praplan, A.P., Platt, S.M., Prévôt, A.S.H., Donahue, N.M., Baltensperger, U.: OH clock determination by proton transfer reaction mass spectrometry at an environmental chamber. Atmos. Meas. Tech. **5**, 647–656 (2012). https://doi.org/10.5194/amt-5-647-2012

Barnes, I., Bastian, V., Becker, K.H., Fink, E.H., Zabel, F.: Reactivity studies of organic substances towards hydroxyl radicals under atmospheric conditions. Atmos. Environ. **1967**(16), 545–550 (1982). https://doi.org/10.1016/0004-6981(82)90163-9

Barnes, I., Bastian, V., Becker, K.H., Tong, Z.: Kinetics and products of the reactions of nitrate radical with monoalkenes, dialkenes, and monoterpenes. J. Phys. Chem. **94**, 2413–2419 (1990). https://doi.org/10.1021/j100369a041

Bianchi, F., Tröstl, J., Junninen, H., Frege, C., Henne, S., Hoyle, C.R., Molteni, U., Herrmann, E., Adamov, A., Bukowiecki, N., Chen, X., Duplissy, J., Gysel, M., Hutterli, M., Kangasluoma, J., Kontkanen, J., Kürten, A., Manninen, H.E., Münch, S., Peräkylä, O., Petäjä, T., Rondo, L., Williamson, C., Weingartner, E., Curtius, J., Worsnop, D.R., Kulmala, M., Dommen, J., Baltensperger, U.: New particle formation in the free troposphere: a question of chemistry and timing. Science **352**, 1109–1112 (2016). https://doi.org/10.1126/science.aad5456

Burkholder, J.B., Mellouki, A., Talukdar, R., Ravishankara, A.R.: Rate coefficients for the reaction of OH with HONO between 298 and 373 K. Int. J. Chem. Kinet. **24**, 711–725 (1992). https://doi.org/10.1002/kin.550240805

Calvert, J.G., Pitts, J.N.J.: Photochemistry, pp. 897. John Wiley and Sons, Inc., New York (1966)

Chan, W.H., Nordstrom, R.J., Calvert, J.G., Shaw, J.H.: Kinetic study of HONO formation and decay reactions in gaseous mixtures of HONO, NO, NO_2, and N_2. Environ. Sci. Technol. **10**, 674–682 (1976)

Cox, R.A.: The photolysis of nitrous acid in the presence of carbon monoxide and sulphur dioxide. J. Photochem. **3**, 291–304 (1974). https://doi.org/10.1016/0047-2670(74)80038-9

Cox, R.A., Ammann, M., Crowley, J.N., Herrmann, H., Jenkin, M.E., McNeill, V.F., Mellouki, A., Troe, J., Wallington, T.J.: Evaluated kinetic and photochemical data for atmospheric chemistry: Volume VII–Criegee intermediates. Atmos. Chem. Phys. **20**, 13497–13519 (2020). https://doi.org/10.5194/acp-20-13497-2020

D'Anna, B., Andresen, Ø., Gefen, Z., Nielsen, C.J.: Kinetic study of OH and NO_3 radical reactions with 14 aliphatic aldehydes. Phys. Chem. Chem. Phys. **3**, 3057–3063 (2001). https://doi.org/10.1039/b103623h

Docherty, K.S., Ziemann, P.J.: Effects of Stabilized Criegee intermediate and OH radical scavengers on aerosol formation from reactions of β-pinene with O 3. Aerosol. Sci. Technol. **37**, 877–891 (2003). https://doi.org/10.1080/02786820300930

Donahue, N.M., Hartz, K.E.H., Chuong, B., Presto, A.A., Stanier, C.O., Rosenhorn, T., Robinson, A.L., Pandis, S.N.: Critical factors determining the variation in SOA yields from terpene ozonolysis: a combined experimental and computational study. Faraday Discuss. **130**, 295–309 (2005)

Dorn, H.P., Apodaca, R.L., Ball, S.M., Brauers, T., Brown, S.S., Crowley, J.N., Dubé, W.P., Fuchs, H., Häseler, R., Heitmann, U., Jones, R.L., Kiendler-Scharr, A., Labazan, I., Langridge, J.M., Meinen, J., Mentel, T.F., Platt, U., Pöhler, D., Rohrer, F., Ruth, A.A., Schlosser, E., Schuster, G., Shillings, A.J.L., Simpson, W.R., Thieser, J., Tillmann, R., Varma, R., Venables, D.S., Wahner, A.: Intercomparison of NO_3 radical detection instruments in the atmosphere simulation chamber SAPHIR. Atmos. Meas. Tech. **6**, 1111–1140 (2013). https://doi.org/10.5194/amt-6-1111-2013

Ehn, M., Thornton, J.A., Kleist, E., Sipilä, M., Junninen, H., Pullinen, I., Springer, M., Rubach, F., Tillmann, R., Lee, B., Lopez, F., Andres, S., Acir, I.-H., Rissanen, M., Jokinen, T., Schobesberger, S., Kangasluoma, J., Kontkanen, J., Nieminen, T., Kurtén, T., Nielsen, L.B., Jørgensen, S., Kjaergaard, H.G., Canagaratna, M., Maso, M.D., Berndt, T., Petäjä, T., Wahner, A., Kerminen, V.-M., Kulmala, M., Worsnop, D.R., Wildt, J., Mentel, T.F.: A large source of low-volatility secondary organic aerosol. Nature **506**, 476–479 (2014). https://doi.org/10.1038/nature13032

Etzkorn, T., Klotz, B., Sørensen, S., Patroescu, I.V., Barnes, I., Becker, K.H., Platt, U.: Gas-phase absorption cross sections of 24 monocyclic aromatic hydrocarbons in the UV and IR spectral ranges. Atmos. Environ. **33**, 525–540 (1999). https://doi.org/10.1016/S1352-2310(98)00289-1

Finlayson-Pitts, B.J., Pitts Jr., J.N.: Atmospheric chemistry: fundamentals and experimental techniques. John Wiley and Sons, New-York (1986). (edited by: Publication, W. I.)

Fouqueau, A.: Reactivity of terpenes with nitrate radical: kinetic and mechanistic studies in atmospheric simulation chambers. Université Paris-Est (2019)

Galloway, M.M., Huisman, A.J., Yee, L.D., Chan, A.W.H., Loza, C.L., Seinfeld, J.H., Keutsch, F.N.: Yields of oxidized volatile organic compounds during the OH initiated oxidation of isoprene, methyl vinyl ketone, and methacrolein under high-NOx conditions. Atmos. Chem. Phys. **11**, 10779–110790 (2011). https://doi.org/10.5194/acp-11-10779-2011

Ghosh, B., Papanastasiou, D.K., Burkholder, J.B.: Oxalyl chloride, ClC(O)C(O)Cl: UV/vis spectrum and Cl atom photolysis quantum yields at 193, 248, and 351 nm. J. Chem. Phys. **137**, 164315 (2012). https://doi.org/10.1063/1.4755769

Gibalov, V.I., Pietsch, G.J.: On the performance of ozone generators working with dielectric barrier discharges. Ozone Sci. Eng. **28**, 119–124 (2006). https://doi.org/10.1080/01919510600559419

Graham, R.A., Winer, A.M., Pitts, J.N.: Temperature dependence of the unimolecular decomposition of pernitric acid and its atmospheric implications. Chem. Phys. Lett. **51**, 215–220 (1977). https://doi.org/10.1016/0009-2614(77)80387-4

Graham, R.A., Winer, A.M., Pitts, J.N., Jr.: Pressure and temperature dependence of the unimolecular decomposition of HO_2NO_2. J. Chem. Phys. **68**, 4505–4510 (1978). https://doi.org/10.1063/1.435554

Gray, J.A., Style, D.W.G.: The photolysis of methyl nitrite. Trans. Faraday Soc. **48**, 1137–1142 (1952). https://doi.org/10.1039/tf9524801137

Hass, S.A., Andersen, S.T., Nielsen, O.J.: Trichloroacetyl chloride, CCl3COCl, as an alternative Cl atom precursor for laboratory use and determination of Cl atom rate coefficients for n-CH2=CH(CH2)xCN (x = 3–4). Environ. Sci. Process Impacts **22**, 1347–1354 (2020). https://doi.org/10.1039/d0em00105h

Henry, K.M., Donahue, N.M.: Effect of the OH radical scavenger hydrogen peroxide on secondary organic aerosol formation from α-pinene ozonolysis. Aerosol. Sci. Technol. **45**, 696–700 (2011). https://doi.org/10.1080/02786826.2011.552926

Henry, K.M., Lohaus, T., Donahue, N.M.: Organic aerosol yields from α-pinene oxidation: bridging the gap between first-generation yields and aging chemistry. Environ. Sci. Technol. **46**, 12347–12354 (2012). https://doi.org/10.1021/es302060y

Johnson, D., Marston, G.: The gas-phase ozonolysis of unsaturated volatile organic compounds in the troposphere. Chem. Soc. Rev. 37, 699–716 (2008). https://doi.org/10.1039/b704260b

Kerdouci, J., Picquet-Varrault, B., Durand-Jolibois, R., Gaimoz, C., Doussin, J.-F.: An experimental study of the gas-phase reactions of NO3 radicals with a series of unsaturated aldehydes: trans-2-Hexenal, trans-2-Heptenal, and trans-2-Octenal. J. Phys. Chem. A **116**, 10135–10142 (2012). https://doi.org/10.1021/jp3071234

Keywood, M.D., Kroll, J.H., Varutbangkul, V., Bahreini, R., Flagan, R.C., Seinfeld, J.H.: Secondary organic aerosol formation from cyclohexene ozonolysis: effect of OH scavenger and the role of radical chemistry. Environ. Sci. Technol. **38**, 3343–3350 (2004). https://doi.org/10.1021/es049725j

Kleffmann, J., Heland, J., Kurtenbach, R., Lörzer, J.C., Wiesen, P.: A new instrument (LOPAP) for the detection of nitrous acid (HONO). Environ. Sci. Pollut. Res. **9**, 48–54 (2002)

McFiggans, G., Mentel, T.F., Wildt, J., Pullinen, I., Kang, S., Kleist, E., Schmitt, S., Springer, M., Tillmann, R., Wu, C., Zhao, D., Hallquist, M., Faxon, C., Le Breton, M., Hallquist, Å.M., Simpson, D., Bergström, R., Jenkin, M.E., Ehn, M., Thornton, J.A., Alfarra, M.R., Bannan, T.J., Percival, C.J., Priestley, M., Topping, D., Kiendler-Scharr, A.: Secondary organic aerosol reduced by mixture of atmospheric vapours. Nature **565**, 587–593 (2019). https://doi.org/10.1038/s41586-018-0871-y

Min, K.E., Washenfelder, R.A., Dubé, W.P., Langford, A.O., Edwards, P.M., Zarzana, K.J., Stutz, J., Lu, K., Rohrer, F., Zhang, Y., Brown, S.S.: A broadband cavity enhanced absorption spectrometer for aircraft measurements of glyoxal, methylglyoxal, nitrous acid, nitrogen dioxide, and water vapor. Atmos. Meas. Tech. **9**, 423–440 (2016). https://doi.org/10.5194/amt-9-423-2016

Mitchell, D.N., Wayne, R.P., Allen, P.J., Harrison, R.P., Twin, R.J.: Kinetics and photochemistry of NO3. Part 1—absolute absorption cross-section. J. Chem. Soc. Faraday Trans. 2 Mol. Chem. Phys. **76**, 785–793 (1980). https://doi.org/10.1039/f29807600785

Nash, T.: Chemical status of nitrogen dioxide at low aerial concentration. Ann. Occup. Hyg. **11**, 235–239 (1968). https://doi.org/10.1093/annhyg/11.3.235

Niki, H., Maker, P.D., Savage, C.M., Breitenbach, L.P.: An FTIR study of mechanisms for the HO radical initiated oxidation of C2H4 in the presence of NO: detection of glycolaldehyde. Chem. Phys. Lett. **80**, 499–503 (1981). https://doi.org/10.1016/0009-2614(81)85065-8

Peeters, J., Müller, J.-F., Stavrakou, T., Nguyen, V.S.: Hydroxyl radical recycling in isoprene oxidation driven by hydrogen bonding and hydrogen tunneling: the upgraded LIM1 mechanism. J. Phys. Chem. A **118**, 8625–8643 (2014). https://doi.org/10.1021/jp5033146

Picquet-Varrault, B.: Etude cinétique et mécanistique de la photooxydation des acétates en atmosphère simulée, Université Denis Diderot Paris 7 (2000)

Rickard, A.R., Johnson, D., McGill, C.D., Marston, G.: OH yields in the gas-phase reactions of ozone with alkenes. J. Phys. Chem. A **103**, 7656–7664 (1999). https://doi.org/10.1021/jp9916992

Riva, M., Heikkinen, L., Bell, D.M., Peräkylä, O., Zha, Q., Schallhart, S., Rissanen, M.P., Imre, D., Petäjä, T., Thornton, J.A., Zelenyuk, A., Ehn, M.: Chemical transformations in monoterpene-derived organic aerosol enhanced by inorganic composition, npj Clim. Atmos. Sci. **2**, 2 (2019). https://doi.org/10.1038/s41612-018-0058-0

Rohrer, F., Bohn, B., Brauers, T., Brüning, D., Johnen, F.J., Wahner, A., Kleffmann, J.: Characterisation of the photolytic HONO-source in the atmosphere simulation chamber SAPHIR. Atmos. Chem. Phys. **5**, 2189–2201 (2005). https://doi.org/10.5194/acp-5-2189-2005

Schlosser, E., Brauers, T., Dorn, H.P., Fuchs, H., Häseler, R., Hofzumahaus, A., Holland, F., Wahner, A., Kanaya, Y., Kajii, Y., Miyamoto, K., Nishida, S., Watanabe, K., Yoshino, A., Kubistin, D., Martinez, M., Rudolf, M., Harder, H., Berresheim, H., Elste, T., Plass-Dülmer, C., Stange, G., Schurath, U.: Technical note: formal blind intercomparison of OH measurements: results from the international campaign HOxComp. Atmos. Chem. Phys. **9**, 7923–7948 (2009). https://doi.org/10.5194/acp-9-7923-2009

Simpson, W.R., Brown, S.S., Saiz, A., Thornton, J.A., von Glasow, R.: Tropospheric halogen chemistry: sources, cycling and impacts. Chem. Rev. **115**, 4035–4062 (2015). https://doi.org/10.1021/cr5006638

Slade, J.H., de Perre, C., Lee, L., Shepson, P.B.: Nitrate radical oxidation of γ-terpinene: hydroxy nitrate, total organic nitrate, and secondary organic aerosol yields. Atmos. Chem. Phys. **17**, 8635–8650 (2017). https://doi.org/10.5194/acp-17-8635-2017

Spittler, M., Barnes, I., Bejan, I., Brockmann, K.J., Benter, T., Wirtz, K.: Reactions of NO_3 radicals with limonene and α-pinene: product and SOA formation. Atmos. Environ. **40**, 116–127 (2006). https://doi.org/10.1016/j.atmosenv.2005.09.093

Stefenelli, G., Jiang, J., Bertrand, A., Bruns, E., Pieber, S., Baltensperger, U., Marchand, N., Aksoyoglu, S., Prevot, A., Slowik, J., El Haddad, I.: Secondary organic aerosol formation from smoldering and flaming combustion of biomass: a box model parametrization based on volatility basis set. Atmos. Chem. Phys. **19**, 11461–11484 (2019). https://doi.org/10.5194/acp-19-11461-2019

Stone, D., Whalley, L.K., Heard, D.E.: Tropospheric OH and HO2 radicals: field measurements and model comparisons. Chem. Soc. Rev. **41**, 6348–6404 (2012). https://doi.org/10.1039/c2cs35140d

Taylor, W.D., Allston, T.D., Moscato, M.J., Fazekas, G.B., Kozlowski, R., Takacs, G.A.: Atmospheric photodissociation lifetimes for nitromethane, methyl nitrite, and methyl nitrate. Int. J. Chem. Kinet. **12**, 231–240 (1980). https://doi.org/10.1002/kin.550120404

Thüner, L., Bardini, P., Rea, G., Wenger, J.: Kinetics of the gas-phase reactions of OH and NO_3 radicals with dimethylphenols. J. Phys. Chem. A **108**, 11019–11025 (2004). https://doi.org/10.1021/jp046358p

Tumbiolo, S., Vincent, L., Gal, J.-F., Maria, P.-C.: Thermogravimetric calibration of permeation tubes used for the preparation of gas standards for air pollution analysis. Analyst **130**, 1369–1374 (2005). https://doi.org/10.1039/b508536e

Wayne, R.P., Barnes, I., Biggs, P., Burrows, J.P., Canosa-Mas, C.E., Hjorth, J., Le Bras, G., Moortgat, G.K., Perner, D., Poulet, G., Restelli, G., Sidebottom, H.: The nitrate radical: physics, chemistry, and the atmosphere. Atmos. Environ. Part A. Gen. Top. **25**, 1–203. https://doi.org/10.1016/0960-1686(91)90192-A, 1991

Zhao, Y., Thornton, J.A., Pye, H.O.T.: Quantitative constraints on autoxidation and dimer formation from direct probing of monoterpene-derived peroxy radical chemistry. Proc Natl Acad Sci USA **115**, 12142–12147 (2018). https://doi.org/10.1073/pnas.1812147115

Chapter 5
Preparation of the Experiment: Addition of Particles

Rami Alfarra, Urs Baltensperger, David M. Bell, Silvia Giulia Danelli,
Claudia Di Biagio, Jean-François Doussin, Paola Formenti,
Martin Gysel-Beer, Dario Massabò, Gordon McFiggans, Rob L. Modini,
Ottmar Möhler, Paolo Prati, Harald Saathoff, and John Wenger

Abstract Atmospheric simulation chambers are often utilized to study the physical properties and chemical reactivity of particles suspended in air. In this chapter, the various approaches employed for the addition of particles to simulation chambers are described in detail. Procedures for the generation of monodispersed seed aerosols, mineral dust, soot particles and bioaerosols are all presented using illustrative examples from chamber experiments. Technical descriptions of the methods used for the addition of whole emissions (gases and particles) from real-world sources such as wood-burning stoves, automobile engines and plants are also included, along with an outline of experimental approaches for investigating the atmospheric processing of these emissions.

5.1 Motivation

During the formation of secondary organic aerosol (SOA) in an atmospheric simulation chamber experiment, deposition of condensable material to the walls of the chamber can be a significant loss term when determining yields of SOA formation (Zhang et al. 2014). As a result, it is now commonplace to conduct experiments with

R. Alfarra · G. McFiggans
University of Manchester, Manchester, UK

U. Baltensperger · D. M. Bell · M. Gysel-Beer · R. L. Modini
Paul Scherrer Institute, Villigen, Switzerland

S. G. Danelli · D. Massabò · P. Prati
Università degli Studi di Genova, Genoa, Italy

C. Di Biagio · J.-F. Doussin · P. Formenti
Centre National de la Recherche Scientifique, Paris, France

O. Möhler · H. Saathoff
Karlsruhe Institute of Technology, Karlsruhe, Germany

J. Wenger (✉)
University College Cork, Cork, Ireland
e-mail: j.wenger@ucc.ie

© The Author(s) 2023
J.-F. Doussin et al. (eds.), *A Practical Guide to Atmospheric Simulation Chambers*,
https://doi.org/10.1007/978-3-031-22277-1_5

unreactive seeds in the chamber to compete with the surfaces of the walls for the condensable material. Many aerosol generators, such as atomizers (typically used in combination with a dryer), produce high aerosol concentrations, which are suitable for out-competing walls for condensable material. Such atomizers are widely used to produce seed aerosols with a wide variety of desired chemical composition, both inorganic and organic, with varying acidity/hygroscopicity and are straightforward in their use (see, e.g. Leskinen et al. 2015; Stirnweis et al. 2017). Inorganic seeds are typically preferred because they allow for easier chemical discrimination of the seed and the formed SOA, e.g. by an aerosol mass spectrometer (AMS). The decrease in the seed aerosol concentration can also be used to correct for wall losses (Zhang et al. 2014).

A downside of the use of atomizers is that they produce rather broad particle size distributions. This results in differences in the wall loss rates for particles of different sizes (see Sect. 2.5), which can complicate the analysis of such experiments. Also, the ratio of SOA to seed aerosol mass will be different for particles of different sizes (Stirnweis et al. 2017). These difficulties can be overcome by using a classifier behind the atomizer/dryer. Such classification can be performed either based on the mobility diameter (differential mobility analyser; DMA), mass (aerosol particle mass analyser; APM) or aerodynamic diameter (aerodynamic aerosol classifier; AAC). The latter device is particularly useful as it delivers truly monomodal aerosol independent of charge distribution. All these classifiers, however, suffer from a substantial reduction in the aerosol concentration compared to the polydispersed aerosol. In the following, two techniques are described that allow for the production of higher concentrations of monodispersed seed aerosols.

5.1.1 Procedure for Generation of Monodispersed Seed Aerosols

Using an electrospray atomizer

During seeded experiments, ammonium sulphate can be generated using an electrospray aerosol generator (TSI-3480). Ammonium sulphate forms approximately spherical seed particles, which is useful when determining total water uptake based on an increase in mobility diameter. An electrospray aerosol generator can produce nearly monodisperse (geom. std. dev. $\sigma_g = 1.3$) aerosol particles at concentrations that are often only achieved for polydisperse seed samples, Fig. 5.1 (Meyer et al. 2008).

Aerosol generation by condensation of heated gases

Aerosol generation by condensation of heated gases has been used in experiments conducted at the cloud chamber at CERN (Hoyle et al. 2016). Experiments were performed in a well-mixed flow chamber mode, with the sample air drawn off by the instruments continually replaced, and the mixing ratio of any added gas-phase

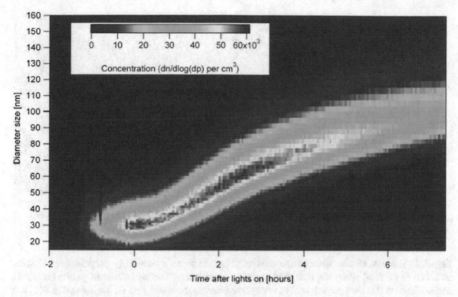

Fig. 5.1 Time series of size distributions measured inside the chamber starting with $(NH_4)_2SO_4$ seed aerosol (seed diameter of approximately 33 nm and $\sigma_g = 1.3$) followed by the growth due to photo-oxidation of α-pinene (Reused with permission from Meyer et al. (2008). Open access under a CC BY 3.0 license, https://creativecommons.org/licenses/by/3.0/)

species being held approximately constant. This results in a dilution lifetime of 2–3 h. O_3 and SO_2 were continually added to the chamber to maintain approximately constant mixing ratios, the dilution lifetime corresponds to the lifetime of the aerosol particles.

Two kinds of seed aerosol were used in these experiments, pure H_2SO_4, and partially to fully neutralized ammonium sulphate aerosol. The pure H_2SO_4 aerosol was formed in an external CCN generator, which comprised a temperature-controlled stainless steel vessel holding a ceramic crucible filled with concentrated H_2SO_4. After heating the vessel to between 150 and 180 °C, depending on the desired characteristics of the aerosol population, a flow of N_2 was passed through the vessel, above the crucible to transport the hot H_2SO_4 vapour into the chamber. In addition, a humidified flow of N_2 was added to the aerosol injection line immediately downstream of the H_2SO_4 vessel, to create more reproducible size distributions. As the vapour cooled in the injection line, H_2SO_4 droplets formed. The partially or fully neutralized aerosol was formed by using the same aerosol generator, and injecting NH_3 directly into the chamber, where it was taken up by the acidic aerosol. The mode diameter of the aerosol distribution produced by this method was approximately 65–75 nm, with a full width at half maximum (FWHM) of approximately 50–70 nm.

In this set-up, the addition of sulphuric acid was stopped at the beginning of an experiment (to avoid the presence of particles with different ageing times. Therefore, the concentration decreased steadily by dilution of the chamber due to the instrument feed, as shown in Fig. 5.2, while the mode of the particles stayed roughly constant.

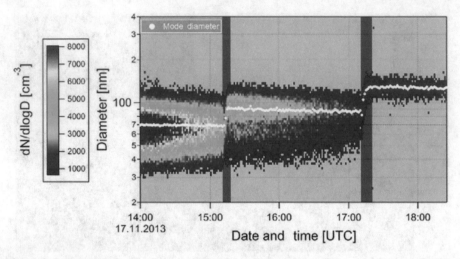

Fig. 5.2 The aerosol size distribution measured by the scanning mobility particle sizer (SMPS) attached to the total sampling line, for a specific experiment. The white line of points shows the mode diameter. Aerosol growth is clearly observed during the cloud periods during which SO₂ was taken up and transformed to sulphuric acid by ozone (marked by the purple vertical bars). (Reused with permission from Hoyle et al. (2016). Open access under a CC BY 3.0 license, https://creativec ommons.org/licenses/by/3.0/)

5.2 Mineral Dust Aerosol and Its Mineral Constituents

5.2.1 Motivation

Mineral dust is one of the dominant aerosol species at the regional and global scales and strongly affects climate via direct and indirect radiative effects and by influencing atmospheric chemistry (Knippertz and Stuut 2014). Chamber experiments are of high relevance to elucidate the properties and processes that drive the climate impacts of mineral dust aerosols. They mostly focus on investigations of the following:

- physicochemical and spectral optical properties such as scattering, absorption, extinction cross section and complex refractive index, i.e. Linke et al. 2006; Wagner et al. 2012; Caponi et al. 2017; Di Biagio et al. 2017a, 2019);
- hygroscopicity properties (Cloud Condensation Nuclei and Ice Nuclei ability; i.e. Czico et al. 2009; Ullrich et al. 2017); and
- heterogeneous chemistry (i.e. Mogili et al. 2006; Chen et al. 2011).

In chamber studies, it is important to ensure that the laboratory-generated dust is similar in particle size, shape and composition to ambient dust aerosols, and free of contamination resulting from the generation process itself.

5.2.2   Generation of Dust Aerosols from Mechanical Agitation and Vibration Devices

These techniques involve putting a certain amount of source soil in a sample holder and then shaking or vibrating it (Lafon et al. 2014; Di Biagio et al. 2017a). Mechanical agitation or vibration provides the source material with the kinetic energy required for breaking the aggregates it contains and to generate dust aerosols which are similar to natural emissions. The mode of energy transfer for these generators is solid–solid given that the aerosols are generated from the abrasion or fracture caused when grains of the source material collide with each other and with the dust holder walls. This is the technique used recently by Utry et al. (2015), Caponi et al. (2017) and Di Biagio et al. (2014, 2017a, 2019) to study the spectral optical properties of mineral dust aerosols.

Experimental procedure

To apply the dust generation protocol described below, it is necessary to use a mechanical shaker or vibrating plate that can be regulated in frequency and amplitude. A glass sample holder with two connections (i.e. conical glass flask-type glass vacuum flask), one for the input of an inert particle-free gas and one for the dust output flow, is required. External connections are needed, i.e. from the gas supplier (gas bottle) to the sample holder (Teflon tubing) and from the sample holder to the chamber. It is recommended that the tubing connecting the sample holder to the chamber is of conductive silicone material to minimize particle loss by electrostatic deposition. There is no specific recommendation for the diameter of the tubing and connections for dust output. A general requirement is that the chamber is equipped with a ventilation system to help the aerosols to remain in suspension, in particular, the super-micron (heaviest) component. The vertical air flux from the ventilation system also allows to homogenize the distribution of the aerosol population within the chamber volume.

Soil preparation

Prior to aerosol generation, the source soil has to be:

− Dry sieved: Source soils can be used in a more or less undisturbed state as they exist in nature or they can be sieved before aerosol generation. Although soil sieving is not deemed to be essential, almost all previous studies using this generation technique have sieved the soils. In order to mimic the generation process and properties of the dust aerosols as they are in nature, a good recommendation is to sieve the soil samples at 1000 μm in order to take into account only the fraction susceptible to erosion, so as to eliminate any non-erodable grains (Lafon et al. 2014; Di Biagio et al. 2017a).
− Dried: water vapour from the soil sample is usually removed to maximize its emission capacity, i.e. to reproduce soil conditions in source dry arid areas. To do so, samples can be heated at more than 100 °C in the oven, held in samples under vacuum, or alternatively, they can be put in a vessel partly filled with silica

gel for a few hours. If a sample shows a particular tendency to retain water, like the mineral montmorillonite, the drying process can require repeated heating and pumping cycles over a few hours or overnight pumping under vacuum conditions in order to better remove residual water (Mogili et al. 2007).

Material preparation

The sample holder is cleaned and dried prior to use. The cleaning procedure may include rinsing with deionized water, 15 min of sonication, and drying under a laminar flow hood.

The silicon tubes and connections used in the system can be flushed with high-speed air to remove residual dust. If needed metallic connections can be additionally rinsed with deionized water, followed by 15 min of sonication and drying under a laminar flow hood.

Aerosol generation and injection

The step-by-step procedure for dust generation is:

1. Place a few grams of soil in the sample holder and fix to the shaker or vibrating plate. The amount of soil depends on the volume of the chamber and the targeted mass or number concentration.
2. One entrance of the sample holder is connected to a source of particle-free inertial gas (N_2) while the other is left open to laboratory air.
3. The sample holder is flushed with inertial gas for 2–3 min to eliminate gaseous impurities within the holder.
4. After flushing, the sample holder is immediately connected to the chamber port but the valve opening into the chamber is keep closed. In this way, the sample holder is closed, i.e. no contamination from ambient air will occur, and the configuration is ready for dust injection in the chamber (it will be necessary only to open the valve to make the generated dust to enter the chamber volume).
5. The shaking or vibration is activated for a few minutes. The operating frequency and amplitude of the shaking/vibration should be regulated depending on the desired number and mass concentration and size distribution of the dust. When the aerosol generation process is effective, it should be possible to see an 'aerosol cloud' within the sample holder. A sensitivity study based on a generation device using a shaking arm (Lafon et al. 2014) showed that shaking at a higher frequency increases the number and mass concentration of aerosol particles and also increases the ratio of the fine to coarse dust. This is because increasing the kinetic energy of the soil aggregates is known to liberate aerosols enriched in fine particles (Sow et al. 2009).
6. Dust injection in the chamber is achieved by flushing the dust aerosol suspension with a particle-free inertial gas (N_2) while continuing to shake or vibrate the soil. The valve between the sample holder and the chamber is opened to allow the aerosol to enter the chamber. The injection typically continues for a few minutes until the desired concentration is obtained.

7. When the injection is finished, the valve connecting the sample holder to the chamber is closed and, at the same time, the inertial flow is stopped.

Example of mineral dust aerosol generation

A picture of the system used in Di Biagio et al. (2014, 2017a, 2019) and Caponi et al. (2017) to generate dust in the 4.2 m^3 CESAM chamber is shown in Fig. 5.3. These studies used 15 g of soil sample sieved at 1000 μm and dried at 100 °C for about 1 h in an oven. The soil was put in a 500 ml glass vacuum flask and the vessel was flushed with N$_2$ at 10 L min^{-1} for about 5 min to remove gaseous impurities. The soil was vibrated for 30 min at a frequency of 100 Hz by means of a sieve shaker (Retsch AS200) operated at an amplitude of 70/100. The dust suspension in the flask was injected into the CESAM chamber by flushing it with N$_2$ at 10 L min^{-1} for about 10–15 min.

The CESAM chamber is equipped with a four-blade stainless steel fan located at the bottom of the chamber that ensures a vertical flux of approximately 10 m s^{-1} and is used to achieve homogeneous conditions within the chamber volume (with a typical mixing time of approximately 10 min).

Figure 5.4 shows the surface size distribution of the suspended dust measured 10 min after injection in the CESAM chamber (Di Biagio et al. 2017a). Different experiments were performed with dust of various origins. Figure 5.4 shows the range of sizes measured in CESAM for experiments with Northern African samples (Tunisia, Morocco, Algeria, Libya, Mauritania, grey-shaded area) in comparison to field observations of the size distribution for dust close to source regions in Northern

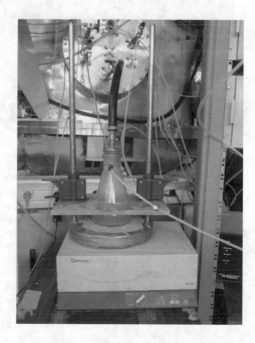

Fig. 5.3 Picture of the system used in Di Biagio et al. (2014, 2017a, 2019) and Caponi et al. (2017) to generate mineral dust aerosols. The vacuum flask is attached to the shaker using a custom-made wood plate. The Teflon white tube connecting the vessel to the N$_2$ gas bottle supplier and the black 0.64 cm silicon tube connecting it to the CESAM chamber are also visible

Africa as measured during AMMA (African Monsoon Multidisciplinary Analyses, Formenti et al. 2011), SAMUM1 (Saharan Mineral Dust Experiment, Weinzerl et al. 2009) and FENNEC (Ryder et al. 2013). The size of the generated dust in CESAM at the beginning of the experiments includes both sub- and super-micron aerosols up to more than 20 μm in diameter and compares well with field observations. This suggests that the aerosol generation procedure is good at reproducing the size of natural dust particles measured close to the source areas.

The dust size distribution in the chamber changes significantly over time due to gravitational settling. In CESAM the lifetime of dust aerosols varies with the size. For particles smaller than 2 μm in diameter the lifetime is >60 min, but for particles larger than about 10 μm in diameter, the lifetime is <10 min (Di Biagio et al. 2017a). The rapid decrease of the coarse mode above 5 μm is due to the much larger settling velocity of the bigger particles (~1 cm s^{-1} at 10 μm, compared to ~0.01 cm s^{-1} at 1 μm).

The range of mass concentrations obtained in CESAM during the two studies by Di Biagio et al. (2017a, 2019) was between 2 and 310 mg m^{-3} at the peak of the injection. The concentration decreased to less than 1 mg m^{-3} after 2 h due to the combined effects of gravitational deposition and dilution caused by instrument sampling. An example of the time evolution of dust mass concentration in CESAM is reported in Fig. 5.5 (Caponi et al. 2017).

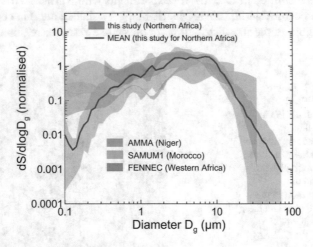

Fig. 5.4 Surface size distribution obtained for dust aerosols from Northern Africa (Figure reused with permission from Di Biagio et al. (2017a). Open access under a CC BY 3.0 license, https://creati vecommons.org/licenses/by/3.0/). The grey-shaded area represents the range of sizes measured in CESAM during experiments with the different Northern African samples. Comparison of CESAM measurements at the peak of the injection with dust size distributions from several airborne field campaigns in Northern Africa is also shown. Data from field campaigns are AMMA (Formenti et al. 2011), SAMUM-1 (Weinzierl et al. 2009) and FENNEC (Ryder et al. 2013). The shaded areas for each dataset correspond to the range of variability observed for the campaigns considered

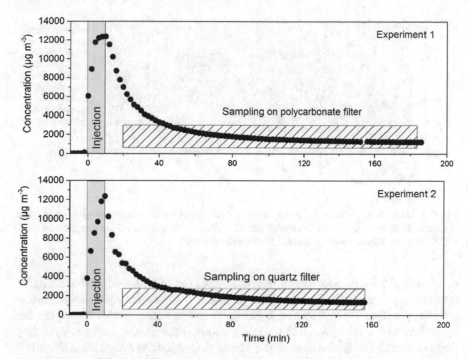

Fig. 5.5 Time series of aerosol mass concentration in the CESAM chamber for two experiments with the same soil sample from Libya. Experiment 1 (top panel) was dedicated to the determination of the chemical composition by sampling on polycarbonate filters. Experiment 2 was dedicated to the determination of the absorption optical properties by sampling on quartz filters (Figure reused with permission from Caponi et al. (2017). Open access under a CC BY 3.0 license, https://creativecommons.org/licenses/by/3.0/)

Figure 5.6 shows the mineralogical composition of the dust aerosols obtained from X-Ray Diffraction analyses of particles collected on filters during experiments. The generated dust contains a wide range of silicates, calcium-rich species, feldspars and iron oxides, as expected for dust samples of varying origins. The proportions of the different minerals are also realistic compared to atmospheric dust samples, which are usually dominated by clays and quartz, with smaller amounts of calcite, dolomite, gypsum, feldspars and iron oxides (i.e. Formenti et al. 2011).

5.2.3   Generation of Dust Aerosols from Fluidization Devices

Fluidization devices simulate the suspension of pre-existing fine particles from a solid surface under the effect of lifting forces or drag, a process that mostly mimics the re-suspension of aerosol deposited on the ground at receptor sites. As a result, the particle size distribution of the aerosol is virtually identical to that produced when the dust is suspended in ambient air. These fluidization devices do not transfer mechanical

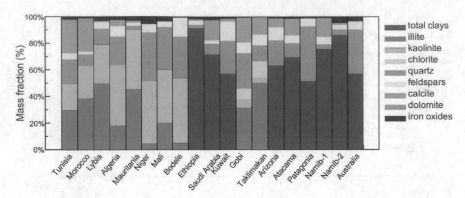

Fig. 5.6 Mineralogical composition for nineteen dust samples with different origins investigated in CESAM (Figure reused with permission from Di Biagio et al. (2017a). Open access under a CC BY 3.0 license, https://creativecommons.org/licenses/by/3.0/)

or kinetic energy to dust source materials and, for this reason, they are usually referred to as 'resuspension chambers'. Indeed, some recent fluidization devices have been designed to transfer kinetic energy to source materials with consequent de-agglomeration of soil material by the application of strong shear forces in expanding flows (nozzles) (Vlasenko et al. 2005; Linke et al. 2006; Mogili et al. 2006, 2007; Wagner et al. 2012) or by the use of high-pressure air 'shots' (Moosmüller et al. 2012; Engelbrecht et al. 2016). The mode of energy transfer for these generators is fluid–solid given that the energy for de-aggregating the dust agglomerates is obtained from shearing stress.

Experimental procedure

Fludization systems require a powder disperser and a nozzle: the disperser feeds the nozzle that de-aggregates the dust agglomerates which then enter the chamber volume. The powder disperser can be a commercial type (as in Linke et al. 2006 or Wagner et al. 2012) or custom-made (as in Mogili et al. 2006, 2007). Commercial dispersers use a rotating belt or a brush to feed dust into the injector nozzle. A custom-made system can be composed of a glass sample holder with two connections similar to the one described in 5.2.2 and a solenoid valve to put between the sample holder and the nozzle. External connections should be of conductive silicone tubing to minimize particle losses. Connections from the gas supply (gas bottle) to the powder disperser are also required (Teflon tubing). There is no specific recommendation for the diameter of the tubing and connections for dust output. A general requirement is that the chamber is equipped with a ventilation system to help the aerosols remain in suspension and to ensure their homogeneous distribution in the chamber volume.

Soil preparation

Prior to use for aerosol generation, the source soil has to be:

Dry sieved: in all previous studies involving fluidization devices, the soils were sieved to keep only a fraction below a certain diameter (usually below 100 μm) for generation. The exact size may depend on the application of the specific study and the desired output size distribution. For instance, Wagner et al. (2012) sieved soils in order to keep only the fraction in the range 20–75 μm, whereas Linke et al. (2006) sieve the soil at 20 μm before generation.

Dried: after sieving the soil samples are dried by heating them at over 100 °C in the oven, holding them under vacuum, or putting them in a vessel partly filled with silica gel for a few hours.

Material preparation

The tubing, connections and (if used) the sample holder are cleaned and dried before utilization. The cleaning procedure may include flushing with high-speed air, or rinsing with deionized water, about 10–15 min of sonication, and drying under a laminar flow hood.

Aerosol generation and injection

The procedure to generate mineral dust aerosols depends on whether a commercial or custom-made disperser is used.

When a commercial disperser is used the procedure is:

1. Fill the disperser with the soil sample (typically a few to hundreds of grams of soil are required depending on the model and set-up).
2. Activate the disperser. The movement of the belt or brush in the disperser ensures a small but constant and reproducible supply of powder to the nozzle. The generated aerosol is available at the output of the nozzle. The resulting particle number concentration can be adjusted by setting the feeding belt or rotating brush speed. Particle production can be stopped without changing the gas flow through the generator by stopping the belt or brush movement.

If a custom-made disperser is used the procedure is:

1. A certain amount of soil sample is placed in the sample holder (the amount of soil will depend on the desired output concentration, but typically is a few to tens of grams).
2. One entrance of the sample holder is connected to a source of a particle-free inertial gas (N_2) while the other is left open.
3. The sample holder is flushed with the inertial gas for 2–3 min to eliminate gaseous impurities within the holder.
4. The sample holder is pressurized to a high level with N_2.
5. The pulsed valve solenoid between the sample holder and the chamber is activated which entrains a certain amount of dispersed soil into the nozzle. The generated aerosol is available at the output of the nozzle.

Other systems are conceived so that the generation is ensured only by the use of a high-pressure air shot without the use of a nozzle (Moosmüller et al. 2012; Engelbrecht et al. 2016). In this case, the procedure is:

1. Place the soil or powder sample in a sample holder.
2. One entrance of the sample holder is connected to a source of a particle-free inertial gas (N_2) while the other is left open.
3. The sample holder is flushed with the inertial gas for 2–3 min to eliminate gaseous impurities within the holder.

The sample holder is connected to the chamber and a pulsed high-speed jet of filtered air is injected. The air jet suspends part of the material and transports it into the chamber.

Example of mineral dust aerosol generation

A good example of the use of this aerosol generation method is provided by Wagner et al. (2012) who generated dust aerosols to study their optical properties in the NAUA chamber (3.7 m^3). In this case, a commercial powder disperser (PALAS RGB 1000) was used followed by a nozzle and a cyclone system (using alternatively one or two stages) to cut the coarse particle size fraction. Both the disperser and the nozzle were operated with dry and particle-free synthetic air and the dispersion pressure of the nozzle was 1.5 bar. The resulting number and volume size distribution measured in the NAUA chamber are shown in Fig. 5.7. Because of the cyclone system, the particle cut-off size is around 2–3 μm in diameter (two or one stages of cyclone, respectively). The initial number concentration is between 860 and 6500 cm^{-3}.

Figure 5.8 shows the size-resolved mineralogical composition for the four dust samples analysed by Wagner et al. (2012). The generated dust contains similar proportions of minerals as the real dust and is dominated by silicates, calcium-rich species and iron oxides as expected for the sub-micron fraction of dust.

A custom-made disperser system was used in Mogili et al. (2006, 2007) to generate aerosols from single synthetic minerals to study their extinction spectra in a 0.151 m^3 environmental chamber. In their system, the mineral dust sample was put in a sample holder (conical glass flask) that was pressurized up to 100 psi (corresponding

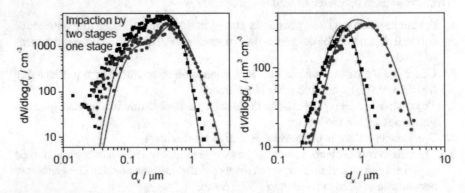

Fig. 5.7 Number size distribution obtained in the NAUA chamber by cutting the dust coarse fraction (Figure reused with permission from Wagner et al. (2012). Open access under a CC BY 3.0 license, https://creativecommons.org/licenses/by/3.0/)

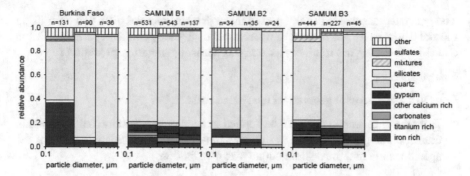

Fig. 5.8 Size-resolved mineralogical composition for four dust samples analysed in the NAUA chamber (Figure reused with permission from Wagner et al. (2012). Open access under a CC BY 3.0 license, https://creativecommons.org/licenses/by/3.0/)

to 6895 mbar) with a particle-free inert gas. A pulsed solenoid valve between the chamber and the sample holder was activated so that the aerosol could enter the chamber through a nozzle and an impactor plate assembly. The impactor cut-off the particles above about 2–3 μm in diameter. The number size distribution obtained in Mogili et al. (2007) was monomodal with diameters between 107 and 357 nm. The number concentration was 10^5–10^6 cm^{-3}.

Some other studies have used fluidization devices to generate dust aerosols or their mineral components (i.e. Vlasenko et al. 2005; Linke et al. 2006; Moosmüller et al. 2012; Engelbrecht et al. 2016). In all these studies, the coarse (super-micron) dust fraction was cut by employing cyclones or impactors. The dust at the output of the nozzle or as entrained by 'air shots' should contain coarse mode particles; however, the aerosols were not added to simulation chambers and the size distribution was not reported.

5.2.4 Generation of Dust Aerosols from Atomization of Liquid Solutions

Atomizers are widely used for studies of dust composed of single minerals (Vlasenko et al. 2006; Hudson et al. 2008a, b; Laskina et al. 2012; Di Biagio et al. 2017b). The principal drawback of this technique is that the particle size of the generated aerosols is usually confined to the accumulation mode range, with the specific diameter depending on the concentration of the solution, but usually limited to around 2.5 μm (Hudson et al. 2008a, b).

Experimental procedure

The generation of aerosols using this approach involves the use of a commercial atomizer to aerosolize a suspension of dust or single minerals in ultrapure water.

The generated aerosol is usually passed through a commercial diffusion drier using inertial particle-free gas to remove water and thus dry the generated particles. Teflon and silicone tubing are routinely used in this aerosol generation procedure.

Soil preparation

Prior to use for aerosol generation, the source soil can be:

Dry sieved: this procedure can be applied to keep only a fraction below a certain diameter for generation. The exact size may depend on the application of the specific study and the desired output size distribution.

Material preparation

The atomizer, tubing and connections should be cleaned and dried before use. The cleaning procedure may include flushing with high-speed air, or rinsing with deionized water, about 10–15 min of sonication, and drying under a laminar flow hood.

Aerosol generation and injection

Aerosol generation will consist of the following steps:

1. The sample is suspended in ultrapure water and the liquid solution is placed in the atomizer bottle.
2. The diffusion drier is connected at the output of the atomizer and the connections to the chamber are set-up.
3. The atomizer is activated by a source of inertial gas (N_2) which starts the aerosolization process. Aerosols enter the chamber by opening the diffusion drier to the chamber.
4. The process is stopped when the desired concentration is achieved in the chamber.

Example of mineral dust aerosol generation

Good examples of number size distributions obtained from the atomization of clays and non-clays components of mineral dust are provided by Hudson et al. (2008a, b). A solution of the minerals in Optima water (Fisher Scientific, W7-4) was atomized with a commercial atomizer (TSI Inc., Model 3076). The generated aerosols were dried to a relative humidity of 15–20% by passing through a diffusion dryer (TSI Inc., Model 3062). The resulting size distribution was shown to be mostly composed of particles <1 μm in diameter, with the peak of the number concentration located between 50 and 500 nm. The size distribution was in most cases well described by a lognormal distribution function and examples of the parameters of the lognormal functions obtained in Hudson et al. (2008a) are provided in Table 5.1. Note that the mineral samples used in the atomized solutions were ground more or less finely by Hudson et al. work. Whether ground or not, the mineral could also affect the size of the generated aerosols. By comparison, Di Biagio et al. (2017b) (not shown) used unground kaolinite mineral and generated aerosols with a larger size spectrum

Table 5.1 Parameters of the lognormal size distribution (Median Diameter, Width) estimated for aerosols produced by atomization of clay components of mineral dust as reported in Hudson et al. (2008a)

Mineral component	Median diameter, D_m (nm)[a]	Width, w[a]
Illite	153.6 ± 2.0	0.88 ± 0.01
Kaolinite	409.6 ± 55.3	0.59 ± 0.02
Montmorillonite	208.8 ± 12.4	1.26 ± 0.09

[a]The lognormal function is defined here as $y = A \exp(-\ln(D/D_m)/\text{width})_2)$

extended to the supermicron range (0.05 M solution, TSI atomizer model 3075, coupled with a diffusion drier TSI model 3062)

5.3 Preparation of Soot Particles for Chamber Experiments

5.3.1 Motivation

Soot and black carbon aerosol particles are often used as synonyms. However, this is incorrect. Black carbon (BC) is an important fraction of the carbonaceous aerosol and is characterized by its strong absorption of visible light and by its resistance to chemical transformation (Petzold et al. 2013). In the ambient atmosphere, it is formed by any incomplete combustion (e.g. gasoline or diesel exhaust, wood and coal combustion). These BC particles typically consist of agglomerates of primary spheres, which are then coated by primary organic aerosol (POA) that is co-emitted during the combustion process and condenses on the agglomerates during the cooling of the exhaust. According to the exact definition, 'soot' includes both the BC core material and the POA; however, it is often also used for the BC material alone. Due to the strong light absorption of black carbon, it exerts the second strongest positive radiative forcing after CO_2. Despite its climatic importance, its radiative forcing is subject to a high degree of uncertainty. This is due partly not only to high uncertainties in the emission inventories but also to large variability in the mass absorption cross section (MAC). In addition, the potential of BC to act as ice nucleating particles (INP) is still under debate. Finally, BC is associated with negative health effects. Sufficient evidence has been found for an association between daily outdoor concentrations of black carbon and mortality, although the causal links have not yet been established. For these reasons, it is important to develop a better method for BC characterization in order to reduce the uncertainties related to climate impact and health effects. Chamber experiments are useful tools for BC characterization, studies of the effects of atmospheric ageing as well as its climate and health impacts.

5.3.2   General Approach

In these experiments, it is important to produce BC samples that are as representative as possible for the source/process under study. Some experiments use emissions from various combustion processes, such as wood combustion or car exhaust (see also Sect. 5.5 on whole emissions from real-world sources). Other experiments use lab-scale burners that produce well-controlled flames to simulate real-world combustion sources. A third type of experiment uses commercial soot samples such as AquaDAG, fullerene soot or Cab-O-Jet which are nebulized and introduced into the chamber. All three procedures are described below.

When real combustion samples are used, the soot particles do not exist of pure black carbon, but are typically coated with POA. The ratio of POA to BC as well as the properties of the BC material itself (size of the primary spherules and agglomerate size) can vary strongly with the combustion conditions (e.g. diesel soot emissions during idling or high load; flaming and smoldering conditions). Concerning the latter, Ward and Hardy (1991) define the flaming and smoldering conditions according to the modified combustion efficiency, $MCE = CO_2/(CO + CO_2)$. Specifically, MCE > 0.9 is identified as flaming conditions, while MCE < 0.85 is identified as smoldering conditions. The actual conditions of the combustion process must therefore be described in detail for reproducible results.

The most commonly used commercial burners for simulating the production of BC are the various different miniCAST burners (Jing Ltd., Zollikofen, Switzerland, www.sootgenerator.com), as well as the recently introduced miniature inverted flame soot generator (Argonaut Scientific, Edmonton, Canada). The fuel and gas flow rates in the latest versions of these burners can be adjusted precisely to produce soot particles of different sizes, concentrations and compositions (i.e. with different ratios of elemental carbon, EC, to organic carbon, OC). The latest prototype miniCAST burner (model 5201) is able to produce polydisperse soot aerosols with high EC/OC ratios over a wide range of geometric mean diameters (~50–170 nm; (Ess and Vasilatou 2019)). The Argonaut inverted burner produces polydisperse soot aerosols that peak at slightly larger diameters, up to 270 nm (Kazemimanesh et al. 2019; Moallemi et al. 2019). It is also possible to use such lab-scale burners to produce soot aerosols with high fractions of 'organic carbon'. However, the operating conditions of these burners need to be chosen carefully to make the produced organic carbon representative of POA formed in engine exhaust (Moore et al. 2014).

When AquaDAG, fullerene soot or Cab-O-Jet samples are used, the batch number has to be specifically noted, as the properties can vary with different batches. Further details about the properties of AquaDAG and fullerene soot particles can be found in Baumgardner et al. (2012), while the properties of Cab-O-Jet particles have been reported by Zangmeister et al. (2019).

Many properties of BC particles vary with the mixing state, i.e. if particles are internally or externally mixed with other aerosol components. An example of externally mixed particles is a mixture of pure BC particles and ammonium sulphate particles. At not-too-high concentrations, the coagulation rate is relatively low, such that the external mixture can be sustained for a few hours. An example of internally mixed particles are BC particles that are coated with secondary organic aerosol (SOA) from a gaseous precursor such as α-pinene. In this case, care has to be taken to keep the concentration of the condensable SOA species sufficiently low to prevent new particle formation, which would result in a combination of external and internal mixtures.

Quantitative determination of the BC concentration is a highly challenging task. Often, absorption measurements are used, however, to retrieve a BC mass concentration the MAC value needs to be known, which can vary substantially depending on the source of the BC and its mixing state. They are therefore not treated here. EC measurements are based on quartz filter samples which undergo thermal treatment with a special protocol where the separation between EC and OC is based on their different volatility and refractoriness. During the heating step, some OC can pyrolyse at high temperature and produce extra EC, which is known as charring. This positive bias can be dealt with by correction for the attenuation of a laser signal by the extra BC, where a thermal optical reflectance (TOR) and a thermal optical transmittance (TOT) are used. In Europe, the so-called EUSAAR_2 method is mostly used (Cavalli et al. 2010). An intercomparison has shown differences by up to a factor of 2 between these different thermal protocols (Chiappini et al. 2014; Wu et al. 2016). In addition, the limited time resolution of this method limits its application in chamber experiments.

Laser-induced incandescence (LII) has been shown to be a promising technique for the determination of refractory black carbon (rBC). The single particle soot photometer (SP2, Droplet Measurement Technologies, CO, USA) facilitates not only the measurement of the mass equivalent diameter of the rBC core but also the coating thickness of non-BC material on a single particle level (Laborde et al. 2013). Moreover, instruments applying a pulsed LII system are available. However, these instruments have mostly been used in the analysis of undiluted combustion aerosols, due to their higher detection limit, and have only recently been used in atmospheric applications.

Since no standards for rBC concentrations are available, calibration of the SP2 is best done with an aerosol particle mass analyser (APM). Using this approach, fullerene soot particles are first mass-selected followed by their rBC mass determination with an SP2. This technique has been shown to provide calibration curves that match the response of the SP2 to BC from diesel exhaust (Laborde et al. 2012a). The SP2 is a single-particle instrument capable of measuring at very high time resolution, which makes it a useful technique for detecting rapidly changing processes in chamber experiments. The main limitation of the technique is its limited detection range in terms of particle size (BC cores with individual masses between ~ 0.5 and 200 fg, which corresponds to mass equivalent diameters between ~ 80 and 600 nm assuming a BC material density of 1.8 g cm^{-3}). This means, for example, that the SP2 is unable to detect small BC particles freshly emitted in diesel exhaust. In addition, at high number concentrations reported by the SP2 can be biased low due to particle coincidence in the instrument. Further details regarding the advantages and limitations of both SP2 and EC/OC measurements can be found in Pileci et al. (2021).

5.3.3 Procedure for Addition of Soot Particles

While several slightly different procedures have successfully been applied (see examples below), it is important to rapidly dilute the exhaust (as it also occurs in the ambient atmosphere), to avoid excessive coagulation at the high concentrations of the undiluted exhaust. To reproduce the same mixture of BC and POA as in the ambient atmosphere, the transfer into the chamber has to be performed via a heated line. If only the BC fraction is to be investigated the POA fraction can in principle be eliminated by a thermal denuder or catalytic stripper. However, it has been shown that a catalytic stripper operated at 350 °C and a residence time of 0.35 s is not able to completely eliminate all the non-refractory aerosol (Yuan et al. 2021).

Experiments with real combustion samples have been performed in various chambers. In the PSI chamber, the ageing of emissions from flaming and smoldering-dominated wood fires has been investigated in different residential stoves, across a wide range of ageing temperatures and emission loads (Bruns et al. 2015; Stefenelli et al. 2019, and references therein). At the ILMARI chamber, different anthropogenic emission sources (small-scale heaters, stoves and boilers, multifuel grate combustion reactor, passenger cars with varying fuel and after-treatment technology) have been used (see e.g. Tiitta et al. 2016). Evolution of straw biomass burning emissions was investigated in the Leipzig Biomass Burning Facility (LBBF). Experiments on the night-time chemical ageing of residential wood combustion emissions have been conducted at the ICE-FORTH environmental chamber and combustion chamber facilities. The NCAS-UMAN aerosol chamber has been coupled to a light-duty diesel engine. The combination of the collapsible 18 m^3 chamber design, a high flow rate clean air source (3 m^{3}min^{-1}) and a three-way valve enabled controlled amounts of the exhaust to be injected in the chamber allowing for a wide range of dilution ratios to be achieved (Pereira et al. 2018). These studies have enabled a focus on both the characterization of POA and the formed SOA during ageing as well as on BC mixing-state and optical properties.

While an extensive campaign was performed on BC at the AIDA chamber in 1999 with the participation of several EUROCHAMP partners (see Saathoff et al. 2003, and the whole corresponding special issue of that journal), only few studies focusing on BC were performed in the chambers of the EUROCHAMP community in more recent years. Soot emissions from a diesel engine test bench, holding a EURO-5 with a 2.0 L series Volkswagen diesel engine was used in an intercomparison study at the AIDA chamber involving six SP2 instruments, each from different research groups (Laborde et al. 2012b). It was shown that the accuracy of the SP2 mass concentration measurement depends on the calibration material chosen. In 2019, a Transnational Activity at the PSI chamber with 12 scientists from 10 different institutions was carried out, with the goal to improve reproducibility of rBC measurements using the LII technique. In these experiments, airborne AquaDAG and fullerene soot (FS) were generated by nebulizing aqueous dispersions (Collison type nebulizer; PSI home-made) followed by drying (using a silica gel-based diffusion dryer) before being injected into a steel cylinder (75 L), which served as buffer volume and, if needed, for particle mixing. In addition, emissions from a diesel-operated, Euro 4 passenger vehicle without particle filter (Opel Combo 1.3 CDTI) were also used as BC test aerosols. The vehicle was operated only under idling conditions, and the emissions passed through a heated sampling line and dilution system before entering the chamber.

The NCAS-UMAN chamber hosted a Transnational Activity at the end of 2019, where a group of scientists from five European and North American institutions quantified the ability of a range of measurement techniques, including LII, to quantify rBC from three different sources. The research also examined the reliability of these measurements in the presence of absorbing and non-absorbing secondary organic aerosol coatings. Black carbon particles were introduced into the chamber from the following sources: (i) light-duty diesel engine (1.9 L, VW), (ii) nebulized AquaDAG and (iii) the miniature inverted flame soot generator (Argonaut Scientific, Edmonton, Canada). In one configuration of the experiments, a Catalytic Stripper (CS015; Catalytic Instruments GmbH) was used upstream of the instruments to remove the semi-volatile fraction for solid particle studies. A second configuration used a combination of a Dekati ejector dilutor, a catalytic stripper and purafil to introduce bare black carbon particles into the Manchester aerosol chamber before coating them with absorbing or non-absorbing SOA material for studies of BC optical properties.

BC particles of different O/C ratios were generated with a CAST burner, diluted and dried before injecting it into the AIDA simulation chamber for studying ice nucleation on flame soot (Möhler et al. 2005a). Information on the potential variation of OC and EC content of these soot particles is given by Schnaiter et al. (2006) and Haller et al. (2019), (Fig. 5.9).

Crawford et al. (2011) studied the heterogeneous ice nucleation on soot with different organic carbon content and with coatings of sulphuric acid. At the AIDA

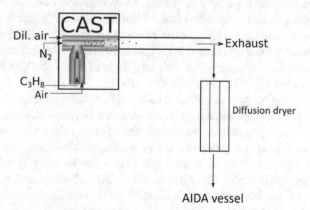

Fig. 5.9 Combustion aerosol standard (CAST) burner for generating soot aerosol particles with different organic carbon content, depending on the C/O-ratio of the propane/ synthetic air mixture (Figure rearranged with permission from Haller et al. (2019). Open access under a CC BY 4.0 license, https://creativecommons.org/licenses/by/4.0/)

simulation chamber, a graphite spark generator (GfG 1000, Palas) was also used to generate model soot particles for aerosol dynamic studies (Saathoff et al. 2003) and ice nucleation studies with and without sulphuric acid coatings (Möhler et al. 2005b). Furthermore, different diesel engines on test stands were used to generate soot for the AIDA chamber, typically connected to the chamber via a denuder system optionally removing humidity, VOCs and NO_x (Saathoff et al. 2003).

It is important to note that, despite recent instrumental developments, there are still considerable uncertainties in the determination of rBC or EC. Pileci et al. (2021) reported systematic discrepancies of up to ~ $\pm50\%$ for the sites investigated. Potential reasons for discrepancies are as follows: a source-specific SP2 response, the possible presence of an additional mode of small BC cores below the lower detection limit of the SP2, differences in the upper cut-off of the SP2 and the inlet line for the EC sampling, or various uncertainties and interferences from co-emitted species in the EC mass measurement. The lack of a traceable reference method or reference aerosols combined with uncertainties in both of the methods make it impossible to clearly quantify the sources of discrepancies or to attribute them to one or the other method, and further research is clearly warranted.

5.4 Bioaerosols

5.4.1 Motivation

Bioaerosols or Primary Biological Aerosol Particles (PBAP) such as pollen, fungal spores and bacteria can affect human health and influence the earth's climate (Després et al. 2012). Among PBAP, bacteria have a crucial role (Bowers et al. 2011). Bacterial viability, including the capability of pathogens to survive in aerosol and maintain their pathogenic potential, depends on the interaction between bacteria and the other organic and inorganic constituents in the atmospheric medium. The interactions of PBAP with other atmospheric constituents such as ozone, nitrogen oxides, volatile organic compounds, became a new, interesting topic of atmospheric science (Amato et al. 2015; Brotto et al. 2015; Massabò et al. 2018).

Atmospheric simulation chambers can provide information on the biological component of atmospheric aerosol and the interaction between bioaerosol and atmospheric conditions. Systematic experiments carried out by chambers give the opportunity to explore bioaerosol behaviour under controlled conditions. The viability of bacteria when they are dispersed in atmosphere can also be investigated, along with their correlation to air quality (Brotto et al. 2015; Massabò et al. 2018). The impact of bacteria on ice nucleation (Amato et al. 2015) is potentially important for climate and could also be relevant for spores, fungi and pollen too.

5.4.2 Generation of Bioaerosols from Liquid Solution

Bioaerosol studies require generators that can provide high particle concentrations with minimal damage to microorganisms. The production of a stable and viable bioaerosol is the first important element for bioaerosol research in a laboratory setting.

Currently, pneumatic nebulizers, such as collison devices, are probably the most frequently used in bioaerosol research (Alsved et al. 2019). The collison nebulizer can not only produce high concentrations of aerosol but can also cause damage to microorganisms due to recirculation of the cell suspension (Zhen et al. 2014). In recent years, several new generators have been designed for bioaerosol research, with the goal of minimizing the damage to microorganisms. Examples are the Blaustein Atomizing Modules (BLAM) and the Sparging Liquid Aerosol Generator (SLAG), both from CH TECHNOLOGIES (Thomas et al. 2011; Zhen et al. 2014). The BLAM unit is an improvement of the pneumatic nebulization without liquid recirculation, aiming to reduce the damage to bacterial culturability and structural integrity. The SLAG is a single-pass bubbling generator designed for low air pressure aerosolization of sensitive and delicate microorganisms: it implements the concept of bursting bubbles to aerosolize particles.

It is recommended to carry out all procedures in a biosafety cabinet or similar structure to ensure operator safety and to avoid contamination of the sample. According to the biosafety level of the microorganism or spore used, all necessary precautions have to be taken in order to protect the operator (attention to leaks in case of over-pressurizing the chamber). It is suggested to limit the experimental research to microorganisms with biosafety levels 1 and 2. Furthermore, the chamber must be equipped with a sterilization system such as germicidal UV lamps, in order to sterilize the entire volume before and after use. On a practical note, one essential requirement is that the chamber has a ventilation system to homogenize the distribution of the aerosol inside the volume and to help the aerosols to remain in suspension.

General procedure

The following equipment are required for the bioaerosol generation protocol described below:

- Teflon tube for the connection between the compressed air source and the nebulizers (typically ¼″ OD tube).
- Mass flow controller to manage the airflow.
- A syringe containing the solution to be sprayed.
- Silicon tube with Luer-lock connection to connect the syringe to the liquid inlet of BLAM or SLAG atomizer.
- A precision pump to feed the BLAM and SLAG atomizer.

It is important to note that each atomizer runs with a different pressure range and aerosolization flow rate. Each nebulizer must have its own adapter, to connect it to the chamber via a gate valve.

The procedure for operating a collision nebulizer is:

1. Connect the collision nebulizer to a source of clean, compressed air (i.e. a cylinder of dry air) with a Teflon tube, Fig. 5.10. The pressure range is 1–6 bar, which corresponds, for a 1-jet model, to an airflow rate from 2 to 7 lpm. It is recommended to use a mass flow meter or a precision pressure gauge to regulate the desired flow.
2. Position the bacteria suspension directly in the glass jar with the liquid level covering the nozzle no more than 1 cm (May 1973; Brotto et al. 2015). When the airflow is switched on, nebulization occurs immediately.

The procedure for operating a BLAM is:

1. Use a ¼″ OD tube to connect the compressed air line to the inlet of the BLAM. Use appropriate pressure and flow controllers to regulate the airflow, Fig. 5.11. The operating air pressure range is from 1 to 6 bar which gives a resulting air flow rate between 1 and 4 lpm.
2. Fill the jar with about 20–30 ml of test solution, taking care not to fill the jar excessively. The solution serves only as a soft impaction surface for the aerosol and will not be used for atomization.

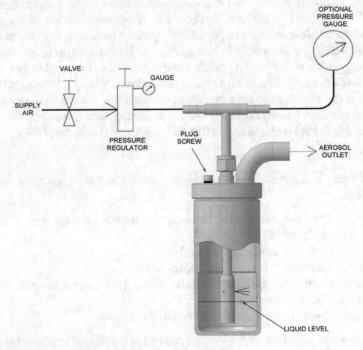

Fig. 5.10 Schematic diagram of Collison nebulizer. Figure extracted from the nebulizer manual by CH Technologies. Collision Nebulizer—User's Manual. Westwood, NJ, USA. https://chtechusa.com/products_tag_lg_collison-nebulizer.php

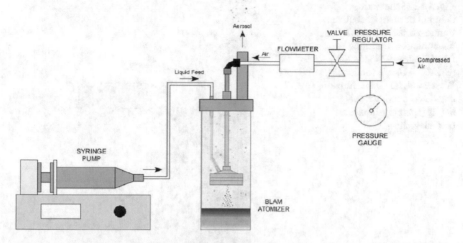

Fig. 5.11 Schematic diagram of BLAM nebulizer. Figure extracted from CH Technologies. Blaustein Atomizer (BLAM) Multi-Jet Model–User's Manual. Westwood, NJ, USA. https://cht echusa.com/products_tag_lg_blaustein-atomizing-modules-blam.php

3. Use a silicon tube to connect the liquid feed port on the nozzle to the liquid feed bulkhead on the lid. Put the bacteria solution in a syringe and use a precision pump for feeding the BLAM (Liquid Feed Rate: 0.1–6 ml/min). When the liquid reaches the liquid feed port of the BLAM, turn on the compressed air supply to the device.
4. Using a mass flow controller, adjust the airflow to about 2 lpm. If a higher output is needed, increase first the upstream pressure of the compressed air line and then increase airflow rate to the atomizer (Massabò et al. 2018).
5. To stop aerosol generation, turn off the air supply to the BLAM and stop operation of the precision pump.

The procedure for operating a SLAG is:

1. Using the air pressure control instrument and a mass flow controller upstream of the SLAG, set the desired air pressure and airflow rate, Fig. 5.12. The standard SLAG model operates between 2 and 6 lpm depending on the input air pressure.
2. Use a precision pump to provide the desired liquid from a syringe filled with the bacteria solution. Use a silicon tube for the connection between the syringe and the SLAG liquid input. The optimal liquid flow rate should be such that there is all the time a thin layer of liquid on the diffusor disc surface.
3. Turn on air supply to the SLAG and in quick sequence turn on the precision pump.
4. To stop aerosol generation, simply turn off air supply to the SLAG. At the same time, stop operation of the precision pump.

Fig. 5.12 Schematic
diagram of SLAG nebulizer.
Figure extracted from CH
Technologies. Sparging
Liquid Aerosol Generator
(SLAG)—User's Manual.
Westwood, NJ, USA. https://
chtechusa.com/products_
tag_lg_sparging-liquid-aer
osol-slag.php

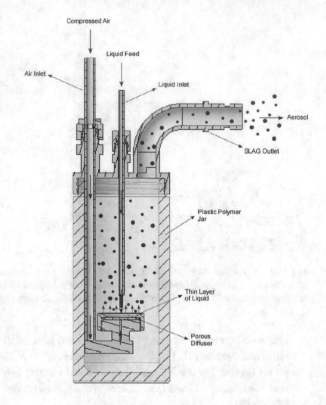

Typically, the injection time to have 10^7–10^8 CFU inside the chamber is a few
minutes for all nebulizers, depending on the airflow, the liquid feed rate and the
desired volume to be injected. A typical value used with the BLAM and SLAG
during chamber experiments is 2 ml of solution sprayed (Massabò et al. 2018). A
particle counter is required to follow the injection inside the chamber. Although,
it should be emphasized that if the microorganisms are suspended in physiological
solution, the particle counter will mainly count the salt particles produced during the
nebulization. When the injection is finished, the valve connecting the nebulizer to the
chamber is closed and at the same time, the airflow and the liquid feed are stopped.

Example of bioaerosol generation

Gram-negative bacteria *Escherichia coli* (ATCC® 25,922™) and the Gram-positive
Bacillus subtilis (ATCC® 6633™) were selected as test bacterial species in
ChAMBRe (Chamber for Aerosol Modelling and Bioaerosol Research, www.lab
fisa.ge.infn.it). These organisms are often used in bioaerosol research as standard
test bacteria (Lee et al. 2002). Prior to experiments, both the strains are cultivated
on Tryptic Soy Broth (TSB) until the mid-exponential phase (Optical Density at λ
= 600 nm around 0.5) and then the bacteria are centrifuged at 4000 g for 10 min.
Afterwards, bacteria are resuspended in a sterile physiological solution (NaCl 0.9%)
to prepare a bacterial solution of approximately 10^7 CFU/ml as verified by standard

dilution plating. From this solution, 2 ml is injected inside the chamber through a flanged connection (Massabò et al. 2018). Since there are few literature available on the efficiency of nebulization of BLAM and SLAG with respect to the most used Collison nebulizer, these injection systems have been extensively character-ized with typical bacterial suspensions (Danelli et al. 2021). Different airflows were tested, using a mass flow controller (Bronkhorst, model F201C-FA), to obtain the best nebulization conditions, in terms of the maximum number of viable aerosolized bacteria at the nebulizer outlet. Sampling has been carried out directly from the output of the nebulizer, through a flanged connection, using an impinging system (liquid impinger by Aquaria srl) filled with 20 mL of sterile physiological solution and operating a constant airflow of 12.5 lpm. The number of cultivable cells inside the liquid impinger was then determined as CFUs, by standard dilution plating: 100 μL of six-fold serial dilutions of the solution was spread on an agar non-selective culture medium (trypticase soy agar, TSA), and incubated at 37 °C for 24 h before the CFU counting. Results are still to be published but, at least with *Escherichia coli* (ATCC® 25,922™), the nebulization efficiency turned out to be well reproducible and in the range of 1% for both the atomizers, with a typical ratio of 3:1 in favor of the BLAM at a fixed inlet airflow (Danelli et al. 2021). At ChAMBRe, considering the range of inlet air flows for the two devices, the typical figure for the ratio between the CFU on Petri dishes (diameter: 10 cm) placed inside the camber to collect the bacteria by a gravitational settling and the injected CFU of *E. coli* (ATCC® 25,922™), is 10^{-5} and 10^{-6}, for BLAM and SLAG, respectively (Danelli et al. 2021).

The injection procedure for bacteria could be updated by adding a real-time monitor of the bioaerosol concentration inside the chamber volume, like the Wide-band Integrated Bioaerosol Sensor (WIBS, University of Hertfordshire, Hertford-shire, UK, now licensed to Droplet Measurement Technologies, Longmont, CO, USA). Furthermore, even if the sequence of operations is well assessed, the need remains to tune each step to the specific bacteria strain under study. Finally, a similar but possibly different approach has to be developed for the injection of spores, fungi or pollens.

5.4.3 Experimental Protocols for Studies on Fungal Spores

Fungal spores are ubiquitous components of air in both indoor and outdoor envi-ronments. They can act as nuclei for water droplets and ice crystals, thereby poten-tially affecting climate and the hydrological cycle (Fröhlich-Nowoisky et al. 2009). Moreover, fungal spores in the respirable fine particle fraction (<3 μm), can impact human health by triggering allergic reactions or causing infectious diseases (Kurup et al. 2000). Measurements of airborne fungal spores are typically performed offline following sampling and collection onto a range of substrates. However, recent devel-opments in this area have seen the introduction of instruments, such as the Waveband Integrated Bioaerosol Sensor (WIBS), for online measurements of PBAP (Healy et al. 2012a).

General procedure

Methods for the addition of various fungal spores to a small ($2\ m^3$) FEP Teflon chamber and use of the WIBS for online characterization of the BPAP have been developed by the University College Cork (Healy et al. 2012b). The general experimental set-up used for testing the addition of PBAP to the FEP Teflon chamber consists of a commercially available small-scale powder disperser (SSPD, Model 3433, TSI Inc.), a condensation particle counter (CPC, Model 3010, TSI Inc.) and a Waveband Integrated Bioaerosol Sensor (WIBS, Model 4), Fig. 5.13. All instruments were connected to the chamber using conductive tubing to minimize particle deposition. The first step in all experiments was to ensure that the chamber was cleaned and flushed with dry purified air (Zander KMA 75). The cleanliness of the chamber was checked using the CPC and deemed to be 'clean' if particle number concentrations were in the range of 0–$50\ cm^{-3}$. The relative humidity in the chamber was increased to 50–60% by gently heating a glass impinger of distilled water in a flow of purified air. The operating temperature in the chamber was in the range 293–295 K. Aerosolization of fungal spores was achieved using the SSPD. Fungal spores were gently brushed onto the surface of a pre-cleaned membrane (Nuclepore Polycarbonate, Whatman) which was subsequently attached to the rotating turntable in the SSPD. Dry purified air (Zander KMA 75) was used to flush the aerosolized spores into the chamber.

Prior to entering the WIBS, the aerosolized fungal spores were diluted at a ratio of 20:1 to safeguard against saturation of the detectors during a sample run. This was achieved using an aerosol diluter (Model 3433, TSI Inc.) and a flow rate of 4.8 l/min generated by supplementing the internal pump of the WIBS (2.4 l/min) with an auxiliary pump controlled by a flow meter, also at 2.4 l/min. The WIBS uses a 635 nm diode laser to detect particles, accompanied by two pulsed xenon UV excitation sources (280 and 370 nm) and three fluorescence detector channels (FL1, FL2 and FL3) which operate over different wavelength ranges (Healy et al. 2012b). The excitation and emission wavelengths are selected to optimize detection of the biological molecules tryptophan and nicotine adenine dinucleotide, NAD(P)H. Ultimately, for each particle, an excitation–emission matrix is recorded along with a measurement of particle size and an index of particle asymmetry, which is used to imply particle shape.

Example of fungal spore aerosolization

The general approach outlined above was used to aerosolize the following fungal spores: *Cladosporium cladosporioides, Cladosporium herbarum, Alternaria notatum, Pencillium notatum* and *Aspergillus fumigatus* (Healy et al. 2012a, b). In general, all of the fungal spore samples gave higher number concentrations in the FL1 channel (Fig. 5.14), suggesting that this may be the best channel for searching for fungal spores in an ambient air. The only exception was for *Aspergillus fumigatus*, which showed similar number concentrations in the same size range for all three fluorescence channels. This observation could provide a basis for distinguishing between *Aspergillus fumigatus* and other fungal spores.

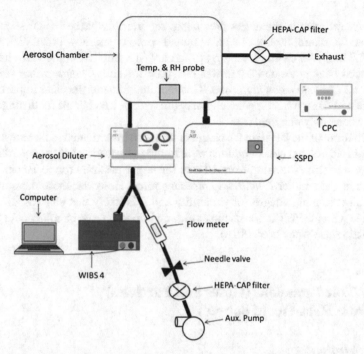

Fig. 5.13 A schematic diagram of the experimental set-up used for the introduction of fungal spores to the FEP Teflon chamber (Healy et al. 2012b)

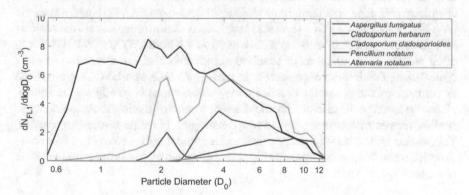

Fig. 5.14 Particle number-size distribution profile for each type of fungal spore measured by the WIBS using fluorescent channels FL1 adapted from Healy et al, 2012b

Two of the spore types—*Pencillium notatum* and *Aspergillus fumigatus*—showed very similar profiles in all three channels. *Pencillium notatum* has by far the broadest size distribution and is the only spore type that reaches the sub-micron range. Aspergillus fumigatus particles are not only observed above 2.5 μm but also have a size distribution that stretches out to 12 μm. These results are in good agreement with the aerodynamic diameters previously reported for these fungal spore types (Baron and Kulkarni 2005). The other fungal spore types show broad distributions in the FL2 and FL3 channels ranging from ca. 3–12 μm, but in the FL1 channel, a pronounced peak at around 2 μm was observed for both *Cladosporium cladosporioides* and *Cladosporium herbarum*. This feature indicates the clear importance of tryptophan in these fungal species and may prove to be another useful distinguishing feature when analysing field data.

The lifetime of the BPAP in the chamber was also investigated in these tests. In all cases, particle number concentrations were found to depend strongly on particle size, resulting in lifetimes ranging from 10 min for larger particles (up to 10 μm) to 3 h for some of the smaller *Pencillium notatum* spores. However, these measurements were subject to a high degree of variability and it is likely that electrostatic effects associated with the FEP Teflon chamber play an important role in influencing particle deposition rates (Wang et al. 2018).

5.5 Whole Emissions (Gases and Particles) from Real-World Sources

5.5.1 Motivation

Due to the importance of volatile organic compounds (VOCs) for atmospheric chemistry and gaps remaining in process-level understanding, both anthropogenic (AVOCs) and biogenic VOCs (BVOCs) and their oxidation processes are the subjects of continuous research. Atmospheric evolution of both AVOCs and BVOCs is often studied using mixtures of standard compounds that are meant to represent typical compounds from each source. In reality, AVOCs and BVOCs are emitted as complex mixtures and the detailed composition depends greatly on the source. Hence, to increase the realism and relevance of the atmospheric simulation chamber studies, measurements using real anthropogenic and biogenic sources are needed. This section will outline the best practices and methodology for coupling whole emissions from combustion sources and biogenic sources with the atmospheric simulation chamber.

5.5.2 Combustion Sources

One difficulty when using real combustion sources is the complexity of varying emission sources, which includes a complicated mixture of VOCs, oxidants (e.g. high concentrations of NO_x), sulphur oxides, CO and CO_2 and water vapour. The concentration of different constituents in the emissions is highly dependent on a number of parameters (e.g. combustion source, operating conditions, type of fuel, etc.). Further complicating matters, the oxidation conditions, VOC-to-NO$_x$ ratios, primary particle concentration to VOC concentration are not easily controllable. In ideal experiments, each of these parameters may be carefully chosen and injected into the chamber. For example, when SOA yields are studied, it is necessary to consider the primary particle concentration in a set of experiments because they will act as seed particles, Fig. 5.15. When using emissions from a real-world source both particulate emissions and gaseous concentrations will vary, making it difficult to prepare a chamber study for a specific source in a reproducible way. As a result, setting a precise ratio between the initial particle concentration and different concentrations of gaseous compounds is difficult.

Wood combustion

In wood combustion studies, the feeding time of the exhaust can be varied so that the desired concentrations in the simulation chamber are achieved. Since the emission characteristics in batch combustion of wood may change remarkably during evolution of the combustion process, the exhaust feeding period must be designed accordingly. For example, in one set of experiments carried out in the ILMARI facility (Tiitta et al. 2016) the study design was to cover the following phases of the sequential batches of wood logs; 'cold ignition', flaming combustion, residual char burning and 'hot ignition'. In the experiments, a middle-European type modern chimney stove (model: Aduro 9.3) fired with dry spruce logs was used as the emission source, and the first batch of wood logs (2.5 kg) was ignited from the top by using sticks of the same wood (0.25 kg) as kindling, and combusted until the residual char burning phase (for 35 min). The sequential batch of wood logs was then ignited by adding the batch on top of the glowing char residue. In this set of experiments, the whole first batch from 'cold start' was injected into the chamber, but it is also possible to inject emissions from any of the above-mentioned burning phases or several of them, depending on the desired aerosol to be studied.

In experiments utilizing pellet boiler emissions with a continuous burning process, a variable feeding time can be used to achieve the desired particle concentration. It is

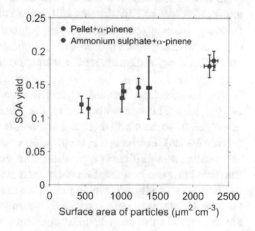

Fig. 5.15 SOA yield as a function of seed surface area using either ammonium sulphate seed or pellet-burning primary aerosol as a seed

recommended that the pellet boiler is operated at its nominal load for at least one hour before injection in order to let the combustion process and emission characteristics stabilize, unless 'cold start' burning is to be studied. In pellet boilers, different kinds of pellets from different woods can be burned by simply loading them into the boiler. Typical wood pellet boilers utilize a device that automatically loads pellets into the fire at a prescribed rate, as shown in Heringa et al. (2012).

The choice of wood stove can depend on the objectives of the study. For instance, in the PSI chamber, three wood stoves were used to probe the formation of SOA from wood combustion (Stefenelli et al. 2019). In this study, 2–3 kg of beech wood was loading into the selected stove and the fire was started with a mixture of paraffin and wood shavings. Flaming and smouldering phases of the fire were investigated. Typically for the smouldering phase the fire was allowed to proceed and the air intake reduced to cool the fire, which transitioned into a smouldering burning phase coupled with a white smoke exhaust. To generate a continuous flaming phase the stove was operated in a high airflow mode to keep flames visible. The emissions were injected into the atmospheric simulation chamber after passing them through a Dekati ejector dilution stage, which dilutes the emissions with purified air at a ratio of 10:1. A final dilution ratio of 100–200:1 was achieved in the chamber. The lines from the stove and the Dekati ejector were all heated to 150 °C to ensure all emissions were injected into the chamber and limit the losses of semi-volatiles and intermediate volatility species. After the emissions were injected, several minutes (5–20 min.) were allowed for the contents of the chamber to equilibrate.

Vehicles and engines

In vehicle emission studies, a variety of engine conditions can be probed depending on the equipment available in each facility. These studies can range from vehicle idling, constant speeds (torque or power), or simulated driving conditions. The studies that are possible depend on the availability and type of a dynamometer. If driving cycles cannot be performed at simulation chamber facilities, portable chambers or dynamometers can be temporarily installed, as described by Platt et al. (2013, 2017).

Engines mounted in a test rig can be used in chamber studies. As mentioned above, the types of experiments will depend on the capability of the test rig and can be conducted using constant or varying engine operation parameters. Here follows the description of a general protocol for coupling the emissions from various engines to atmospheric simulation chambers, adapted from Platt et al. (2013, 2017), and Pereira et al. (2018).

An example experimental set-up from the University of Manchester is provided in Fig. 5.16. The warm-up time of the engine depends significantly on the type of emissions to be studied (cold start, constant operational conditions, or standard driving cycles). For instance, if cold start experiments are the aim of the study then there will be no significant preparation of the engine. Injections of cold start emissions into the atmospheric simulation chamber must occur on a very fast time scale, within the first 60 s of starting the engine. This time scale ensures the engine and after-treatment systems are not sufficiently warm and will capture most of the VOCs that are emitted. On the other hand, if constant conditions or driving cycles are desired then a warm-up time will be necessary and this will vary according to engine type. However, in general, the warm-up time to reach a steady temperature in the engine is ~10 min. After the warm-up time has lapsed, the emissions can be injected into the chamber for as long as required, depending on the aim of the study.

Depending on the study design, the emission can be diluted before injecting it into the chamber, or the emission can be injected directly into the chamber. The latter procedure can be used in cases where rapid changes in driving or combustion conditions take place (e.g. standard driving cycles), however, it is not exclusively used in these circumstances (see Platt et al. 2017). In typical chamber experiments with combustion exhaust in ILMARI, the sample is diluted in a two-stage dilution system with purified air at room temperature (Fig. 5.17), and the total dilution ratio (DR) is determined by measuring the CO_2 concentration in the raw emission and in the chamber. One requirement is that the sample transport line before the dilution system is heated to 150–250 °C, depending on the temperature at the sampling point. It is also recommended that the sample transport line between the first and second diluter is heated to approximately 80–120 °C in order to avoid condensation and thermophoretic sample losses.

If the emission is injected directly into the chamber, the best practice procedures for the transfer of engine emissions into a reaction chamber include:

Use of a high flow (0.1–3 m^3 min^{-1}) of clean air to mix and dilute the engine emissions into the chamber at ambient temperature.
– Introduction of raw exhaust emissions directly into the chamber while cooling and diluting into clean air, which closely represents combustion emissions in the atmosphere.

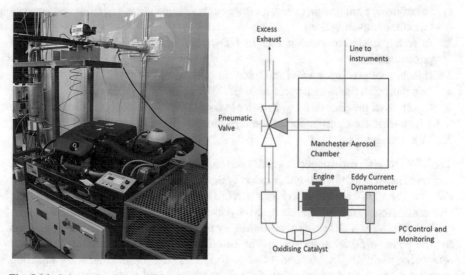

Fig. 5.16 Schematic of engine injection set-up at the University of Manchester

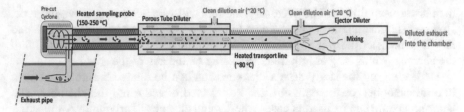

Fig. 5.17 Schematic of the dilution system at the ILMARI chamber used to inject complex emissions from an engine into the chamber. The emissions first pass through a heated cyclone to remove large particulates, then through a heated sampling line to a porous diluter to dilute the emission by up to a factor of 10. The diluted emissions pass through a second dilution stage and into the chamber at high flow rates ($\sim$0.1–3 m^3 min^{-1}). Figure by Olli Sippula, UEF

To verify that the gas phase emissions have been effectively transferred to the atmospheric simulation chamber, a comparison of the emissions directly from the source engine should be compared to the gas phase concentrations in the chamber. This can be accomplished by comparing the normalized CO_2 concentrations to other relevant measured VOCs and total hydrocarbons. For instance, in Platt et al. (2013), the emissions of all relevant gas phase species, including total hydrocarbons, were within 20% of their values directly emitted by the source, thus confirming the effectiveness of the transfer process.

Photochemical ageing experiments on combustion emissions

The current procedure for performing ageing experiments on a combination of exhaust emissions and single precursors in ILMARI is (Kari et al. 2017):

1. Injection of combustion exhaust, either from a single source or from two sources (simultaneous injection).
2. Injection of O_3 to convert NO to NO_2, thus enabling a faster start for the photochemistry.
3. Injection of precursor VOC and tracer (e.g. butanol-d9).
4. Injection of oxidant or its precursor (HONO or H_2O_2 for OH, or O_3).
5. Injection of propene or NO_2, in order to adjust the VOC-to-NO_x ratio, if needed.
6. Allow time for stabilization of the injected compounds.
7. Turn the lights on.

The VOC-to-NO_x ratio depends greatly on the type of sources. For example, in diesel engine exhaust, the VOC-to-NO_x ratio is typically very low, while in gasoline engine exhaust, the VOC-to-NO_x ratio is often in the atmospherically relevant range, which enables branching of the different reaction pathways occurring in the atmosphere. The critical values for VOC-to-NO_x ratio ranging between 3 and 15 have been suggested for the point of 50:50 branching of the reaction pathways (Hoyle et al. 2011, and references therein). The desired VOC-to-NO_x ratio can be increased by injecting propene or decreased by injecting NO_2. It must be noted that if the VOC-to-NO_x ratio is very low, photochemical oxidation of VOCs is very slow, because the OH produced in the chamber is consumed by NO_2.

When several emission sources are used, the best practice is to inject the emissions simultaneously, if possible. Simultaneous feeding has been regarded as the best practice in ILMARI because the injection time from a single emission source is relatively long (e.g. 50 min), depending on the desired concentration in the chamber. If the emissions are injected sequentially, the first injected emissions could already start transforming during the injection of the second (and later) emission(s). In ILMARI the simultaneous feeding practice has been used in experiments with emissions from a diesel engine and wood-burning stove.

5.5.3   Plant Emissions

Due to the importance of biogenic volatile organic compounds (BVOC) for atmospheric chemistry (Guenther 2002) and the large remaining gaps in process-level understanding, BVOC are a subject of continuous research. One concern with investigations of BVOC and their impact on atmospheric chemistry arises from the fact that under natural conditions BVOCs are emitted as complex mixtures, whereas many simulation chamber experiments use single BVOC or simple combinations of BVOC to explore atmospheric chemistry processes. The use of direct emissions from plants is a way of progressing towards more realistic experimental simulations. To improve our understanding on the influence of BVOC emissions on atmospheric processes it is important to be able to study the complex plant emissions from different species under a large variety of different conditions ranging from normal to extreme conditions for the plants. Since BVOC emissions are significantly different between different plant species and can vary significantly with environmental conditions, it is important to have stable environments for the enclosed plants. It is also important to ensure a quantitative and reproducible mechanism for transferring emissions into the simulation chamber so that the complex mixtures and the emission patterns remain unchanged.

Experimental procedure

The addition of real plant emissions into the atmospheric simulation chamber SAPHIR is achieved by coupling SAPHIR with a PLant chamber Unit for Simulation (PLUS) (Hohaus et al. 2016), Fig. 5.18.

The number of trees needed to reach sufficiently high concentrations in a simulation chamber depends on the volume of the chamber and concentration levels at which users want to work. For experiments to be conducted at atmospheric concentrations of organic compounds in SAPHIR, six trees placed in a sea container beneath the chamber are sufficient. In order to avoid that interactions of the atmosphere with soil influence, the mixture of trace gases that is transferred, the canopies of the trees need to be housed in one Teflon bag (the gas exchange volume), in which emissions of the plants are released. To maximize the transfer and to avoid possible specific compound losses all surfaces inside the gas exchange volume should be chemically inert. In the Jülich chambers, all transfer lines and surfaces are either

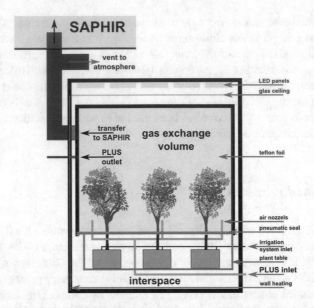

Fig. 5.18 Schematic of the PLUS plant container as an example of a plant chamber acting as a source for realistic tree emissions. The green lines show the inflow of synthetic air and other gases into PLUS (PLUS inlet) and the red line are the outflows of PLUS to either instrumentation (PLUS outlet), SAPHIR or the atmosphere (Figure reused with permission from Hohaus et al. (2016). Open access under a CC BY 3.0 license, https://creativecommons.org/licenses/by/3.0)

made of PFA or have a PFA cover. Also, all cables and connectors inside the gas exchange volume are Teflon-covered. Operation of the plant chamber as a turbulently mixed continuous flow-through reactor is recommended to ensure a homogeneous mixture of plant emissions. Environmental conditions such as light, air and soil humidity, CO_2 concentrations and temperature in the plant chamber need to be controlled and possibly varied to represent normal or more extreme (e.g. drought) environmental conditions. Tree emissions can be transported using a high flow of air (up to 30 m^3/h for the SAPHIR chamber) in order to reach representative concentrations of organic compounds. It is recommended that the ingoing and outgoing flows are continuously monitored to access potential leakage and the outgoing flow is measured contactless with an ultra-sonic flowmeter. Plant conditions and their emissions are recommended to be continuously monitored inside the gas exchange volume by measuring the temperature of the leaves, CO_2 and water vapour by cavity ring-down spectroscopy (for example, using a CRDS, Picarro, Model G2301, intsrument) and emission strength and patterns by, for example, gas chromatography–mass spectrometry (GC–MS) and/or proton reaction mass spectrometry (PTR-MS).

The transfer of organic compounds has been proven to be quantitative by using best practice procedures:

- Use of inert surfaces for the transfer line (PFA).
- Use of a high flow (up to 30 m³/h) to minimize the residence time in the transfer line (order of seconds) compared to the residence time of air in the plant chamber (10–50 min).
- Monitoring of the transfer flow rate by an ultra-sonic flowmeter that measures the flow without contact with the air.

Protocol for transfer efficiency and installation and preparation of the plants

The transfer efficiency of the complete system can be tested using gas standards. As an example, the gas standard used for tests at SAPHIR consisted of acetone, isoprene, α-pinene, nopinone and methyl salicylate (MeSa) in N_2 (99,999% purity). The compounds should be chosen to represent typical BVOC emissions (isoprene, α-pinene, MeSa), while also possessing a significant variety in molecular mass, boiling point and solubility in water (see Table 5.2).

The transfer efficiencies between inlet and outlet of the plant chamber and between the plant chamber outlet and simulation chamber inlet can be determined by the ratio of the measured VOC and CO_2 concentration divided by the calculated concentration. Throughout the experiment, the CO_2 can be used as an inert tracer. Relative humidity inside the plant chamber should be varied between 25 and 100% in order to determine any humidity effect on the transmission efficiency. Results of the transfer efficiency as measured for the Jülich chamber are shown in Fig. 5.19. No significant difference in the transfer efficiencies for different VOC can be observed, indicating that within the range of vapour pressure and polarity investigated, the VOC mixtures emitted from trees enclosed in the gas exchange volume are transferred to SAPHIR without changes to the relative composition of the VOC mixture. Transfer is furthermore independent of relative humidity in the range of 25%–100% for both transfer between PLUS inlet and outlet and transfer between PLUS and SAPIIIR. VOC mixtures were shown to be quantitatively transferred to SAPHIR, ensuring the emission pattern remains unchanged.

Table 5.2 Summary of physical and chemical properties of VOC (acetone, isoprene, α-pinene, nopinone, and methyl salicylate) used in gas standard for transfer efficiency characterization as an example for a suitable mixture of species for testing the transfer efficiency

VOC	Molecular formula	Molar mass	Boiling point (K)
Acetone	C_3H_6O	58	329.3
Isoprene	C_5H_8	68	307
α-pinene	$C_{10}H_{16}$	136	430
Nopinone	$C_9H_{14}O$	138	482.15
Methyl salicylate	$C_8H_8O_3$	152	495.2–496.5

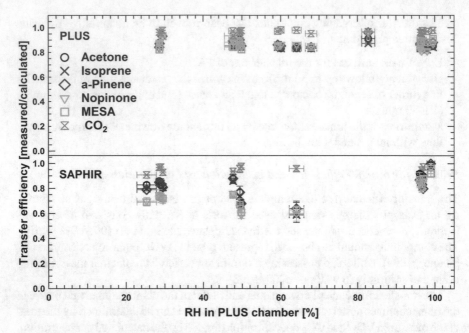

Fig. 5.19 Example of a test on the transfer efficiency between plant chamber and simulation chamber giving measured/calculated averaged mixing ratios of VOC gas standard compounds. The upper panel shows the transfer efficiency between the Jülich PLUS inlet and PLUS outlet. Lower panel shows the transfer efficiency between PLUS outlet and the SAPHIR chamber. Error bars shown are the standard deviation (NPLUS = 18, NSAPHIR = 37) (Figure reused with permission from Hohaus et al. (2016). Open access under a CC BY 3.0 license, https://creativecommons.org/licenses/by/3.0)

Coupling of the plant chamber with the simulation chamber for transfer of BVOC emitted from trees can be achieved by either continuously transferring air from PLUS to SAPHIR or by a short-pulsed coupling (time scale several minutes to hours). For example, Fig. 5.20 displays the temporal evolution of monoterpenes emitted from six Quercus ilex trees during transfer of emissions into the SAPHIR chamber. The efficiency of the transfer can be checked by calculating the expected concentration in the simulation chamber from the measured concentrations of organic compounds in the plant chamber and the transfer flow. Apart from the very beginning of the experiment, where calculated concentration of monoterpenes exceeds the observed value, this calculation confirms the high transfer efficiency for the SAPHIR chamber.

Plants should be installed at least 48 h before the start of experiments to allow for the trees to adjust to the new environment. Also, possible changes in the emissions due to damage of leaves or branches during installation need this time to return to normal. Respiration rate, transpiration rate and emission pattern should be monitored. The emission strength of plants is temperature-dependent. This can be tested by

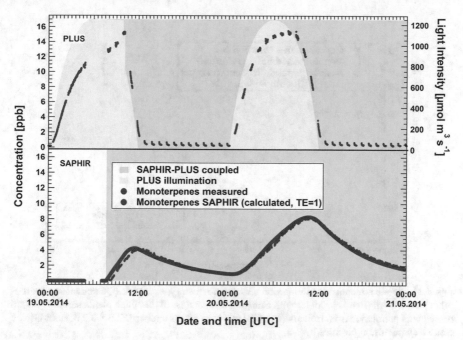

Fig. 5.20 Illustration of monoterpene transfer from PLUS to SAPHIR when continuously coupling the chambers with a flow of 30 m³/h. Yellow shaded areas indicate time periods and intensity of light in the plant chamber, green shaded area indicates coupling of PLUS to SAPHIR (Figure reused with permission from Hohaus et al. (2016). Open access under a CC BY 3.0 license, https://creativecommons.org/licenses/by/3.0)

measuring the sum of monoterpenes with a PTR-MS for different types of trees (oak and birch) inside the plant chamber. The emission strengths can be also fitted to a parameterization developed by Guenther et al. (1993) as a consistency check. For the SAPHIR chamber, this resulted in a good description for several tested trees, Fig. 5.21.

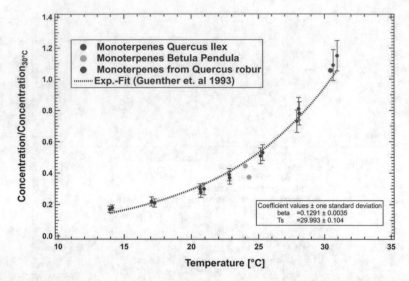

Fig. 5.21 Relative monoterpene emission strength for two types of oak (Quercus ilex, Quercus robur) and a birch (Betula pendula) with changing temperature of the Jülich plant chamber (Figure reused with permission from Hohaus et al. (2016). Open access under a CC BY 3.0 license, https://creativecommons.org/licenses/by/3.0)

References

Alsved, M., Bourouiba, L., Duchaine, C., Löndahl, J., Marr, L.C., Parker, S.T., Prussin, A.J., II, Thomas R.J.: Natural sources and experimental generation of bioaerosols: challenges and perspectives. Aerosol Sci. Technol. **54**, 547–571 (2019). https://doi.org/10.1080/02786826.2019.1682509

Amato, P., Joly, M., Schaupp, C., Attard, E., Möhler, O., Morris, C.E., Brunet, Y., Delort, A.M.: Survival and ice nucleation activity of bacteria as aerosols in a cloud simulation chamber. Atmos. Chem. Phys. **15**, 6455–6465 (2015). https://doi.org/10.5194/acp-15-6455-2015

Baron, P.A., Kulkarni, P. (eds.): Aerosol Measurement: Principles, Techniques, and Applications. John Wiley and Sons, West Sussex, UK (2005)

Baumgardner, D., Popovicheva, O., Allan, J., Bernardoni, V., Cao, J., Cavalli, F., Cozic, J., Diapouli, E., Eleftheriadis, K., Genberg, P.J., Gonzalez, C., Gysel, M., John, A., Kirchstetter, T.W., Kuhlbusch, T.A.J., Laborde, M., Lack, D., Müller, T., Niessner, R., Petzold, A., Piazzalunga, A., Putaud, J.P., Schwarz, J., Sheridan, P., Subramanian, R., Swietlicki, E., Valli, G., Vecchi, R., Viana, M.: Soot reference materials for instrument calibration and intercomparisons: a workshop summary with recommendations. Atmos. Meas. Tech. **5**, 1869–1887 (2012). https://doi.org/10.5194/amt-5-1869-2012

Bowers, R.M., McLetchie, S., Knight, R., Fierer, N.: Spatial variability in airborne bacterial communities across land-use types and their relationship to the bacterial communities of potential source environments. ISME J. **5**, 601–612 (2011). https://doi.org/10.1038/ismej.2010.167

Brotto, P., Repetto, B., Formenti, P., Pangui, E., Livet, A., Bousserrhine, N., Martini, I., Varnier, O., Doussin, J.F., Prati, P.: Use of an atmospheric simulation chamber for bioaerosol investigation: a feasibility study. Aerobiologia **31**, 445–455 (2015). https://doi.org/10.1007/s10453-015-9378-2

Bruns, E.A., Krapf, M., Orasche, J., Huang, Y., Zimmermann, R., Drinovec, L., Močnik, G., El-Haddad, I., Slowik, J.G., Dommen, J., Baltensperger, U., révôt, A.S.H.: Characterization of primary and secondary wood combustion products generated under different burner loads. Atmos. Chem. Phys. **15**, 2825–2841 (2015). https://doi.org/10.5194/acp-15-2825-2015

Caponi, L., Formenti, P., Massabo, D., Di Biagio, C., Cazaunau, M., Pangui, E., Chevaillier, S., Landrot, G., Andreae, M.O., Kandler, K., Piketh, S., Saeed, T., Seibert, D., Williams, E., Balkanski, Y., Prati, P., Doussin, J.F.: Spectral- and size-resolved mass absorption efficiency of mineral dust aerosols in the shortwave spectrum: a simulation chamber study. Atmos. Chem. Phys. **17**, 7175–7191 (2017). https://doi.org/10.5194/acp-17-7175-2017

Cavalli, F., Viana, M., Yttri, K.E., Genberg, J., Putaud, J.-P.: Toward a standardised thermal-optical protocol for measuring atmospheric organic and elemental carbon: the EUSAAR protocol. Atmos. Meas. Tech. **3**, 79–89 (2010). https://doi.org/10.5194/amt-3-79-2010

CH Technologies: Blaustein Atomizer (BLAM) Single-Jet Model—User's Manual. Westwood, NJ, USA. https://chtechusa.com/products_tag_lg_blaustein-atomizing-modules-blam.php

CH Technologies: Collision Nebulizer—User's Manual. Westwood, NJ, USA. https://chtechusa.com/products_tag_lg_collison-nebulizer.php

CH Technologies: Sparging Liquid Aerosol Generator (SLAG)—User's Manual. Westwood, NJ, USA. https://chtechusa.com/products_tag_lg_sparging-liquid-aerosol-slag.php

Chiappini, L., Verlhac, S., Aujay, R., Maenhaut, W., Putaud, J.P., Sciare, J., Jaffrezo, J.L., Liousse, C., Galy-Lacaux, C., Alleman, L.Y., Panteliadis, P., Leoz, E., Favez, O.: Clues for a standardised thermal-optical protocol for the assessment of organic and elemental carbon within ambient air particulate matter. Atmos. Meas. Tech. **7**, 1649–1661 (2014). https://doi.org/10.5194/amt-7-1649-2014

Chen, H., Navea, J.G., Young, M.A., Grassian, V.H.: Heterogeneous photochemistry of trace atmospheric gases with components of mineral dust aerosol. J. Phys. Chem. A **115**, 490–499 (2011). https://doi.org/10.1021/jp110164j

Crawford, I., Möhler, O., Schnaiter, M., Saathoff, H., Liu, D., McMeeking, G., Linke, C., Flynn, M., Bower, K.N., Connolly, P.J., Gallagher, M.W., Coe, H.: Studies of propane flame soot acting as heterogeneous ice nuclei in conjunction with single particle soot photometer measurements. Atmos. Chem. Phys. **11**, 9549–9561 (2011). https://doi.org/10.5194/acp-11-9549-2011

Cziczo, D., Froyd, K., Gallavardin, S., Möhler, O., Benz, S., Saathoff, H., Murphy, D.: Deactivation of ice nuclei due to atmosphercally relevant surface coatings. Environ. Res. Lett. **4**, 044013 (2009). https://doi.org/10.1088/1748-9326/4/4/044013

Danelli, S.G., Brunoldi, M., Massabò, D., Parodi, F., Vernocchi, V., Prati, P.: Comparative characterization of the performance of bio-aerosol nebulizers in connection with atmospheric simulation chambers. Atmos. Meas. Tech. **14**, 4461–4470 (2021). https://doi.org/10.5194/amt-14-4461-2021

Després, V., Huffman, J.A., Burrows, S.M., Hoose, C., Safatov, A., Buryak, G., Fröhlich-Nowoisky, J., Elbert, W., Andreae, M., Pöschl, U., Jaenicke, R.: Primary biological aerosol particles in the atmosphere: a review. Tellus B: Chem. Phys. Meteorol. **64**, 15598 (2012). https://doi.org/10.3402/tellusb.v64i0.15598

Di Biagio, C., Formenti, P., Styler, S.A., Pangui, E., Doussin, J.F.: Laboratory chamber measurements of the longwave extinction spectra and complex refractive indices of African and Asian mineral dusts. Geophys. Res. Lett. **41**, 6289–6297 (2014). https://doi.org/10.1002/2014gl060213

Di Biagio, C., Formenti, P., Balkanski, Y., Caponi, L., Cazaunau, M., Pangui, E., Journet, E., Nowak, S., Caquineau, S., Andreae, M.O., Kandler, K., Saeed, T., Piketh, S., Seibert, D., Williams, E., Doussin, J.F.: Global scale variability of the mineral dust long-wave refractive index: a new dataset of in situ measurements for climate modeling and remote sensing. Atmos. Chem. Phys. **17**, 1901–1929 (2017a). https://doi.org/10.5194/acp-17-1901-2017

Di Biagio, C., Formenti, P., Balkanski, Y., Caponi, L., Cazaunau, M., Pangui, E., Journet, E., Nowak, S., Andreae, M.O., Kandler, K., Saeed, T., Piketh, S., Seibert, D., Williams, E., Doussin, J.-F.: Complex refractive indices and single-scattering albedo of global dust aerosols in the shortwave spectrum and relationship to size and iron content. Atmos. Chem. Phys. **19**, 15503–15531 (2019). https://doi.org/10.5194/acp-19-15503-2019

Di Biagio, C., Formenti, P., Cazaunau, M., Pangui, E., Marchand, N., Doussin, J.F.: Aethalometer multiple scattering correction Cref for mineral dust aerosols. Atmos. Meas. Tech. **10**, 2923–2939 (2017b). https://doi.org/10.5194/amt-10-2923-2017

Engelbrecht, J.P., Moosmüller, H., Pincock, S., Jayanty, R.K.M., Lersch, T., Casuccio, G.: Technical note: Mineralogical, chemical, morphological, and optical interrelationships of mineral dust re-suspensions. Atmos. Chem. Phys. **16**, 10809–10830 (2016). https://doi.org/10.5194/acp-16-10809-2016

Ess, M.N., Vasilatou, K.: Characterization of a new mini-CAST with diffusion flame and premixed flame options: Generation of particles with high EC content in the size range 30 nm to 200 nm. Aerosol Sci. Technol. **53**, 29–44 (2019). https://doi.org/10.1080/02786826.2018.1536818

Formenti, P., Rajot, J.L., Desboeufs, K., Saïd, F., Grand, N., Chevaillier, S., Schmechtig, C.: Airborne observations of mineral dust over western Africa in the summer Monsoon season: spatial and vertical variability of physico-chemical and optical properties. Atmos. Chem. Phys. **11**, 6387–6410 (2011). https://doi.org/10.5194/acp-11-6387-2011

Fröhlich-Nowoisky, J., Pickersgill, D.A., Després, V.R., Pöschl, U.: High diversity of fungi in air particulate matter. Proc. Natl. Acad. Sci. **106**, 12814–12819 (2009). https://doi.org/10.1073/pnas.0811003106

Guenther, A.: The contribution of reactive carbon emissions from vegetation to the carbon balance of terrestrial ecosystems. Chemosphere **49**, 837–844 (2002). https://doi.org/10.1016/S0045-6535(02)00384-3

Guenther, A.B., Zimmerman, P.R., Harley, P.C., Monson, R.K., Fall, R.: Isoprene and monoterpene emission rate variability: model evaluations and sensitivity analyses. J. Geophys. Res.: Atmos. **98**, 12609–12617 (1993). https://doi.org/10.1029/93JD00527

Haller, T., Rentenberger, C., Meyer, J.C., Felgitsch, L., Grothe, H., Hitzenberger, R.: Structural changes of CAST soot during a thermal–optical measurement protocol. Atmos. Meas. Tech. **12**, 3503–3519 (2019). https://doi.org/10.5194/amt-12-3503-2019

Healy, D., O'Connor, D., Sodeau, J.: Measurement of the particle counting efficiency of the "Waveband Integrated Bioaerosol Sensor" model number 4 (WIBS-4). J. Aerosol Sci. **47**, 94–99 (2012a). https://doi.org/10.1016/j.jaerosci.2012.01.003

Healy, D.A., O'Connor, D.J., Burke, A.M., Sodeau, J.R.: A laboratory assessment of the Waveband Integrated Bioaerosol Sensor (WIBS-4) using individual samples of pollen and fungal spore material. Atmos. Environ. **60**, 534–543 (2012b). https://doi.org/10.1016/j.atmosenv.2012.06.052

Heringa, M.F., DeCarlo, P.F., Chirico, R., Lauber, A., Doberer, A., Good, J., Nussbaumer, T., Keller, A., Burtscher, H., Richard, A., Miljevic, B., Prevot, A.S.H., Baltensperger, U.: Time-resolved characterization of primary emissions from residential wood combustion appliances. Environ. Sci. Technol. **46**, 11418–11425 (2012). https://doi.org/10.1021/es301654w

Hohaus, T., Kuhn, U., Andres, S., Kaminski, M., Rohrer, F., Tillmann, R., Wahner, A., Wegener, R., Yu, Z., Kiendler-Scharr, A.: A new plant chamber facility, PLUS, coupled to the atmosphere simulation chamber SAPHIR. Atmos. Meas. Tech. **9**, 1247–1259 (2016). https://doi.org/10.5194/amt-9-1247-2016

Hoyle, C.R., Boy, M., Donahue, N.M., Fry, J.L., Glasius, M., Guenther, A., Hallar, A.G., Huff Hartz, K., Petters, M.D., Petäjä, T., Rosenoern, T., Sullivan, A.P.: A review of the anthropogenic influence on biogenic secondary organic aerosol. Atmos. Chem. Phys. **11**, 321–343 (2011). https://doi.org/10.5194/acp-11-321-2011

Hoyle, C.R., Fuchs, C., Järvinen, E., Saathoff, H., Dias, A., El Haddad, I., Gysel, M., Coburn, S.C., Tröstl, J., Bernhammer, A.K., Bianchi, F., Breitenlechner, M., Corbin, J.C., Craven, J., Donahue, N.M., Duplissy, J., Ehrhart, S., Frege, C., Gordon, H., Höppel, N., Heinritzi, M., Kristensen, T.B., Molteni, U., Nichman, L., Pinterich, T., Prévôt, A.S.H., Simon, M., Slowik, J.G., Steiner, G., Tomé, A., Vogel, A.L., Volkamer, R., Wagner, A.C., Wagner, R., Wexler, A.S., Williamson, C., Winkler, P.M., Yan, C., Amorim, A., Dommen, J., Curtius, J., Gallagher, M.W., Flagan, R.C., Hansel, A., Kirkby, J., Kulmala, M., Möhler, O., Stratmann, F., Worsnop, D.R., Baltensperger, U.: Aqueous phase oxidation of sulphur dioxide by ozone in cloud droplets. Atmos. Chem. Phys. **16**, 1693–1712 (2016). https://doi.org/10.5194/acp-16-1693-2016

Hudson, P.K., Gibson, E.R., Young, M.A., Kleiber, P.D., Grassian, V.H.: Coupled infrared extinction and size distribution measurements for several clay components of mineral dust aerosol. J. Geophys. Res.: Atmos. **113** (2008a). https://doi.org/10.1029/2007JD008791

Hudson, P.K., Young, M.A., Kleiber, P.D., Grassian, V.H.: Coupled infrared extinction spectra and size distribution measurements for several non-clay components of mineral dust aerosol (quartz, calcite, and dolomite). Atmos. Environ. **42**, 5991–5999 (2008b)

Kari, E., Hao, L., Yli-Pirilä, P., Leskinen, A., Kortelainen, M., Grigonyte, J., Worsnop, D.R., Jokiniemi, J., Sippula, O., Faiola, C.L., Virtanen, A.: Effect of pellet boiler exhaust on secondary organic aerosol formation from α-Pinene. Environ. Sci. Technol. **51**, 1423–1432 (2017). https://doi.org/10.1021/acs.est.6b04919

Kazemimanesh, M., Moallemi, A., Thomson, K., Smallwood, G., Lobo, P., Olfert, J.S.: A novel miniature inverted-flame burner for the generation of soot nanoparticles. Aerosol Sci. Technol. **53**, 184–195 (2019). https://doi.org/10.1080/02786826.2018.1556774

Knippertz, P., Stuut, J.B.: Mineral Dust—A Key Player in the Earth System. Springer, Dordrecht–Heidelberg–New York–London (2014)

Kulkarni, P., Baron, P.A., Willeke, K.: Aerosol Measurement: Principles, Techniques and Applications. Wiley (2011)

Kurup, V.P., Shen, H.-D., Banerjee, B.: Respiratory fungal allergy. Microbes Infect. **2**, 1101–1110 (2000). https://doi.org/10.1016/S1286-4579(00)01264-8

Laborde, M., Mertes, P., Zieger, P., Dommen, J., Baltensperger, U., Gysel, M.: Sensitivity of the single particle soot photometer to different black carbon types. Atmos. Meas. Tech. **5**, 1031–1043 (2012a). https://doi.org/10.5194/amt-5-1031-2012

Laborde, M., Schnaiter, M., Linke, C., Saathoff, H., Naumann, K.H., Möhler, O., Berlenz, S., Wagner, U., Taylor, J.W., Liu, D., Flynn, M., Allan, J.D., Coe, H., Heimerl, K., Dahlkötter, F., Weinzierl, B., Wollny, A.G., Zanatta, M., Cozic, J., Laj, P., Hitzenberger, R., Schwarz, J.P., Gysel, M.: Single particle soot photometer intercomparison at the AIDA chamber. Atmos. Meas. Tech. **5**, 3077–3097 (2012b). https://doi.org/10.5194/amt-5-3077-2012

Laborde, M., Crippa, M., Tritscher, T., Jurányi, Z., Decarlo, P.F., Temime-Roussel, B., Marchand, N., Eckhardt, S., Stohl, A., Baltensperger, U., Prévôt, A.S.H., Weingartner, E., Gysel, M.: Black carbon physical properties and mixing state in the European megacity Paris. Atmos. Chem. Phys. **13**, 5831–5856 (2013). https://doi.org/10.5194/acp-13-5831-2013

Lafon, S., Alfaro, S.C., Chevaillier, S., Rajot, J.L.: A new generator for mineral dust aerosol production from soil samples in the laboratory: GAMEL. Aeol. Res. **15**, 319–334 (2014). https://doi.org/10.1016/j.aeolia.2014.04.004

Laskina O., Young M.A., Kleiber P.D., Grassian V.H.: Infrared extinction spectra of mineral dust aerosol: single components and complex mixtures. J. Geophys. Res.: Atmos. **117**, D18210 (2012). https://doi.org/10.1029/2012JD017756

Lee, B., Kim, S., Kim, S.S.: Hygroscopic growth of E. coli and B. subtilis bioaerosols. J. Aerosol. Sci. **33**, 1721–1723 (2002). https://doi.org/10.1016/s0021-8502(02)00114-3

Leskinen, A., Yli-Pirilä, P., Kuuspalo, K., Sippula, O., Jalava, P., Hirvonen, M.R., Jokiniemi, J., Virtanen, A., Komppula, M., Lehtinen, K.E.J.: Characterization and testing of a new environmental chamber. Atmos. Meas. Tech. **8**, 2267–2278 (2015). https://doi.org/10.5194/amt-8-2267-2015

Linke, C., Möhler, O., Veres, A., Mohácsi, A., Bozóki, Z., Szabó, G., Schnaiter, M.: Optical properties and mineralogical composition of different Saharan mineral dust samples: a laboratory study. Atmos. Chem. Phys. **6**, 3315–3323 (2006)

Massabò, D., Danelli, S.G., Brotto, P., Comite, A., Costa, C., Di Cesare, A., Doussin, J.F., Ferraro, F., Formenti, P., Gatta, E., Negretti, L., Oliva, M., Parodi, F., Vezzulli, L., Prati, P.: ChAMBRe: a new atmospheric simulation chamber for aerosol modelling and bio-aerosol research. Atmos. Meas. Tech. **11**, 5885–5900 (2018). https://doi.org/10.5194/amt-11-5885-2018

May, K.R.: The collison nebulizer: description, performance and application. J. Aerosol Sci. **4**, 235–243 (1973). https://doi.org/10.1016/0021-8502(73)90006-2

Meyer, N.K., Duplissy, J., Gysel, M., Metzger, A., Dommen, J., Weingartner, E., Alfarra, M.R., Fletcher, C., Good, N., Mcfiggans, G., Jonsson, Å.M., Hallquist, M., Baltensperger, U., Ristovski, Z.D.: Analysis of the hygroscopic and volatile properties of ammonium sulphate seeded and un-seeded SOA particles. Atmos. Chem. Phys. Discuss. **8**, 8629–8659 (2008)

Moallemi, A., Kazemimanesh, M., Corbin, J.C., Thomson, K., Smallwood, G., Olfert, J.S., Lobo, P.: Characterization of black carbon particles generated by a propane-fueled miniature inverted soot generator. J. Aerosol Sci. **135**, 46–57 (2019). https://doi.org/10.1016/j.jaerosci.2019.05.004

Mogili, P.K., Kleiber, P.D., Young, M.A., Grassian, V.H.: N_2O_5 hydrolysis on the components of mineral dust and sea salt aerosol: comparison study in an environmental aerosol reaction chamber. Atmos. Environ. **40**, 7401–7408 (2006). https://doi.org/10.1016/j.atmosenv.2006.06.048

Mogili, P.K., Yang, K.H., Young, M.A., Kleiber, P.D., Grassian, V.H.: Environmental aerosol chamber studies of extinction spectra of mineral dust aerosol components: broadband IR-UV extinction spectra. J. Geophys. Res.: Atmos. **112** (2007). https://doi.org/10.1029/2007JD008890

Möhler, O., Linke, C., Saathoff, H., Schnaiter, M., Wagner, R., Mangold, A., Krämer, M., Schurath, U.: Ice nucleation on flame soot aerosol of different organic carbon content. Meteorologische Zeitschrift **14**, 477–484 (2005a). https://doi.org/10.1127/0941-2948/2005/0055

Möhler, O., Büttner, S., Linke, C., Schnaiter, M., Saathoff, H., Stetzer, O., Wagner, R., Kramer, M., Mangold, A., Ebert, V., Schurath, U.: Effect of sulfuric acid coating on heterogeneous ice nucleation by soot aerosol particles. J. Geophys. Res. **110**, D11210 (2005b). https://doi.org/10.1029/2004JD005169

Moore, R.H., Ziemba, L.D., Dutcher, D., Beyersdorf, A.J., Chan, K., Crumeyrolle, S., Raymond, T.M., Thornhill, K.L., Winstead, E.L., Anderson, B.E.: Mapping the operation of the miniature combustion aerosol standard (Mini-CAST) soot generator. Aerosol Sci. Technol. **48**, 467–479 (2014). https://doi.org/10.1080/02786826.2014.890694

Moosmüller, H., Engelbrecht, J.P., Skiba, M., Frey, G., Chakrabarty, R.K., Arnott, W.P.: Single scattering albedo of fine mineral dust aerosols controlled by iron concentration. J. Geophys. Res.: Atmos. **117** (2012). https://doi.org/10.1029/2011JD016909

Pereira, K.L., Dunmore, R., Whitehead, J., Alfarra, M.R., Allan, J.D., Alam, M.S., Harrison, R.M., McFiggans, G., Hamilton, J.F.: Technical note: use of an atmospheric simulation chamber to investigate the effect of different engine conditions on unregulated VOC-IVOC diesel exhaust emissions. Atmos. Chem. Phys. **18**, 11073–11096 (2018). https://doi.org/10.5194/acp-18-11073-2018

Petzold, A., Ogren, J.A., Fiebig, M., Laj, P., Li, S.-M., Baltensperger, U., Holzer-Popp, T., Kinne, S., Pappalardo, G., Sugimoto, N., Wehrli, C., Wiedensohler, A., Zhang, X.-Y.: Recommendations for reporting "black carbon" measurements. Atmos. Chem. Phys. **13**, 8365–8379 (2013). https://doi.org/10.5194/acp-13-8365-2013

Pileci, R.E., Modini, R.L., Bertò, M., Yuan, J., Corbin, J.C., Marinoni, A., Henzing, B., Moerman, M.M., Putaud, J.P., Spindler, G., Wehner, B., Müller, T., Tuch, T., Trentini, A., Zanatta, M., Baltensperger, U., Gysel-Beer, M.: Comparison of co-located refractory black carbon (rBC) and elemental carbon (EC) mass concentration measurements during field campaigns at several European sites. Atmos. Meas. Tech. **14**, 1379–1403 (2021). https://doi.org/10.5194/amt-14-1379-2021

Platt, S.M., El Haddad, I., Zardini, A.A., Clairotte, M., Astorga, C., Wolf, R., Slowik, J.G., Temime-Roussel, B., Marchand, N., Ježek, I., Drinovec, L., Močnik, G., Möhler, O., Richter, R., Barmet, P., Bianchi, F., Baltensperger, U., Prévôt, A.S.H.: Secondary organic aerosol formation from gasoline vehicle emissions in a new mobile environmental reaction chamber. Atmos. Chem. Phys. **13**, 9141–9158 (2013). https://doi.org/10.5194/acp-13-9141-2013

Platt, S.M., El Haddad, I., Pieber, S.M., Zardini, A.A., Suarez-Bertoa, R., Clairotte, M., Daellenbach, K.R., Huang, R.J., Slowik, J.G., Hellebust, S., Temime-Roussel, B., Marchand, N., de Gouw, J., Jimenez, J.L., Hayes, P.L., Robinson, A.L., Baltensperger, U., Astorga, C., Prévôt, A.S.H.: Gasoline cars produce more carbonaceous particulate matter than modern filter-equipped diesel cars. Sci. Rep. **7**, 4926 (2017). https://doi.org/10.1038/s41598-017-03714-9

Ryder, C.L., Highwood, E.J., Rosenberg, P.D., Trembath, J., Brooke, J.K., Bart, M., Dean, A., Crosier, J., Dorsey, J., Brindley, H., Banks, J., Marsham, J.H., McQuaid, J.B., Sodemann, H., Washington, R.: Optical properties of Saharan dust aerosol and contribution from the coarse mode as measured during the Fennec 2011 aircraft campaign. Atmos. Chem. Phys. **13**, 303–325 (2013). https://doi.org/10.5194/acp-13-303-2013

Saathoff, H., Naumann, K.H., Schnaiter, M., Schock, W., Möhler, O., Schurath, U., Weingartner, E., Gysel, M., Baltensperger, U.: Coating of soot and $(NH_4)_2SO_4$ particles by ozonolysis products of α-pinene. J. Aerosol. Sci. **34**, 1297–1321 (2003). https://doi.org/10.1016/S0021-8502(03)003 64-1

Schnaiter, M., Gimmler, M., Llamas, I., Linke, C., Jäger, C., Mutschke, H.: Strong spectral dependence of light absorption by organic carbon particles formed by propane combustion. Atmos. Chem. Phys. **6**, 2981–2990 (2006). https://doi.org/10.5194/acp-6-2981-2006

Sow, M., Alfaro, S.C., Rajot, J.L., Marticorena, B.: Size resolved dust emission fluxes measured in Niger during 3 dust storms of the AMMA experiment. Atmos. Chem. Phys. **9**, 3881–3891 (2009). https://doi.org/10.5194/acp-9-3881-2009

Stefenelli, G., Jiang, J., Bertrand, A., Bruns, E.A., Pieber, S.M., Baltensperger, U., Marchand, N., Aksoyoglu, S., Prévôt, A.S.H., Slowik, J.G., El Haddad, I.: Secondary organic aerosol formation from smoldering and flaming combustion of biomass: a box model parametrization based on volatility basis set. Atmos. Chem. Phys. **19**, 11461–11484 (2019). https://doi.org/10.5194/acp-19-11461-2019

Stirnweis, L., Marcolli, C., Dommen, J., Barmet, P., Frege, C., Platt, S.M., Bruns, E.A., Krapf, M., Slowik, J.G., Wolf, R., Prévôt, A.S.H., Baltensperger, U., El-Haddad, I.: Assessing the influence of NO_x concentrations and relative humidity on secondary organic aerosol yields from α-pinene photo-oxidation through smog chamber experiments and modelling calculations. Atmos. Chem. Phys. **17**, 5035–5061 (2017). https://doi.org/10.5194/acp-17-5035-2017

Thomas, R.J., Webber, D., Hopkins, R., Frost, A., Laws, T., Jayasekera, P.N., Atkins, T.: The cell membrane as a major site of damage during aerosolization of Escherichia coli. Appl Environ Microbiol **77**, 920–925 (2011). https://doi.org/10.1128/aem.01116-10

Tiitta, P., Leskinen, A., Hao, L., Yli-Pirilä, P., Kortelainen, M., Grigonyte, J., Tissari, J., Lamberg, H., Hartikainen, A., Kuuspalo, K., Kortelainen, A.M., Virtanen, A., Lehtinen, K.E.J., Komppula, M., Pieber, S., Prévôt, A.S.H., Onasch, T.B., Worsnop, D.R., Czech, H., Zimmermann, R., Jokiniemi, J., Sippula, O.: Transformation of logwood combustion emissions in a smog chamber: formation of secondary organic aerosol and changes in the primary organic aerosol upon daytime and nighttime aging. Atmos. Chem. Phys. **16**, 13251–13269 (2016). https://doi.org/10.5194/acp-16-13251-2016

Ullrich, R., Hoose, C., Möhler, O., Niemand, M., Wagner, R., Höhler, K., Hiranuma, N., Saathoff, H., Leisner, T.: A new ice nucleation active site parameterization for desert dust and soot. J. Atmos. Sci. **74**, 699–717 (2017). https://doi.org/10.1175/jas-d-16-0074.1

Utry, N., Ajtai, T., Pintér, M., Tombácz, E., Illés, E., Bozóki, Z., Szabó, G.: Mass-specific optical absorption coefficients and imaginary part of the complex refractive indices of mineral dust components measured by a multi-wavelength photoacoustic spectrometer. Atmos. Meas. Tech. **8**, 401–410 (2015). https://doi.org/10.5194/amt-8-401-2015

Vlasenko, A., Sjögren, S., Weingartner, E., Gäggeler, H.W., Ammann, M.: Generation of submicron Arizona test dust aerosol: chemical and hygroscopic properties. Aerosol. Sci. Technol. **39**, 452–460 (2005). https://doi.org/10.1080/027868290959870

Vlasenko, A., Sjogren, S., Weingartner, E., Stemmler, K., Gäggeler, H.W., Ammann, M.: Effect of humidity on nitric acid uptake to mineral dust aerosol particles. Atmos. Chem. Phys. **6**, 2147–2160 (2006). https://doi.org/10.5194/acp-6-2147-2006

Wagner, R., Ajtai, T., Kandler, K., Lieke, K., Linke, C., Müller, T., Schnaiter, M., Vragel, M.: Complex refractive indices of Saharan dust samples at visible and near UV wavelengths: a laboratory study. Atmos. Chem. Phys. **12**, 2491–2512 (2012). https://doi.org/10.5194/acp-12-2491-2012

Wang, N., Jorga, S.D., Pierce, J.R., Donahue, N.M., Pandis, S.N.: Particle wall-loss correction methods in smog chamber experiments. Atmos. Meas. Tech. **11**, 6577–6588 (2018). https://doi.org/10.5194/amt-11-6577-2018

Ward, D.E., Hardy, C.C.: Smoke emissions from wildland fires. Environ. Int. **17**, 117–134 (1991). https://doi.org/10.1016/0160-4120(91)90095-8

Weinzierl, B., Petzold, A., Esselborn, M., Wirth, M., Rasp, K., Kandler, K., SchüTz, L., Koepke, P., Fiebig, M.: Airborne measurements of dust layer properties, particle size distribution and mixing state of Saharan dust during SAMUM 2006. Tellus B: Chem. Phys. Meteorol. **61**, 96–117 (2009). https://doi.org/10.1111/j.1600-0889.2008.00392.x

Wu, C., Huang, X.H.H., Ng, W.M., Griffith, S.M., Yu, J.Z.: Inter-comparison of NIOSH and IMPROVE protocols for OC and EC determination: implications for inter-protocol data conversion. Atmos. Meas. Tech. **9**, 4547–4560 (2016). https://doi.org/10.5194/amt-9-4547-2016

Yuan, J., Modini, R.L., Zanatta, M., Herber, A.B., Müller, T., Wehner, B., Poulain, L., Tuch, T., Baltensperger, U., Gysel-Beer, M.: Variability in the mass absorption cross section of black carbon (BC) aerosols is driven by BC internal mixing state at a central European background site (Melpitz, Germany) in winter. Atmos. Chem. Phys. **21**, 635–655 (2021). https://doi.org/10.5194/acp-21-635-2021

Zangmeister, C.D., Grimes, C.D., Dickerson, R.R., Radney, J.G.: Characterization and demonstration of a black carbon aerosol mimic for instrument evaluation. Aerosol Sci. Technol. **53**, 1322–1333 (2019). https://doi.org/10.1080/02786826.2019.1660302

Zhang, X., Cappa, C.D., Jathar, S.H., McVay, R.C., Ensberg, J.J., Kleeman, M.J., Seinfeld, J.H.: Influence of vapor wall loss in laboratory chambers on yields of secondary organic aerosol. Proc. Natl. Acad. Sci. **111**, 5802–5807 (2014). https://doi.org/10.1073/pnas.1404727111

Zhen, H., Han, T., Fennell, D.E., Mainelis, G.: A systematic comparison of four bioaerosol generators: affect on culturability and cell membrane integrity when aerosolizing Escherichia coli bacteria. J. Aerosol Sci. **70**, 67–79 (2014). https://doi.org/10.1016/j.jaerosci.2014.01.002

Chapter 6
Sampling for Offline Analysis

Esther Borrás, Hartmut Herrmann, Markus Kalberer, Amalia Muñoz, Anke Mutzel, Teresa Vera, and John Wenger

Abstract The detailed chemical characterization of gas and particle phase species is essential for interpreting the results of atmospheric simulation chamber experiments. Although the application of online techniques has advanced significantly over the last two decades, offline analytical methods such as GC–MS and LC–MS are still frequently used. In this chapter, the approaches commonly employed for gas and particle sampling prior to subsequent offline analysis are described in detail. Methods involving the use of cartridges, canisters, bags and sorbent tubes for gas sampling are described with the support of examples reported in the literature. Technical descriptions related to the application of different types of filters, inertial classifiers and particle-into-liquid samplers for the collection of particles are also provided.

Although online techniques to characterize gas and particle phase chemical composition from chamber experiments have advanced significantly over the last two decades, offline analytical methods are still frequently used.

One main reason for the continued use of offline chemical analysis for characterizing gas and particle composition is the possibility to use a wide range of techniques and instruments which are not suitable for continuous-flow sample analysis. Chromatography and mass spectrometry methods are frequently used to analyse the particle and gas composition in chamber experiments with unprecedented molecular

E. Borrás · A. Muñoz · T. Vera
Fundación Centro de Estudios Ambientales del Mediterráneo, Valencia, Spain

H. Herrmann
Leibniz-Institut für Troposphärenforschung e.V. (TROPOS), Leipzig, Germany

M. Kalberer
University of Basel, Basel, Switzerland

A. Mutzel
Eurofins, Food & Feed Testing, Leipzig GmbH, Leipzig, Germany

J. Wenger (✉)
University College Cork, Cork, Ireland
e-mail: j.wenger@ucc.ie

© The Author(s) 2023
J.-F. Doussin et al. (eds.), *A Practical Guide to Atmospheric Simulation Chambers*,
https://doi.org/10.1007/978-3-031-22277-1_6

detail and accuracy. While methods typically require non-continuous samples, fast chromatographic methods are available to perform analysis cycles with a few minutes of time resolution, which is often enough to capture important time trends in chamber experiments. On the other hand, a number of powerful analytical techniques such as NMR (nuclear magnetic resonance) spectroscopy or ESR (electron spin resonance) spectroscopy can only be used with offline samples.

In addition to the greater choice of analytical techniques available for offline analysis, they are often also less expensive compared to online techniques either due to the exclusive use of many online instruments for atmospheric analysis applications or the possibility to share offline techniques with other users.

New analytical methods for particle or gas-phase characterization are often established as offline techniques to assess their suitability and sensitivity in chamber experiments, before online instruments are developed. One example is the methods to quantify the oxidative potential of particles, which were originally developed using offline analysis, but have recently been adapted to create dedicated online instruments (e.g. Wragg et al. 2016; Puthussery et al. 2018).

This chapter contains detailed descriptions of procedures commonly used for gas and particle sampling prior to subsequent offline analysis.

6.1 Gas-Phase Sampling

Gas-phase sampling is routinely performed for offline analysis of volatile organic compounds (VOCs). The choice of air sampling method depends on the volatility and polarity range of the target VOCs (Woolfenden 2010a, b). During sampling, it is very important to know the airflow rate as it enables the exact volume of air collected to be determined. The flow rate must be kept constant in order to obtain reliable measurements. There are different ways of doing this, with the most common methods involving a critical orifice (a restrictor placed in the sampling line that is equivalent to a certain flow rate) or the use of a mass flow controller.

Another issue that has to be considered when an offline sample is collected is the tubing material (Deming et al. 2019). Deming and co-workers have studied different tubing materials, classified as absorbent (such as PFA, FEP Teflon and PTFE among others) or adsorbent (such as electropolished steel, glass or silonite among others). In studies of the absorbent materials, PEEK, PTFE and conductive PTFE demonstrated a higher retention capability (longer delays) than PFA and FEP Teflon probably because both materials have shorter polymer chain lengths and increased chain entanglements compared with PTFE. Therefore, Deming et al. (2019) recommend the use of PFA or FEP Teflon for collecting air samples of VOCs. On the other hand, measurements made using adsorptive, metal-like, tubing materials were strongly affected by humidity, with the longest measured delay times found for aluminium tubing and aluminium tubing treated with hexavalent chromate. Besides humidity, the measured tubing delay also depends on the VOC concentration and researchers are advised to condition the sampling lines in order to reduce memory effects and delays. If adsorbent tubing must be used, it is recommended that the relative humidity is maintained above 20%. The best tubing adsorbent materials are conductive PFA tubing and Silonite. However, even though it was not studied, Deming et al. (2019) recommended the use of conductive FEP Teflon instead of conductive PFA since it can combine good gas and particle transmission at nearly half the price. It should be noted that further studies are needed to improve our knowledge of the role of different tubing materials for different types of functionalized organic compounds, concentration and other parameters (temperature and relative humidity) during air sampling.

6.1.1 Cartridge Sampling

A range of different cartridges containing solid sorbents are used to collect VOCs in simulation chamber experiments and used for subsequent analysis in LC, LC–MS, GC and GC–MS. Cartridge sampling is an active sampling technique and it is important that the airflow rate and sampling duration are known. In order to prevent breakthrough, an estimation of the expected concentration of target compounds is recommended. In some cases, two cartridges or solid sorbents can be connected in series to determine the extent of breakthrough.

DNPH-silica cartridges

DNPH-silica cartridges trap aldehydes and ketones in air by allowing them to react with 2,4-dinitrophenylhydrazine (DNPH) in the cartridge to form stable hydrazone derivatives. The methodology is based on US EPA Methods TO-11A and TO-5 (US EPA 2022) which have been updated in order to analyse samples by LC–MS. The derivatization reaction (Fig. 6.1) takes place during sample collection. The derivatives are later eluted and analysed.

The US EPA recommends using pre-coated silica DNPH cartridges. However, users can coat the cartridges themselves following the instructions detailed in Method TO-11A. Among the advantages of using pre-coated DNPH cartridges is the lower and more consistent background concentration of carbonyls. The main disadvantage of the pre-coated cartridges is the price and the fact that they are discarded after use. C18 cartridges coated with acidic DNPH solution can also be utilized. However, there are very few references using this sampling methodology.

The main manufacturers of DNPH-Silica coated cartridges are Waters and Sigma-Aldrich. Both cartridges are very similar, Fig. 6.2 (Tejada 1986; Winberry et al. 1990; Sirju and Shepson 1995).

Ozone has been shown to interfere with the analysis of carbonyl compounds in air samples that have been drawn through cartridges containing silica coated with 2,4-dinitrophenylhydrazine (Tejada 1986, Arnts 1989). Ozone Scrubber cartridges are designed to remove this ozone interference, while scrubber stainless steel coils filled with KI can be used too. These disposable devices are intended for use in series with the DNPH-Silica cartridges. Each Ozone Scrubber cartridge contains granular

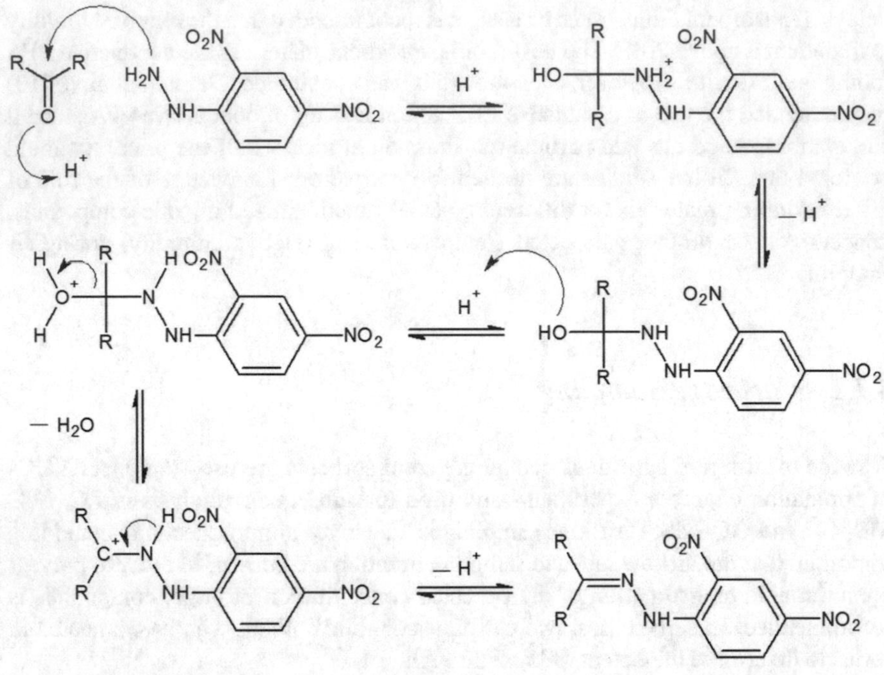

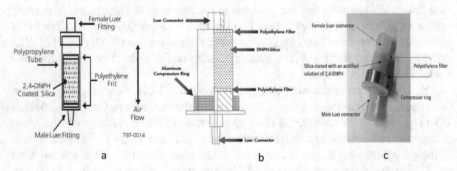

Fig. 6.1 Derivatization of carbonyl compounds by reaction with 2,4-dinitrophenylhydrazine (DNPH) to form stable hydrazones (DNPH-derivatives)

Fig. 6.2 DNPH silica coated cartridges: **a** LpDNPH S10L from Sigma-Aldrich; **b** Sep-Pak DNPH-silica cartridge from Waters; **c** a photograph of a DNPH cartridge. © EUPHORE

potassium iodide. When air containing ozone is drawn through this packed bed, iodide is oxidized to iodine, consuming the ozone. The purity of acetonitrile used for eluting the samples is very important since it can affect the carbonyl background level in the cartridge.

Fig. 6.3   C18 cartridge, © EUPHORE

C18 cartridges

C18 is an octadecylsilane-bonded silica sorbent with the surface passivated by non-polar paraffinic groups which make it hydrophobic and relatively inert. Due to these properties, C18 is regularly used as an adsorbent trap for trace organics in environmental samples. C18 cartridges can be used for a wider group of compounds, although tests have to be made before (Fig. 6.3).

Sampling procedure for DNPH-silica and C18 cartridges

- Measurement of the sampling airflow at the beginning and the end of the sampling period. Flow rate should be between 1 and 2 L/min.
- Connection of the cartridge in the Teflon sampling line with the thinner end in the upper position (most of the cartridges are bidirectional, however, read the instructions from the manufacturer).
- Connection of the Luer end at the pump using silicone tubing.
- Usually, 30 min of sampling at 1 L/min is sufficient when working at ppb level. If the expected concentrations are lower, the sampling time could be longer.
- When using DNPH-silica cartridges, if the ozone concentration is 70 ppb or higher, an ozone scrubber has to be connected to prevent artefacts.
- When sampling is completed, the cartridge has to be removed, capped, labelled and stored at 4 °C in dark conditions. Samples have to be analysed as soon as possible (storage time is set by the manufacturer).
- To quantify both carbonyl and VOC compounds, external calibrations must be performed.

Examples of applications in the literature

Small DNPH-coated C18 cartridges have been successfully used for the sampling of carbonyls in air since the 1990s (Druzik et al. 1990; Sirju and Shepson 1995). The recovery of carbonyls by cartridge elution is typically over 95% efficient and analysis has generally been performed using liquid chromatography. This technique has been widely used in the fields of atmospheric chemistry, indoor and outdoor air quality research.

Application of the cartridge sampling technique to a simulation chamber study was demonstrated by Brombacher et al. (2001), who collected air samples during experiments on the OH radical-initiated oxidation of cis-3-acetyl-2,2-dimethylcyclobutylethanal (pinonal) and *cis*-3-acetyl-2,2-dimethylcyclobutylcarbaldehyde (nor-pinonal). High-performance liquid chromatography combined with ion trap mass spectrometry (online HPLC-MS*n*) was used to identify carbonyl oxidation products at the picogram level.

6.1.2   Canister Sampling

Canisters can be used to collect gaseous compounds during chamber studies for subsequent offline analysis by GC or GC–MS. This approach is most appropriate for highly volatile, non-polar compounds (Cardin and Noad 2018) and typically involves the use of evacuated stainless steel canisters with electro-polished inner surfaces, called SUMMA canisters. These canisters are widely used for sampling VOCs in ambient air (US EPA Methods TO-14A and TO-15, US EPA 2022) and have been tested on a range of volatile species, including aliphatic and aromatic hydrocarbons, as well as chlorinated compounds (Sin et al. 2001). Canisters offer the following advantages; a sampling pump is not needed, problems associated with collection efficiency and analyte recovery when using sorbents and filters are avoided, repeat injections or dilutions can be made during analysis.

Samples are collected by opening an evacuated stainless steel canister to the air. Prior to analysis, the canisters are pressurized using nitrogen and aliquots of the air sample are withdrawn, cryofocused and analysed. The canister volume can vary from 400 mL to several litres. Most compounds are stable in canister samples for around 30 days and in some cases up to 4 months (Sin et al. 2001). Canisters can be re-used after a cleaning process.

Air samples are collected through a sampling orifice which can either be a simple open/close set-up or pressure regulated to allow for sampling times of a few minutes at a desired flow rate.

Sampling begins immediately, and is completed when the pressure inside the canister is equal to the atmospheric pressure on the outside, or when the sampling orifice is detached from the canister. In some cases, a sampling orifice with regulator is attached to the inlet of the canister, and a length of inert tubing leading from the chamber is connected to the inlet of the orifice. A flow controller can also be connected to the canister. Small samples can be collected by attaching a sampling orifice to the inlet of a MiniCan.

Canisters are recommended for sampling VOCs up to approximately C12 and permanent gases. As explained above, air may be collected as grab samples (instantaneous fill) or time-integrated samples (using a flow controller or a critical orifice assembly). Canisters exposed to high vapour concentrations can require extensive cleaning post-analysis, particularly if the contaminants are polar or have a higher boiling point than toluene. Canister cleaning typically involves a sequence of evacuations and air purges, often at elevated temperatures, followed by an analysis of zero air from the cleaned canister to confirm that all contamination has been removed.

Procedure for canister sampling

- Choose the canister (6 L canister, 2 L canister or MiniCans) appropriate for the desired application.
- Holding the canister, slide back the knurled collar, remove the protective end cap and connect the canister tip to the sampling regulator (flow controller, critical orifice assembly…).
- Insert the canister tip into the sampling regulator and release the knurled collar.
- Sampling begins immediately, write down the initial time.
- When sampling is complete, reverse the above steps to disengage the canister from the regulator and separate canister.
- Put the protective end cap onto the canister and seal it. Label the canister with the information needed to identify the sample.
- Write down the end time.
- If the canisters are assured to be cleaned at the outset of sampling, no blank is needed.
- In the laboratory, the canister is pressurized with nitrogen, and the contents are analysed by gas chromatography/mass spectrometry.
- To be applicable, it is critical that the canisters are cleaned and tested to assure inertness. Be careful with the canister valves, do not over-tighten them. Label all the samples taken.
- As a prerequisite, it is useful to have a rough idea about the expected concentrations in order to calculate the sampling time and volume.
- Depending on the type of canister, a wrench might be needed, together with a flow controller.

Examples of applications in the literature

Spicer et al. (1994) studied the composition and photochemical reactivity of a turbine engine exhaust to establish the environmental impact of the organic compounds emitted from aircraft turbine engines. Authors wanted to identify and quantify the VOCs present in gaseous emissions from jet engines and to study the photochemical reactivity of those compounds. For studying the photochemical reactivity, exhaust fumes were introduced into two 8.5 m^3 outdoor Teflon simulation chambers. Among all the compounds sampled and quantified, there were carbonyls (using DNPH derivatization reaction), sorbent tubes filled with XAD-2 and canisters. Specially passivated aluminium cylinders were used for collecting air samples that were analysed by GC and also by GC–MS.

Some decades after, Miracolo et al. (2011) studied the secondary organic aerosol (SOA) formation from photochemicalageing of aircraft exhaust in a smaller Teflon chamber (7 m^3). Despite the fact that the objective of the study was SOA formation, gas-phase VOCs were also collected using SUMMA canisters and analysed using GC–MS. In total, 94 volatile organic gases were identified and quantified.

Wang et al. (2012) carried out a study on environmental tobacco smoke generated by adding smoke from different brands of cigarettes to a simulation chamber. The identified and quantified pollutants were both inorganic compounds and organic compounds. The test chamber was an 18.26 m^3 stainless steel chamber with temperature and RH maintained at 23 °C and 50%, respectively, to simulate the typical indoor air conditions. Air samples were collected through a sampling port to different samplers or analysers connected in series. Carbonyl compounds were sampled using DNPH-coated cartridges and VOC samples were collected using SUMMA canisters at 4.0–6.0 L/min, using mass flow controllers. Chemical analysis was performed by GC using procedures based on the US EPA Method TO-14 (US EPA 2022).

6.1.3 Bag Sampling

Bag sampling is a convenient and accurate means of collecting gases and vapours when concentrations are expected to be higher than the detection limits of common analytical instruments. Sampling bags are typically made of Tedlar®, FEP Teflon foil or other inert materials (SamplePro FlexFilm, FlexFoil). They are inexpensive, simple to use and available in a range of sizes, from around 0.5 L–100 L in volume. The bags can be reused after several cycles of cleaning with pure air or nitrogen and evacuating using a pump. The main disadvantage of sampling bags is that some of the collected chemical species may not remain stable for more than 1–3 days (Wang and Austin 2006; Kumar and Víden 2007; Ras et al. 2009).

Tedlar® is the most popular material used for sampling bags because it retains the quality of the collected air sample and also provides the best options for storage and transport. Tedlar bags are generally made from polyvinyl fluoride (PVF) film, which has the following beneficial properties:

- High level of inertness to a wide range of chemicals,
- Resistant to corrosion,
- High tensile strength and abrasion-free,
- Low absorption rate,
- High resistance to gas permeability and
- High resistance to increases in temperature.

The sampling bag has a valve fitting, which can be made of polypropylene (PP), polytetrafluoroethylene (PTFE or Teflon) or Stainless Steel (SS). The fittings connect easily to a tube for air sampling and many of them are also fitted with a silicone septum to allow syringe samples to be injected into the bag directly. This is a strong and reliable fitting system as the silicone septum acts as a barrier between the two parts of the bag and can also be easily detached if required.

Procedure for bag sampling

Usually, air sample bags are only used for short periods of time. Some of them can be re-used, while others are designed for single-use only. In the case of re-usable bags, it is very important to ensure the bag is properly cleaned to avoid contamination. Although taking air samples using Tedlar bags is a quite efficient and straightforward procedure, the following points need to be considered:

- Do not fill the Tedlar bag completely. Only fill to about half of the bag's total capacity. This helps to ensure that the container maintains an ideal temperature even with a change in ambient air pressure, such as while being transported in an airplane.
- Although Tedlar bags are highly durable, unforeseen circumstances may result in leakage. The use of two bags to collect the same sample provides adequate back-up.
- Ensure prompt shipping arrangements as the Tedlar bags can only hold air samples effectively for around 72 h. Try to ship the bag the same day as sample collection to ensure on-time and intact delivery of the sample.

In order to sample with a plastic bag, a pump capable of operating at the recommended flow rate is required. An airflow calibrator is also needed to confirm the flow rate. The user has to choose between a bag with single fittings (a hose/valve for flushing and filling the bag and sealing it off after sampling or a syringe port with a septum for removing the sample for analysis) or dual-fitted bags (with separate hose/valve and syringe port fittings).

When sampling directly from the air, the procedure is:

- Attach a piece of flexible PTFE tubing to the valve on the bag.
- Connect the other end of the tubing to the sampling pump.
- To begin sampling, open the valve on the bag, turn on the pump and note the start time.
- Gently fill the bag until it is approximately half full and close the valve securely before disconnecting the bag.
- Store the bag out of direct sunlight and away from heat to prevent the contents from reacting or degrading.

When collecting an air sample using an air-tight syringe, the procedure is:

- Insert the syringe into the septum of the port on the Tedlar bag and slowly push the plunger in.
- Fill until the bag is approximately half full.
- Slowly remove the syringe from the port on the bag.

Examples of applications in the literature

Some literature references for the sampling of air by using plastic bags are Cariou and Guillot (2006), Wang and Austin (2006), Guo et al. (2007), Kumar and Víden (2007), Wang et al. (2012), Chang et al. (2018) among others.

6.1.4   Sorbent Tube Sampling

Sorbent tubes are widely used for sampling gas-phase species in air. The collected species can be extracted from the sorbent by using a solvent or thermal desorption. Extraction into a solvent makes the sample amenable to chemical analysis by either liquid or gas chromatography. In thermal desorption, high-temperature gas streams are used to remove the compounds from the sorbent and inject them, often with cryofocusing, into an instrument, such as GC–MS for analysis. Sorbent tubes are generally good for sampling both polar and non-polar compounds but not suitable for highly volatile species. A range of materials can be used in sorbent tubes and the user should choose the material that is the most appropriate for the compounds of interest.

Some of the key advantages of sorbent tubes are:

- Small, portable and light weight.
- The availability of a large selection of sorbents to match the target compounds, which can be polar and non-polar VOCs. If there is no commercial combination that matches the target compounds, it is easy to produce home-made combinations.
- The commercial availability of thermal desorption systems to release compounds from the sorbent and into the analytical system.
- The possibility of dealing with water using a combination of hydrophobic sorbents.
- Sample tubes used in thermal desorption can usually be reused at least 100 times before the sorbent needs to be replaced.

It is important to know the concentration range of target VOCs in the air samples, since the tube dimensions selected must facilitate these two essential functions without introducing their own practical limitations. Caution must be exercised in order to avoid sample breakthrough. Representative samples are obtained when the correct air volume and sorbent size are employed. Therefore, the total volume of sample collected must be known. The amount of VOCs retained on a sorbent is determined to a large extent by the sorbent bed length and sorbent mass. Typically, a sorbent tube has a length of 90 mm and an outer diameter of 6 mm, containing 0.1–1 g of the sorbent.

It is very important when choosing the most appropriate sorbent to consider the following parameters: hydrophobicity, thermostability and loadability. For example, the less water is retained by the sorbent, the less interference is experienced during analysis. When a single sorbent is not sufficient to capture a range of target compounds, a combination of sorbents can be employed.

Sorbent types

The sorbent is placed in a glass or stainless steel tube and VOCs present in the air are collected onto one or more sorbent tubes using a sampling pump. The use of sorbent tubes for sampling VOCs in ambient air followed by thermal desorption GC and GC–MS has been the subject of several reviews (Woolfenden 1997, 2010a, b).

The different kinds of sorbents that are regularly used include:

Tenax

Tenax tubes contain the polymer p-phenylene oxide packed in glass or stainless steel tubes. They are used in the US EPA Methods T-O1 and VOST for the collection of non-polar VOCs, as well as some polar VOCs and some lighter semi-volatile organics. Tenax is not suitable for organic compounds with high volatility, e.g. those with a vapour pressure greater than approximately 250 mbar.

Carbon Molecular Sieves

Carbon molecular sieves (CMS) are commercially available carbon polymers packed in stainless-steel sampling tubes. They contain tiny crystals of graphite that are cross-linked to yield a microporous structure with high surface area. Tubes containing CMS are used in the US EPA Method TO-2 for sampling and analysis of highly volatile non-polar organic compounds.

Mixed Sorbent Tubes

Mixed sorbent tubes contain two or more types of sorbents. The advantages of each sorbent combine to increase the range of compounds that can be sampled. The use of mixed sorbent tubes can also reduce the chance of highly volatile compounds breaking through the sorbent media. Tenax and CMS are a good combination for a mixed sorbent tube as the former material efficiently collects a wide range of organic compounds, while the latter is effective for the species with high volatility.

Chemically Treated Silica Gel

Silica gel can be treated or coated with chemical species to facilitate sampling of specific compounds in air. One of the most widely used examples of this approach is the DNPH-coated silica gel cartridge used with US EPA Method TO-11.

XAD-2 Polymer

Amberlite® XAD-2 polymers are hydrophobic, cross-linked polystyrene copolymer resins used for the collection of semi-volatile polar and non-polar organic compounds. The XAD-2 polymer is usually packed in tubes along with polyurethane foam and used with US EPA Method TO-13 or the semi-VOST method. The compounds collected on the XAD-2 polymer are chemically extracted for analysis.

Charcoal Cartridges

Charcoal cartridges contain two sections for adsorbing compounds from air. The adsorbed compounds are usually extracted into a solvent and analysed by GC or GC–MS. Quantitative sample collection is demonstrated when target chemicals are detected on the first charcoal section but not on the second. Flow rates and sample volumes can be adjusted to minimize the breakthrough of compounds from the first to the second section.

A summary of types, properties and most suitable target compounds for various sorbents for use in Method TO-17 is shown in Table 6.1.

Table 6.1 Summary of types, properties and most suitable target compounds for various sorbents for use in US EPA Method TO-17 (US EPA 2022)

Sample tube sorbent[c, d]	Approx. analyte volatility range	Hydrophobic (?)	Max. temp (°C)	Specific surface area (m²/g)	Example analytes
CarbotrapC® CarbopackC® Anasorb® GCB2	n-C_8 to n-C_{10}	Yes	>400	12	Alkyl benzenes and aliphatics ranging in volatility from n-C to n-C
Tenax® TA	bp 100–400 °C n-C_7 to n-C_{16}	Yes	350	35	Aromatics except benzene. Non-polar components (bp > 100 °C) and less volatile polar components (bp > 150 °C)
Tenax GR	bp 100–450 °C n-C_7 to n-C_{10}	Yes	350	35	Alkyl benzenes, vapour phase PAHs and PCBs and as above for Tenax TA
Carbotrap® CarbopackB® Anasorb® GCB1	(n-C_4)n-C_3 to n-C_{14}	Yes	>400	100	Wide range of VOCs inc., ketones, alcohols, and aldehydes (bp > 75 °C) and all non-polar compounds within the volatility range specified. Plus perfluorocarbon tracer gases
Chromosorb® 102	bp 50–200 °C	Yes	250	350	Suits a wide range of VOCs incl. oxygenated compounds and haloforms less volatile than methylene chloride

(continued)

Table 6.1 (continued)

Sample tube sorbent[c, d]	Approx. analyte volatility range	Hydrophobic (?)	Max. temp (°C)	Specific surface area (m^2/g)	Example analytes
Chromosorb® 106	bp 50–200 °C	Yes	250	750	Suits a wide range of VOCs incl. hydrocarbons from n-C to n-C. Also good for volatile oxygenated compounds
Porapak Q	bp 50–200 °C n-C_4 to n-C_{12}	Yes	250	550	Suits a wide range of VOCs including oxygenated compounds
Porapak N	bp 50–150 °C n-C_3 to n-C_8	Yes	180	300	Specifically selected for volatile nitriles: acrylonitrile, acetonitrile and propionitrile. Also good for pyridine, volatile alcohols from EtOH, MEK, etc.
Spherocarb[a]	−30–150 °C	No	>400	1,200	Good for very volatile compounds such as VCM, ethylene oxide, CS and CH Cl. Also good for volatile polar organics, e.g. methanol. Ethanol and acetone
Carbosieve SIII[a]® Carboxen 1000[a]® Anasorb CMS[a]	−60–80 °C	No	400	800	Good for ultra volatile compounds such as C C hydrocarbons, volatile haloforms and freons

(continued)

Table 6.1 (continued)

Sample tube sorbent[c, d]	Approx. analyte volatility range	Hydrophobic (?)	Max. temp (°C)	Specific surface area (m²/g)	Example analytes
Zeolite Molecular Sieve 13X[b]	−60–80 °C	No	350		Used specifically for 1,3-butadiene and nitrous oxide
Coconut Charcoal[a] (Coconut Charcoal is rarely used)	−80–50 °C		>400	>1,000	Rarely used for thermal desorption because metal content may catalyse analyte degradation. Petroleum charcoal and Anasorb® 747 are used with thermal desorption in the EPA's volatile organic sampling train (VOST), Methods 0030 and 0031

[a]These sorbents exhibit some water retention. Safe sampling volumes should be reduced by a factor of 10 if sampling a high (>90%) relative humidity

[b]Significantly hydrophilic. Do not use in high-humidity atmospheres unless silicone membrane caps can be fitted for diffusive monitoring purposes

[c]CarbotrapC™, CarbopackC™, CarbopackB™, Carboxen™ and Carbosieve SIII™ are all trademarks of Supeloo, Inc.; Tenax® is a trademark of Enka Research Institute

[d]Chromosorb® is a trademark of Manville Corp.; Ansaorb® is a trademark of SKC, Inc.; Porapak® is a trademark of Waters Corporation

Procedure for sorbent tube sampling

The main factors to consider before sampling onto sorbent tubes are:

- Selection of the tube and sorbent packing for the sampling application (using Table 6.1).
- Selection of the sampling volume, considering the breakthrough characteristics of the sorbents.
- Selection of sampling time taking into account expected concentration and breakthrough.
- Ensure that the tubes are properly conditioned—bear in mind that newly packed tubes have to be conditioned for at least 2 h at 350 °C passing at least 50 mL/min of pure helium carrier gas through them. After that, the tubes have to be sealed and stored at 4 °C until use.
- All appropriate equipment is available—selected sorbent tubes, calibrated pump and flow controller, tubing to connect the tubes to the chamber and to the pump.
- If the expected concentrations are close to the breakthrough of the first sorbent tube, a second tube could be connected to ensure complete collection of the target compounds.

The step-by-step procedure is:

1. Using clean gloves, remove the sorbent tube caps and attach them to the sampling lines.
2. Set the flow rates of the pump using a mass flow monitor and adjust the flow rate to the decided value for sampling.
3. Sample for the selected period. Recheck the sampling flow rates at the end of the monitoring.
4. Make notes of all relevant sampling parameters (sampling time, flow rates, sample code/number/identification).
5. Remove the sampling tubes using clean gloves, recap the tubes with their fittings, wrap the tubes (for example, with uncoated Al foil) and place them in a clean, opaque airtight container or envelope adequately labelled.
6. Store the containers/envelopes adequately labelled in a clean, cool (4 °C) organic solvent-free environment until time for analysis.

Examples of applications in the literature

Miracolo et al. (2011) studied the aircraft exhaust fumes in a chamber using both online instruments and off-line sampling techniques. Tenax sorbent tubes were among the offline techniques used. Something similar was studied by Presto et al. (2011).

Riemer et al. (1994) studied terpene and related compounds in semi-urban air, nevertheless the applied offline techniques are also suitable to be used in chamber experiments.

Composition of SOA together with the gas phase composition has been studied by Nordin et al. (2013). VOC gas phase samples were collected on sorbent tubes filled with Tenax-TA and Carbopack-B.

Tenax has been widely used for determining VOC composition in air monitoring activities. Srivastava and Devotta (2007) used this offline sampling technique to determine the indoor air quality of public places in India; therefore, it is suitable for use in chamber experiments.

As an example of a combination of sorbents, Kuntasal et al. (2005) determined VOCs in different environments using sorbent tubes, among other sampling techniques.

6.2 Particle Sampling

The collection of particles produced during chamber experiments is routinely carried out for offline analysis of their chemical and physical properties. The detailed chemical composition of particles produced from VOC oxidation is often studied to understand SOA formation mechanisms that are used in atmospheric models and simulations. Offline chemical analysis allows the identification and quantification of target species, as well as the determination of more general parameters such as total organic carbon, water-soluble organic carbon and carbon oxidation state. Physical properties of SOA particles, such as the UV–visible absorption, are also determined to further our understanding of the impacts of secondary aerosol formation and chemistry on radiation balance in the troposphere.

Filter sampling, inertial classification, gravitational sedimentation, centrifugation and thermal precipitation are the most important techniques used to collect particles in different environments. For chamber investigations, filter sampling, as well as inertial classification, are the most important techniques. While inertial classifiers are often applied in field studies, their use in chamber experiments is limited due to the high sampling volume needed. The great advantage of inertial classifiers is the size segregation, which is usually not possible with standard filter sampling. In the following sections, filter sampling techniques as well as inertial classification in chamber experiments will be discussed.

6.2.1 Filter-Based Particle Collection

The collection of particles by filter is based on the interaction of five different mechanisms (Raynor et al. 2011).

(a) **Interception**: particles in an air stream contact the filter surface. Relevant for those particles that are larger than the filter pores.

(b) **Impaction**: flow direction of an air stream transporting particles changes and the inertia of the particles lead to collision with the surface. Most important for particles larger than 1 μm. Process becomes more important as the density, velocity and diameter of the particle increase.

(c) **Diffusion**: Collision of particles with the surface due to Brownian motion. Most likely for particles of ≤0.1 μm.

(d) **Electrostatic attraction**: electrostatic charge causes attraction between particles and filter. Charged filter can attract neutral particles and vice versa.

(e) **Sedimentation**: Particles fall onto filter due to gravitational forces. Very likely for large particles or slow flow velocities. Only relevant for smaller particles if air is moving downward onto the filter.

Filter material

The size, shape, density and electrostatic charge of particles, as well as the chemical and physical properties, can all affect the filtration mechanism. Available filters are made of different materials, coatings and sizes. The majority of the filters belong to one of the following groups:

(a) **Fibrous filter**: composed of a deep mesh of fibres with a random orientation, e.g. glass fibre filter (Fig. 6.5)

(b) **Membrane filter**: complex structure which enhances circuitous travel routes for particles, e.g. mixed cellulose ester (MCE, Fig. 6.5) or polytetrafluoroethylene (PTFE, Fig. 6.5)

(c) **Capillary pore filter**: circular pores, e.g. polycarbonate or polyethylene terephthalate (PET, Fig. 6.5).

The material affects the pore size and with this the collection efficiency and artefact vulnerability. Therefore, a decision on the type of filter used for chamber experiments should take into account various technical requirements including particle size and the chemical identity of the target compounds.

Particles collected by a fibrous or porous membrane are forced to "travel" through the filter via circuitous routes that increase the interaction of particles with the filter and enhance the collection efficiency dramatically. Capillary pore filters often show a lower collection efficiency than fibrous or porous membrane filters of the same pore size (or pore diameter). Thus, it can be stated that the pore size or pore diameter does not reflect the size of particles collected by this filter. Therefore, special effort should be spent selecting the filter material to collect chamber-generated SOA (Burton et al. 2006).

Furthermore, as shown in Fig. 6.4, the collection efficiency for a polycarbonate filter is lowest between 40 and 60 nm. This is caused by the fact that the impaction mechanism is less efficient for particles smaller than 100 nm. In the range ≤100 nm, diffusion is more important but less efficient for collection efficiency. This is defined as "most penetrating particle size" and describes the smallest particle size collected by a filter. According to previous studies, this size is affected by flow rate, charge, filter material and loading (Lee and Liu 1980; Martin and Moyer 2000).

Additionally, filters are available with different types of coatings, binder or additional content to enhance their collection efficiency. Each filter type has their own optimal set-up and flow rate which typically depends on filter type and pore size. While performing filter sampling on a chamber, two aspects need be considered loading effects and pressure drop.

Loading effects and pressure drop

Aside from the filter material and the pore size (or equivalent pore diameter), some additional parameters should be kept in mind while selecting the proper filter material. These include the pressure drop and loading effects. The **pressure drop** describes the loss of static pressure from the front surface of the filter to the rear side. This needs to be considered not only in terms of the lifetime of the pump but also with regards to the filter thickness, solidity and face velocity. Suitable filters are characterized by a low-pressure drop combined with a high collection efficiency. Special care should be taken if the pressure drop is very small or changes rapidly. This usually indicates an improper seal in the holder and the air stream carrying particles inadvertently passes the filter.

Loading effects need to be considered if a large mass is loaded onto the filter. Particles loaded on a filter tend to form dendrites that can be seen as chains emanating from the filter surface. On one hand, these dendrites increase the collection efficiency as particles can be collected additionally at the end of the chain. On the other hand, dendrites lead to a larger pressure drop and they can break down during collection, filter storage and sample preparation for offline analysis. Consequently, this will lead to a loss of collected material. Therefore, massive filter loadings should be avoided as well as folding of loaded filters.

A large loading on the filter directly affects the pressure drop. If the loading is too large, the flow through the filter changes and the calculation of sampling volume is

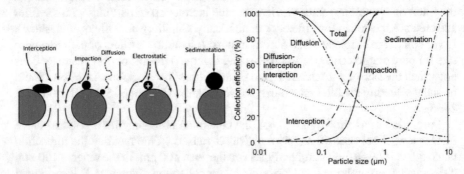

Fig. 6.4 Left: Mechanisms for collection of aerosol particles on filters (Lindsley, NIOSH 2016). Right: theoretical collection efficiencies for each collection mechanism as a function of particle size (Lindsley, NIOSH 2016)

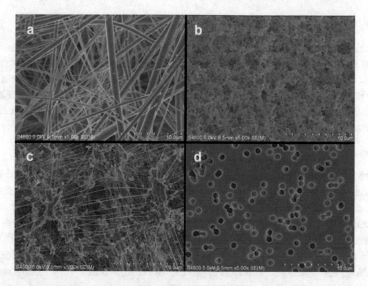

Fig. 6.5 Electron microscope picture from a glass fibre filter (**a**, 1 μm equivalent pore diameter), a mixed cellulose ester filter (**b**, 0.8 μm equivalent pore diameter), a polytetrafluoroethylene (**c**, 3 μm equivalent pore diameter) and a polycarbonate capillary core filter (**d**, 1 μm equivalent pore size). Picture taken from Lindsley, NIOSH (2016)

no longer accurate. This problem can be overcome by using a flow controller unit. Furthermore, it is very helpful for an estimation of the formed particle mass can be given. This should be considered together with the mass needed for offline analysis.

Artefacts

The limitations of filters are mainly caused by positive artefacts due to adsorption of water and organics or negative artefacts due to evaporation of collected material. Filter artefacts lead to inaccuracy while determining chemical composition, loading and physical properties. The artefacts are caused by chemical and physical properties of the particles as well as of the filter material and collection method. Generally, filter artefacts can be split into the following groups, (a) volatilization of collected material, (b) particle bounce, (c) moisture effect and (d) non-aqueous adsorption.

Volatilization of collected material

Volatile or semi-volatile compounds in the particle phase evaporate back into the airflow and pass through the filter. According to Raoult's and Henry's law, this mechanism is most likely if the partial pressure of a compound at the particle surface is greater than the particle pressure in the air passing through the filter. Consequently, this leads to a lower-than-expected concentration of target species in the particle phase. The most important factors influencing the volatilization are particle concentration, filter face velocity, pressure drop and the gas–particle equilibrium (e.g. Cheng and Tsai 1997; Zhang and McMurry 1991; Ashbaugh and Eldred 2004).

Particle bounce

When collecting small particles at low flow rates, the adhesive forces capturing the particles greatly exceed their kinetic energy. If the particle size is increasing as well as the air velocity, larger particles "bounce off" the filter surface. Consequently, this process leads to an underestimation of target species on the filter. On the other hand, if the air is going downwards into the filter, particles can "bounce in" the filter surface. This process can therefore result in an overestimation of target species.

Moisture effects

Moisture greatly influences particle collection on filters, in particular the gravimetry. In general, water can adsorb or desorb from a filter before, during and after sampling. The importance of moisture differs greatly from the filter material. For example, when comparing membrane filters made of Teflon and MCE, it was found that Teflon was less affected by moisture whereas the MCE filter showed a massive effect (Tsai et al. 2002). In particular, if the filter was not equilibrated to environmental conditions prior to weighing, the MCE showed non-reproducible results and a very strong moisture adsorption. Therefore, it is strongly recommended to consider carefully, the filter material for collection under high RH as well as a re-equilibration to environmental conditions prior to sample treatment (Tsai et al. 2002).

Non-aqueous vapour adsorption

Besides water, VOCs and in particular OVOCs can be adsorbed to filter surfaces and/or onto collected material. This lowers the gas-phase concentration while increasing the particle-phase concentration. Sulfates, nitrates and semi-volatile compounds are also very prone to this behaviour. Another aspect to be considered is the artefact formation due to the interaction of collected organics with reactive gases such as ozone, SO_2, etc. It has been demonstrated that high levels of ozone during sampling can lead to the decomposition of particulate OVOCs. This has been intensively reported for PAHs (Liu et al. 2006; Schauer et al. 2003, see review by Menichini 2009 and citations therein), but only rarely for single compounds (Limbeck et al. 2001; Yao et al. 2002; Kerminen et al. 1999; Warnke et al. 2006).

Intensive studies have been carried out to investigate artefacts caused by evaporation and adsorption on the determination of OC/EC (Turpin et al. 1994; Mader and Pankow 2001; Mader et al. 2003; Kim et al. 2001; Subramanian et al. 2004; Kirchstetter et al. 2001).

In particular, quartz fibre filters are very prone to artefacts due to their active surface. Even though, they need to be used for OC/EC analysis as they can be heated prior to analysis up to 800 °C. Studies investigating artefact effects on OC determination revealed an error of -80% up to $+50\%$ (Turpin et al. 2000).

Artefacts can be identified either by a complex sampling set-up or by using a backup filter which is placed behind the target filter (Mader et al. 2003; Warnke et al. 2006). Applying the back-up filter method, it was shown that pinonic acid, an important α-pinene oxidation product, caused a massive artefact due to adsorption. Therefore, great care needs to be taken when quantifying semi-volatile organics.

Denuders

Artefacts due to non-aqueous vapour adsorption can largely be avoided by placing a denuder in front of the filter. So-called **denuder-filter devices** are widely used.

Various types of denuders are used to remove OVOCs as well as inorganic and basic gases. An activated charcoal denuder can be used if OVOCs need to be removed (Eatough et al. 2001). In addition, an annular denuder (e.g. Temime et al. 2007; Healy et al. 2008; Kahnt et al. 2011), coiled denuder (Pui et al. 1990), honeycomb denuder (Koutrakis et al. 1993) and porous-metal denuder (Huang et al. 2001; Tsai et al. 2001) have also been successfully applied. In general, these denuders fulfil two important tasks; (i) avoid artefacts due to adsorption of organics and inorganics, (ii) provide information about gas-phase chemical composition.

In the simplest approach, the denuder contains charcoal to adsorb all gas-phase compounds that could affect filter sampling. If information about the gas-phase chemical composition is needed, the denuder can be coated with an appropriate type of resin (e.g. XAD-4, XAD-2), extracted and subsequently analysed. The coating will directly influence the type of organics trapped on the denuder and can enhance the adsorption potential of the denuder. Many of the resins are effective at trapping non-polar species, but they are much less efficient for removing polar organics. To improve the collection of polar compounds, the denuder can also be coated with an appropriate derivatizing agent. For example, Temime et al. (2007) used a denuder coated with XAD-4 resin and the derivatizing agent O-(2,3,4,5,6-pentafluorobenzyl)hydroxylamine (PFBHA) to enable the on-tube conversion of gas-phase carbonyls to their oxime derivatives which were extracted and identified by GC–MS. This technique not only prevented carbonyls from depositing on a filter but also allowed for their quantification in the gas phase. Using a similar approach, Kahnt et al. (2011) used DNPH on XAD-4 coated denuders to successfully trap a range of gas-phase carbonyl compounds, including methylglyoxal, glyoxal, benzaldehyde, formaldehyde and acetone. As shown in Table 6.2, the use of the derivatizing agent proved to be very effective in reducing the amount of carbonyls detected on the filter.

It should be remembered that the use of a denuder disturbs the gas-particle equilibrium and enhances the re-evaporation of collected species (Zhang and McMurry 1991), in particular for semi-volatile species. Therefore, impregnated filters can also be used to enhance the capture of semi-volatile species, such as carbon-impregnated glass fibre filter (Eatough et al. 2001), XAD-impregnated quartz fibre filter (Swartz et al. 2003), nylon filter (Tsai and Perng 1998) and citric acid-coated filter (Tsai et al. 2000).

General procedure for filter sampling

The following equipment and materials are required:

- Air pump with a flow controller to set the desired rate. Pumps and flow controllers have to be calibrated before use.
- Correct filter material with the corresponding pore size.
- Gloves to handle the samples, opaque envelopes and suitable labels.

Table 6.2 Results from tests using an annular denuder coated with XAD-4 resin and the derivatization reagent 2,4-dintrophenylhydrazine (DNPH) to reduce carbonyl artefacts on PTFE filters (data from Kahnt et al. 2011)

Compound		Fraction detected on filter (%)		Break-through potential (%)	
		<3% RH 10 L min^{-1}	<50% RH 10 L min^{-1}	<3% RH 10 L min^{-1}	<50% RH 10 L min^{-1}
Formaldehyde	XAD-4	0	0.6	49	67
	XAD-4/DNPH	0.5	0.2	44	9.5
Acetone	XAD-4	19	8.7	95	96
	XAD-4/DNPH	0	0	6.4	2.1
Acetaldehyde	XAD-4	0.9	0	16	35
	XAD-4/DNPH	0	0	24	20
Hydroxyacetone	XAD-4	1.9	0	19	98
	XAD-4/DNPH	0	0	1.4	0
Methyl vinyl ketone	XAD-4	1.6	4.9	92	98
	XAD-4/DNPH	0	0.1	11	0.9
Methacrolein	XAD-4	0	0	100	100
	XAD-4/DNPH	0	0	3.1	0
Glyoxal	XAD-4	5.9	36	42	91
	XAD-4/DNPH	1.3	0.9	35	23
Methylglyoxal	XAD-4	2.2	12	8.9	100
	XAD-4/DNPH	7.1	0.6	0	0
Benzaldehyde	XAD-4	0	36	0	20
	XAD-4/DNPH	0	0	0.2	0.4
Campholenic aldehyde	XAD-4	0	0	0	0
	XAD-4/DNPH	0	0	0	0
Nopinone	XAD-4	0	0	0	0

A schematic of a general set-up for filter sampling is shown in Fig. 6.6. A denuder can also be placed in front of the filter if desired. The general procedure is as follows:

- Place the filter in the filter holder.
- Connect the sampling line to the chamber just before sample collection begins. The filter should not be connected for hours without sampling.
- Start sampling and use the flow controller to control the sampling rate.
- The chosen sampling time should take into account the expected concentration of species, safe sampling volumes and potential for breakthrough.
- Stop sampling while closing the flow controller.

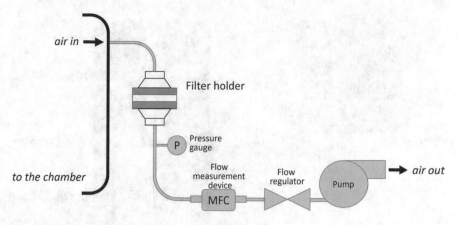

Fig. 6.6 Schematic of a filter collection system

- Take the filter out of the holder, place it in a suitable petri dish and store at a minimum of −20 °C until analysis. If the sample contains semi-volatile components, filters can be wrapped in aluminium foil, placed in sealed plastic bag and stored in a freezer until analysis.

6.2.2 Inertial Classifiers

The term "inertial classifier" includes impactors, virtual impactors and cyclones (see review by Marple and Olson (2011) and citations therein). The collection and the size classification follow the inertia of the particle. The sample flow is transported through the classifier and while turning the flow direction, particles of sufficient inertia are captured. Particles of less inertia remain in the gas flow and are transported to the next stage or escape the classifier. Various inertial classifiers are available for different kind of purposes and technical requirements. Criteria for selection include the number of stages, flow rate and cut points. Particle collection with classifiers is often limited in chamber experiments due to the high flow rates needed for sampling. Several classifiers apply around 70–100 L min^{-1} to ensure sufficient loading on the single stages (high-flow cascade impactor, high-volume virtual impactor, gravimetric impactor). In addition to these high-volume samplers, some classifiers apply a flow between 10 and 30 L min^{-1} (e.g. Moudi, nano-Moudi, Dekati Mass monitor). Only a few chamber studies report the use of inertial classifiers for chamber use (e.g. Palen et al. 1993; Sax et al. 2005; Jain and Petrucci 2015; Hosny et al. 2016).

Impactors are most common in atmospheric science, in particular conventional and cascade impactors. **Conventional impactors** collect particles larger than the cut size. Particles smaller than the cut size remain in the airflow and are not impacted at the substrate surface. Ideal impactors are characterized by a sharp collection efficiency

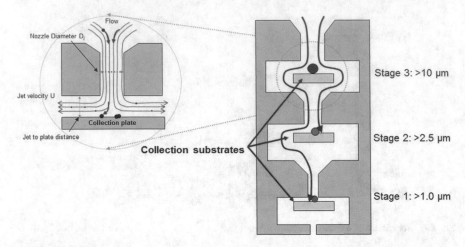

Fig. 6.7 Schematic of a DLPI+ (Dekati® Low-Pressure Impactor, Dekati Ltd), illustrating the collection principle of a cascade impactor

curve. If several stages of impactor plates are connected in a row, this is referred to as a **cascade impactor**, Fig. 6.7.

In the cascade impactor, particle collection can be described using the Stokes number which characterizes the velocity increase of the sample flow throughout successive stages impacting smaller particles down to the lowest plates. Glass plates, membrane filters or foils can be used as substrates.

Virtual impactors are similar to conventional impactors with the exception that the impaction plate is removed by collection probe. Within the collection probe, classification takes place whereby larger particles penetrate more into the probe than smaller ones. Particles larger than the cut size are transported through the probe via the minor flow, whereas particles smaller than the cut size are transported via the major flow (exists at the top of the probe). Both flows can be further transported into other devices.

Cyclones use a cyclonic sample flow which swirls downwards into a conical section and spirals upwards around the cyclone axis to the upper end. Particles are deposited on the surface walls and in the cone. A grit pot is installed downwards to collect particles that settle down.

Similar to filter sampling, particle collection by impaction has limitations that are highly related to artefact formation including particle bounce, evaporation of organics, overloading and interstage losses. Particle bounce, evaporation processes and the effects of overloading are described in the section below. Interstage losses occur when material impacts internal surfaces other than the collection substrate. These loss processes are more pronounced under turbulent conditions.

General procedure for inertial classifier sampling

The following equipment and materials are required:

- Air pump with a flow controller to set the desired rate. Pumps and flow controllers have to be calibrated before use.
- Chosen substrate (filter, metal foil or other).
- Gloves to handle the samples, opaque envelopes and suitable labels.

The cut size of the inertial classifier should be selected carefully. If too little mass is collected for the different size ranges, no further analysis is possible. The general procedure is as follows:

- Place the substrate(s) on the plate(s).
- If applicable, turn on the conditioning system.
- Connect the sampling line to the chamber just before sample collection begins. The filter should not be connected for hours without sampling.
- Start sampling and use the flow controller to control the sampling rate (for low-volume sampler, large-volume samplers often have no flow controller unit). Large-volume samplers usually contain an internal flow meter or measure at least the pressure drop via a pressure gauge.
- The chosen sampling time should take into account the expected concentration of species.
- Stop sampling.
- Take the substrates from the plates, taking care not to damage them, especially if they need to be weighed.
- Place the substrates in a suitable petri dish and store at a minimum of −20 °C until analysis. If the sample contains semi-volatile components, filters can be wrapped in aluminium foil, placed in sealed plastic bag and stored in a freezer until analysis.
- Clean plate(s) and control the nozzles (Fig. 6.8).

Fig. 6.8 Loaded impaction plate (stage 2) from a 5-stage Berner-type cascade impactor (*Picture source* ©TROPOS)

6.2.3 Particle-into-Liquid Sampler

A particle-into-liquid sampler (PILS) is an aqueous-solution-based online technique for determining the bulk chemical composition of aerosol particles (Weber et al. 2001). Particles are collected by impaction in the instrument after condensational growth by supersaturated water vapour typically at elevated temperatures of around 100 °C. This collection technique allows the sampling of about 10–20 L/min of air into small liquid flows in the microliter/min range, Fig. 6.9. Instruments using this collection method have been developed that utilize a sample flow of up to 2500 L/min (Demokritou et al. 2002).

The effluent of a PILS or similar instrument can in principle be combined with any analysis technique suitable for liquid samples and has been used to collect liquid fractions for further detailed offline analyses such as high-pressure liquid chromatography (HPLC) mass spectrometry, which allows to characterize the detailed particle composition with a time resolution of a few minutes (e.g. Zhang et al. 2016; Bateman et al. 2011). A good example of the use of highly time-resolved PILS in simulation

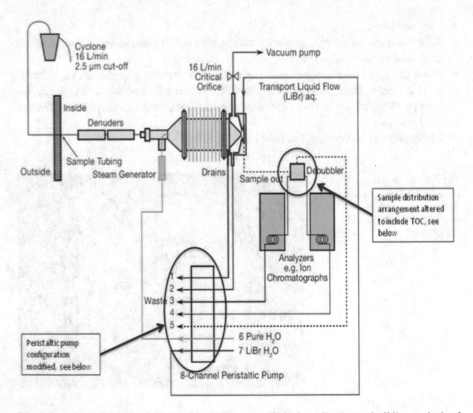

Fig. 6.9 Schematic of a Particle-into-Liquid Sampler (PILS) for collection and off-line analysis of atmospheric aerosol particles (Watson 2016)

chamber studies is provided by Pereira et al. (2015), who performed state-of-the-art mass spectrometry analysis on samples collected by PILS to follow the formation and evolution of individual compounds in SOA produced by the photooxidation of aromatic compounds. In addition to the high time resolution for particle collection, PILS-type devices allow for automated sampling from atmospheric simulation chambers or to couple the liquid effluent directly to analytical instruments, such as an ion chromatograph (e.g. Sorooshian et al. 2006) for inorganic ions, small carboxylic acids or total water-soluble organic carbon (e.g. Peltier et al. 2007). Overall, the number of studies applying PILS to simulation chamber experiments is rather limited (Sierau et al. 2003; Nakao et al. 2011; Zhang et al. 2016). However, an interesting recent study which combined PILS with the Dithiothreitol (DTT) assay to evaluate the oxidizing potential of particles generated in a simulation chamber shows that this sampling technique still has potential applications (Jiang et al. 2016).

An alternative design for condensational growth of particles is the o-MOCA instrument, where particles are passing through a supersaturated laminar flow wet-walled tube at room temperature to incorporate particles into droplets, where supersaturation conditions are more accurately controlled compared to PILS-type instruments. Droplets are then impacted into a small liquid volume for further offline or semi-continuous online analysis. The advantage of this technique compared to typical PILS designs is the milder droplet growth conditions (room temperature), although sampling flow rates are typically lower than for PILS instruments, e.g. a few L/min which results in higher detection limits (Eiguren-Fernandez et al. 2017).

Other instruments, which utilize condensational growth of particles and impaction, have been developed over the last two decades. Examples are the Particle Collection System (PCS) developed by Simon and Dasgupta (1995) with a 10 L/min sample flow, the Steam Jet Aerosol Collector (SJAC, Slanina et al. 2001) or the condensation growth and impaction system (C-GIS, Sierau et al. 2003) which was coupled to a differential mobility analyser to allow particle size-dependent compositional analyses of the impacted liquid samples, Fig. 6.10.

Similar instruments, such as the MARGA (Monitor for AeRosols and GAses) use steam jet aerosol collectors to trap particle components and wet rotating denuders to absorb water-soluble gases (Rumsey et al. 2014). The aqueous aerosol and gas extracts are then typically coupled with ion chromatography and analysed for major inorganic ions or low molecular weight organic acids (Chen et al. 2017; Stieger et al. 2019) in a semi-online manner. The advantage of this instrument is the simultaneous collection and characterization of the gas and particle phase to study partitioning and particle growth processes.

There are a number of limitations associated with the use of PILS in atmospheric simulation chamber experiments. Firstly, the rather high flow rates at which PILS collectors are usually operated are potentially an issue for smaller chambers. In addition, most PILS designs operate by mixing hot water vapour into the aerosol sample flow, which might cause some undesired reactions in the droplets or evaporation of semi-volatile particle components. Furthermore, gas/particle partitioning of particle components in an aqueous droplet (i.e. governed by Henry's Law coefficients) should be considered to assess possible sampling artefacts of these particle collection devices

Fig. 6.10 Schematic of the condensation growth and impaction system (C-GIS) for collection and of-line analysis of atmospheric aerosol particles

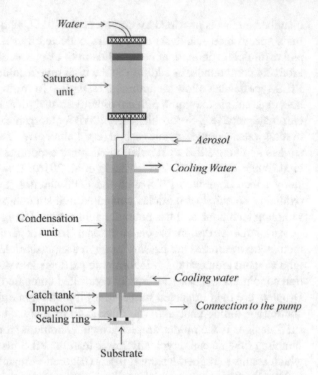

(Zhang et al. 2016). Such potential artefacts need to be characterized to establish if they are relevant for a particular set of target analytes. Finally, it should be kept in mind that PILS is only suitable for water-soluble compounds and that the samples are usually highly diluted which requires highly sensitive analytical techniques or enrichment procedures applied after sampling.

General procedure for particle-into-liquid sampling

The following equipment and materials are required:

- Air pump with a flow controller to set the desired rate. Pumps and flow controllers have to be calibrated before use.
- Chosen vials or substrates for sample collection.
- Gloves to handle the samples, opaque envelopes and suitable labels.

The general procedure is as follows:

- Place the vials or substrates into the sampler.
- Connect the set-up to the chamber just before sample collection begins.
- Start sampling and use the flow controller to control the sampling rate.
- The chosen sampling time should taking into account the expected concentration of species.
- Stop sampling.

Take the samples, label them and store at a minimum of $-20\,°C$ until analysis.

References

Arnts, R.R.: 2,4-dinitrophenylhydrazine-coated silica gel cartridge method for determination of formaldehyde in air: identification of an ozone interference. Environ. Sci. Technol. **23**, 1428–1430, **1423**(1411) (1989). https://doi.org/10.1021/es00069a018

Ashbaugh, L., Eldred, R.: Loss of particle nitrate from teflon sampling filters: effects on measured gravimetric mass in California and in the IMPROVE network. J. Air Waste Manag. Assoc. **1995**(54), 93–104 (2004). https://doi.org/10.1080/10473289.2004.10470878

Bateman, A.P., Nizkorodov, S.A., Laskin, J., Laskin, A.: Photolytic processing of secondary organic aerosols dissolved in cloud droplets. Phys. Chem. Chem. Phys. **13**, 12199–12212 (2011). https://doi.org/10.1039/c1cp20526a

Brombacher, S., Oehme, M., Beukes, J.A.: HPLC combined with multiple mass spectrometry (MSn): an alternative for the structure elucidation of compounds and artefacts found in smog chamber samples. J. Environ. Monit. **3**, 311–316 (2001). https://doi.org/10.1039/b101204p

Burton, N.C., Grishpun, S.A., Reponen, T.: Physical collection efficiency of filter materials for bacteria and viruses. Ann. Occup. Hyg. **51**, 143–151 (2006). https://doi.org/10.1093/annhyg/mel073

Cardin, D.B., Noad, V.L.: The combination of 3 new sampling techniques paired with GCMS for determination of uptake rates and accurate monitoring of SVOC endocrine disruptors in indoor air, entech instruments (2018). https://www.entechinst.com/download/combination-of-3-newsampling/

Cariou, S., Guillot, J.-M.: Double-layer Tedlar bags: a means to limit humidity evolution of air samples and to dry humid air samples. Anal. Bioanal. Chem. **384**, 468–474 (2006). https://doi.org/10.1007/s00216-005-0177-4

Chen, X., Walker, J.T., Geron, C.: Chromatography related performance of the Monitor for AeRosols and GAses in ambient air (MARGA): laboratory and field-based evaluation. Atmos. Meas. Tech. **10**, 3893–3908 (2017). https://doi.org/10.5194/amt-10-3893-2017

Cheng, Y.-H., Tsai, C. J.: Evaporation loss of ammonium nitrate particles during filter sampling. J. Aerosol Sci. **28**, 1553–1567 (1997). https://doi.org/10.1016/S0021-8502(97)00033-5

Deming, B.L., Pagonis, D., Liu, X., Day, D.A., Talukdar, R., Krechmer, J.E., de Gouw, J.A., Jimenez, J.L., Ziemann, P.J.: Measurements of delays of gas-phase compounds in a wide variety of tubing materials due to gas–wall interactions. Atmos. Meas. Tech. **12**, 3453–3461 (2019). https://doi.org/10.5194/amt-12-3453-2019

Demokritou, P., Gupta, T., Koutrakis, P.: A high volume apparatus for the condensational growth of ultrafine particles for inhalation toxicological studies. Aerosol Sci. Technol. **36**, 1061–1072 (2002). https://doi.org/10.1080/02786820290092230

Druzik, C.M., Grosjean, D., Van Neste, A., Parmar, S.S.: Sampling of atmospheric carbonyls with small DNPH-coated C18 cartridges and liquid chromatography analysis with diode array detection. Intern. J. Environ. Anal. Chem. **38**, 495–512 (1990). https://doi.org/10.1080/03067319008026952

Eatough, D.J., Eatough, N.L., Obeidi, F., Pang, Y., Modey, W., Long, R.: Continuous determination of PM2.5 mass, including semi-volatile species. Aerosol. Sci. Technol. **34**, 1–8 (2001). https://doi.org/10.1080/02786820121229

Eiguren-Fernandez, A., Kreisberg, N., Hering, S.: An online monitor of the oxidative capacity of aerosols (o-MOCA). Atmos. Meas. Tech. **10**, 633–644 (2017). https://doi.org/10.5194/amt-10-633-2017

Guo, H., So, K.L., Simpson, I., Barletta, B., Meinardi, S., Blake, D.R.: C1–C8 volatile organic compounds in the atmosphere of Hong Kong: overview of atmospheric processing and source apportionment. Atmos. Environ. **41**, 1456–1472 (2007). https://doi.org/10.1016/j.atmosenv.2006. 10.011

Healy, R.M., Wenger, J.C., Metzger, A., Duplissy, J., Kalberer, M., Dommen, J.: Gas/particle partitioning of carbonyls in the photooxidation of isoprene and 1,3,5-trimethylbenzene. Atmos. Chem. Phys. **8**, 3215–3230 (2008). https://doi.org/10.5194/acp-8-3215-2008

Hosny, N.A., Fitzgerald, C., Vyšniauskas, A., Athanasiadis, A., Berkemeier, T., Uygur, N., Pöschl, U., Shiraiwa, M., Kalberer, M., Pope, F.D., Kuimova, M.K.: Direct imaging of changes in aerosol particle viscosity upon hydration and chemical aging. Chem. Sci. **7**, 1357–1367 (2016). https://doi.org/10.1039/c5sc02959g

Huang, C.-H., Tsai, C.-J., Shih, T.-S.: Particle collection efficiency of an inertial impactor with porous metal substrates. J. Aerosol Sci. **32**, 1035–1044 (2001). https://doi.org/10.1016/s0021-8502(01)00038-6

Jain, S., Petrucci, G.A.: A new method to measure aerosol particle bounce using a cascade electrical low pressure impactor. Aerosol Sci. Technol. **49**, 390–399 (2015). https://doi.org/10.1080/027 86826.2015.1036393

Jiang, H., Jang, M., Sabo-Attwood, T., Robinson, S.E.: Oxidative potential of secondary organic aerosols produced from photooxidation of different hydrocarbons using outdoor chamber under ambient sunlight. Atmos. Environ. **131**, 382–389 (2016). https://doi.org/10.1016/j.atmosenv. 2016.02.016

Kahnt, A., Iinuma, Y., Böge, O., Mutzel, A., Herrmann, H.: Denuder sampling techniques for the determination of gas-phase carbonyl compounds: a comparison and characterisation of in situ and ex situ derivatisation methods. J. Chromatogr. B **879**, 1402–1411 (2011). https://doi.org/10. 1016/j.jchromb.2011.02.028

Kerminen, V.-M., Teinilä, K., Hillamo, R., Mäkelä, T.: Size-segregated chemistry of particulate dicarboxylic acids in the Arctic atmosphere. Atmos. Environ. **33**, 2089–2100 (1999). https://doi. org/10.1016/S1352-2310(98)00350-1

Kim, B., Cassmassi, J., Hogo, H., and Zeldin, M.: Positive organic carbon artifacts on filter medium during PM2.5 sampling in the South Coast air basin. Aerosol. Sci. Technol. **34**, 35–41 (2001). https://doi.org/10.1080/02786820118227

Kirchstetter, T.W., Corrigan, C.E., Novakov, T.: Laboratory and field investigation of the adsorption of gaseous organic compounds onto quartz filters. Atmos. Environ. **35**, 1663–1671 (2001)

Koutrakis, P., Sioutas, C., Ferguson, S.T., Wolfson, J.M., Mulik, J.D., Burton, R.M.: Development and evaluation of a glass honeycomb denuder/filter pack system to collect atmospheric gases and particles. Environ. Sci. Technol. **27**, 2497–2501 (1993). https://doi.org/10.1021/es00048a029

Kumar, A., Víden, I.: Volatile organic compounds: sampling methods and their worldwide profile in ambient air. Environ. Monit. Assess. **131**, 301–321 (2007). https://doi.org/10.1007/s10661-006-9477-1

Kuntasal, Ö.O., Karman, D., Wang, D., Tuncel, S.G., Tuncel, G.: Determination of volatile organic compounds in different microenvironments by multibed adsorption and short-path thermal desorption followed by gas chromatographic–mass spectrometric analysis. J. Chromatogr. A **1099**, 43–54 (2005). https://doi.org/10.1016/j.chroma.2005.08.093

Lee, K.W., Liu, B.Y.H.: On the minimum efficiency and the most penetrating particle size for fibrous filters. J. Air Pollut. Control Assoc. **30**, 377–381 (1980). https://doi.org/10.1080/00022470.1980. 10464592

Limbeck, A., Puxbaum, H., Otter, L., Scholes, M.: Semivolatile behavior of dicarboxylic acids and other polar organic species at a rural background site (Nylsvley, RSA). Atmos. Environ. **35**, 1853–1862 (2001)

Lindsley, W.G.: Filter pore size and aerosol sample collection. National Institute for Occupational Safety and Health (2016). https://www.cdc.gov/niosh/docs/2014-151/pdfs/chapters/chapter-fp. pdf

Liu, Y., Sklorz, M., Schnelle-Kreis, J., Orasche, J., Ferge, T., Kettrup, A., Zimmermann, R.: Oxidant denuder sampling for analysis of polycyclic aromatic hydrocarbons and their oxygenated derivates in ambient aerosol: evaluation of sampling artefact. Chemosphere **62**, 1889–1898 (2006). https://doi.org/10.1016/j.chemosphere.2005.07.049

Mader, B.T., Pankow, J.F.: Gas/solid partitioning of semivolatile organic compounds (SOCs) to air filters. 3. An analysis of gas adsorption artifacts in measurements of atmospheric SOCs and organic carbon (OC) when using teflon membrane filters and quartz fiber filters. Environ. Sci. Technol. **35**, 3422–3432 (2001). https://doi.org/10.1021/es0015951

Mader, B., Schauer, J., Seinfeld, J., Flagan, R., Yu, J., Yang, H., Lim, H.-J., Turpin, B., Deminter, J., Heidemann, G., Bae, M., Quinn, P., Bates, T., Eatough, D., Huebert, B., Bertram, T., Howell, S.: Sampling methods used for the collection of particle-phase organic and elemental carbon during ACE-Asia. Atmos. Environ. **37**, 1435–1449 (2003). https://doi.org/10.1016/s1352-2310(02)01061-0

Marple, V.A., Olson, B.A.: Sampling and measurement using inertial, gravitational, centrifugal, and thermal techniques. In: Aerosol Measurement: Principles, Techniques, and Applications. Wiley (2011)

Martin, S.B., Moyer, E.S.: Electrostatic respirator filter media: filter efficiency and most penetrating particle size effects. Appl. Occup. Environ. Hyg. **15**, 609–617 (2000). https://doi.org/10.1080/10473220050075617

Menichini, E.: On-filter degradation of particle-bound benzo[a]pyrene by ozone during air sampling: a review of the experimental evidence of an artefact. Chemosphere **77**, 1275–1284 (2009). https://doi.org/10.1016/j.chemosphere.2009.09.019

Miracolo, M.A., Hennigan, C.J., Ranjan, M., Nguyen, N.T., Gordon, T.D., Lipsky, E.M., Presto, A.A., Donahue, N.M., Robinson, A.L.: Secondary aerosol formation from photochemical aging of aircraft exhaust in a smog chamber. Atmos. Chem. Phys. **11**, 4135–4147 (2011). https://doi.org/10.5194/acp-11-4135-2011

Nakao, S., Clark, C., Tang, P., Sato, K., Cocker Iii, D.: Secondary organic aerosol formation from phenolic compounds in the absence of NO_x. Atmos. Chem. Phys. **11**, 10649–10660 (2011). https://doi.org/10.5194/acp-11-10649-2011

Nordin, E.Z., Eriksson, A.C., Roldin, P., Nilsson, P.T., Carlsson, J.E., Kajos, M.K., Hellén, H., Wittbom, C., Rissler, J., Löndahl, J., Swietlicki, E., Svenningsson, B., Bohgard, M., Kulmala, M., Hallquist, M., Pagels, J.H.: Secondary organic aerosol formation from idling gasoline passenger vehicle emissions investigated in a smog chamber. Atmos. Chem. Phys. **13**, 6101–6116 (2013). https://doi.org/10.5194/acp-13-6101-2013

Palen, E.J., Allen, D.T., Pandis, S.N., Paulson, S., Seinfeld, J.H., Flagan, R.C.: Fourier transform infrared analysis of aerosol formed in the photooxidation of 1-octene. Atmos. Environ. a. Gen. Top. **27**, 1471–1477 (1993). https://doi.org/10.1016/0960-1686(93)90133-J

Peltier, R.E., Weber, R.J., Sullivan, A.P.: Investigating a liquid based method for online organic carbon detection in atmospheric particles. Aerosol Sci. Technol. **41**, 1117–1127 (2007). https://doi.org/10.1080/02786820701777465

Pereira, K.L., Hamilton, J.F., Rickard, A.R., Bloss, W.J., Alam, M.S., Camredon, M., Ward, M.W., Wyche, K.P., Muñoz, A., Vera, T., Vázquez, M., Borrás, E., Ródenas, M.: Insights into the formation and evolution of individual compounds in the particulate phase during aromatic photooxidation. Environ. Sci. Technol. **49**, 13168–13178 (2015). https://doi.org/10.1021/acs.est.5b03377

Presto, A.A., Nguyen, N.T., Ranjan, M., Reeder, A.J., Lipsky, E.M., Hennigan, C.J., Miracolo, M.A., Riemer, D.D., Robinson, A.L.: Fine particle and organic vapor emissions from staged tests of an in-use aircraft engine. Atmos. Environ. **45**, 3603–3612 (2011). https://doi.org/10.1016/j.atmosenv.2011.03.061

Pui, D.Y.H., Lewis, C.W., Tsai, C.J., Liu, B.Y.H.: A compact coiled denuder for atmospheric sampling. Environ. Sci. Technol. **24**, 307–312 (1990). https://doi.org/10.1021/es00073a003

Puthussery, J.V., Zhang, C., Verma, V.: Development and field testing of an online instrument for measuring the real-time oxidative potential of ambient particulate matter based on dithiothreitol assay. Atmos. Meas. Tech. **11**, 5767–5780 (2018). https://doi.org/10.5194/amt-11-5767-2018

Ras, M., Borrull, F., Marcé, R.: Sampling and preconcentration techniques for determination of volatile organic compounds in air samples. TrAC, Trends Anal. Chem. **28**, 347–361 (2009). https://doi.org/10.1016/j.trac.2008.10.009

Raynor, P.C., Leith, D., Lee, K.W., Mukund, R.: Sampling and analysis using filters. In: Aerosol Measurement: Principles, Techniques, and Applications. Wiley (2011)

Riemer, D.D., Milne, P.J., Farmer, C.T., Zika, R.G.: Determination of terpene and related compounds in semi-urban air by GC-MSD. Chemosphere **28**, 837–850 (1994). https://doi.org/10.1016/0045-6535(94)90235-6

Rumsey, I.C., Cowen, K.A., Walker, J.T., Kelly, T.J., Hanft, E.A., Mishoe, K., Rogers, C., Proost, R., Beachley, G.M., Lear, G., Frelink, T., Otjes, R.P.: An assessment of the performance of the monitor for AeRosols and GAses in ambient air (MARGA): a semi-continuous method for soluble compounds. Atmos. Chem. Phys. **14**, 5639–5658 (2014). https://doi.org/10.5194/acp-14-5639-2014

Sax, M., Zenobi, R., Baltensperger, U., Kalberer, M.: Time resolved infrared spectroscopic analysis of aerosol formed by photo-oxidation of 1,3,5-Trimethylbenzene and α-Pinene. Aerosol. Sci. Technol. **39**, 822–830 (2005). https://doi.org/10.1080/02786820500257859

Schauer, C., Niessner, R., Pöschl, U.: Polycyclic aromatic hydrocarbons in urban air particulate matter: decadal and seasonal trends. Chem. Degradat. Sampling Artif. Environ. Sci. Technol. **37**, 2861–2868 (2003). https://doi.org/10.1021/es034059s

Sierau, B., Stratmann, F., Pelzing, M., Neusüss, C., Hofmann, D., Wilck, M.: A condensation-growth and impaction method for rapid off-line chemical-characterization of organic submicrometer atmospheric aerosol particles. J. Aerosol Sci. **34**, 225–242 (2003). https://doi.org/10.1016/s0021-8502(02)00159-3

Simon, P.K., Dasgupta, P.K.: Continuous automated measurement of the soluble fraction of atmospheric particulate matter. Anal. Chem. **67**, 71–78 (1995). https://doi.org/10.1021/ac00097a012

Sin, D., Wong, Y.-C., Sham, W.-C., Wang, D.: Development of an analytical technique and stability evaluation of 143 C-3-C-12 volatile organic compounds in Summa((R)) canisters by gas chromatography-mass spectrometry. Analyst **126**, 310–321 (2001). https://doi.org/10.1039/b008746g

Sirju, A.-P., Shepson, P.B.: Laboratory and field investigation of the DNPH cartridge technique for the measurement of atmospheric carbonyl compounds. Environ. Sci. Technol. **29**, 384–392 (1995). https://doi.org/10.1021/es00002a014

Slanina, J., ten Brink, H., Otjes, R., Even, A., Jongejan, P., Khlystov, A., Waijers-Ijpelaan, A., Hu, M., Lu, Y.: The continuous analysis of nitrate and ammonium in aerosols by the steam jet aerosol collector (SJAC): extension and validation of the methodology. Atmos. Environ. **35**, 2319–2330 (2001). https://doi.org/10.1016/s1352-2310(00)00556-2

Sorooshian, A., Brechtel, F.J., Ma, Y., Weber, R.J., Corless, A., Flagan, R.C., Seinfeld, J.H.: Modeling and characterization of a particle-into-liquid sampler (PILS). Aerosol. Sci. Technol. **40**, 396–409 (2006). https://doi.org/10.1080/02786820600632282

Spicer, C.W., Holdren, M.W., Riggin, R.M., Lyon, T.F.: Chemical composition and photochemical reactivity of exhaust from aircraft turbine engines. Ann. Geophys. **12**, 944–955 (1994). https://doi.org/10.1007/s00585-994-0944-0

Srivastava, A., Devotta, S.: Ind0aces in Mumbai, India in terms of volatile organic compounds. Environ. Monit. Assess. **133**, 127–138 (2007). https://doi.org/10.1007/s10661-006-9566-1

Stieger, B., Spindler, G., van Pinxteren, D., Grüner, A., Wallasch, M., Herrmann, H.: Development of an online-coupled MARGA upgrade for the 2 h interval quantification of low-molecular-weight organic acids in the gas and particle phases. Atmos. Measure. Tech. **12**, 281–298 (2019). https://doi.org/10.5194/amt-12-281-2019

Subramanian, R., Khlystov, A.Y., Cabada, J.C., Robinson, A.L.: Positive and negative artifacts in particulate organic carbon measurements with denuded and undenuded sampler configurations special issue of aerosol science and technology on findings from the fine particulate matter supersites program. Aerosol. Sci. Technol. **38**, 27–48 (2004). https://doi.org/10.1080/027868203 90229354

Swartz, E., Stockburger, L., Vallero, D.: Polycyclic aromatic hydrocarbons and other semivolatile organic compounds collected in New York City in response to the events of 9/11. Environ. Sci. Technol. **37**, 3537–3546 (2003). https://doi.org/10.1021/es0303561

Tejada, S.B.: Evaluation of silica gel cartridges coated in situ with acidified 2, 4-dinitrophenylhydrazine for sampling aldehydes and ketones in air. Int. J. Environ. Anal. Chem. **26**, 167–185 (1986). https://doi.org/10.1080/03067318608077112

Temime, B., Healy, R.M., Wenger, J.C.: A denuder-filter sampling technique for the detection of gas and particle phase carbonyl compounds. Environ. Sci. Technol. **41**, 6514–6520 (2007). https://doi.org/10.1021/es070802v

Tsai, C.-J., Perng, S.-N.: Artifacts of ionic species for hi-vol PM10 and PM10 dichotomous samplers. Atmos. Environ. **32**, 1605–1613 (1998). https://doi.org/10.1016/S1352-2310(97)00387-7

Tsai, C.-J., Perng, S.-B., Chiou, S.-F.: Use of two different acidic aerosol samplers to measure acidic aerosols in Hsinchu, Taiwan. J. Air Waste Manag. Assoc. **50**, 2120–2128 (2000). https://doi.org/10.1080/10473289.2000.10464237

Tsai, C.-J., Huang, C.-H., Wang, S.-H., Shih, T.-S.: Design and testing of a personal porous-metal denuder. Aerosol Sci. Technol. **35**, 611–616 (2001). https://doi.org/10.1080/02786820117809

Tsai, C.-J., Chang, C.-T., Shih, B.-H., Aggarwal, S.G., Li, S.-N., Chein, H.M., Shih, T.-S.: The effect of environmental conditions and electrical charge on the weighing accuracy of different filter materials. Sci. Total Environ. **293**, 201–206 (2002). https://doi.org/10.1016/s0048-9697(02)000 15-3

Turpin, B.J., Huntzicker, J.J., Hering, S.V.: Investigation of organic aerosol sampling artifacts in the los angeles basin. Atmos. Environ. **28**, 3061–3071 (1994). https://doi.org/10.1016/1352-231 0(94)00133-6

Turpin, B., Saxena, P., Andrews, E.: Measuring and simulating particulate organics in the atmosphere: problems and prospects. Atmos. Environ. **34**, 2983–3013 (2000). https://doi.org/10.1016/s1352-2310(99)00501-4

US EPA. Compendium of methods for the determination of toxic organic compounds in ambient air (2022). https://www.epa.gov/amtic/compendium-methods-determination-toxic-organic-compounds-ambient-air

Wang, D.K.W., Austin, C.C.: Determination of complex mixtures of volatile organic compounds in ambient air: an overview. Anal. Bioanal. Chem. **386**, 1089–1098 (2006). https://doi.org/10.1007/s00216-006-0475-5

Wang, B., Ho, S.S.H., Ho, K.F., Huang, Y., Chan, C.S., Feng, N., Ip, S.H.S.: An environmental chamber study of the characteristics of air pollutants released from environmental tobacco smoke. Aerosol. Air Qual. Res. **12**, 1269–1281 (2012). https://doi.org/10.4209/aaqr.2011.11.0221

Warnke, J., Bandur, R., Hoffmann, T.: Capillary-HPLC-ESI-MS/MS method for the determination of acidic products from the oxidation of monoterpenes in atmospheric aerosol samples. Anal. Bioanal. Chem. **385**, 34–45 (2006). https://doi.org/10.1007/s00216-006-0340-6

Watson, T.B.: Particle-into-Liquid Sampler (PILS) Instrument Handbook, United States (2016). https://www.osti.gov/servlets/purl/1251405

Weber, R.J., Orsini, D., Daun, Y., Lee, Y.N., Klotz, P.J., Brechtel, F.: A particle-into-liquid collector for rapid measurement of aerosol bulk chemical composition. Aerosol. Sci. Technol. **35**, 718–727 (2001). https://doi.org/10.1080/02786820152546761

Winberry, W.T., Forehand, L., Murphy, N.T., Ceroli, A., Phinney, B.: Compendium of methods for the determination of air pollutants in indoor air; Engineering-Science, Inc., Cary, NC (USA)PB-90-200288/XAB United States NTIS, PC A99/MF E06 GRA English, Medium: X; Size, 845 p (1990)

Woolfenden, E.: Monitoring VOCs in air using sorbent tubes followed by thermal desorption-capillary GC analysis: summary of data and practical guidelines. J. Air Waste Manag. Assoc. **47**, 20–36 (1997). https://doi.org/10.1080/10473289.1997.10464411

Woolfenden, E.: Sorbent-based sampling methods for volatile and semi-volatile organic compounds in air. Part 1: Sorbent-based air monitoring options. J. Chromatgr. A **1217**, 2674–2684 (2010a). https://doi.org/10.1016/j.chroma.2009.12.042

Woolfenden, E.: Sorbent-based sampling methods for volatile and semi-volatile organic compounds in air. Part 2. Sorbent selection and other aspects of optimizing air monitoring methods. J. Chromatgr. A **1217**, 2685–2694 (2010b). https://doi.org/10.1016/j.chroma.2010.01.015

Wragg, F.P.H., Fuller, S.J., Freshwater, R., Green, D.C., Kelly, F.J., Kalberer, M.: An automated online instrument to quantify aerosol-bound reactive oxygen species (ROS) for ambient measurement and health-relevant aerosol studies. Atmos. Meas. Tech. **9**, 4891–4900 (2016). https://doi.org/10.5194/amt-9-4891-2016

Yao, X., Fang, M., Chan, C.: Size distributions and formation of dicarboxylic acids in atmospheric particles. Atmos. Environ. **36**, 2099–2107 (2002). https://doi.org/10.1016/s1352-2310(02)00230-3

Zhang, X., McMurry, P.H.: Theoretical analysis of evaporative losses of adsorbed or absorbed species during atmospheric aerosol sampling. Environ. Sci. Technol. **25**, 456–459 (1991). https://doi.org/10.1021/es00015a012

Zhang, X., Dalleska, N.F., Huang, D.D., Bates, K.H., Sorooshian, A., Flagan, R.C., Seinfeld, J.H.: Time-resolved molecular characterization of organic aerosols by PILS + UPLC/ESI-Q-TOFMS. Atmos. Environ. **130**, 180–189 (2016). https://doi.org/10.1016/j.atmosenv.2015.08.049

Chapter 7
Analysis of Chamber Data

Paul Seakins, Arnaud Allanic, Adla Jammoul, Albelwahid Mellouki,
Amalia Muñoz, Andrew R. Rickard, Jean-François Doussin,
Jorg Kleffmann, Juha Kangasluoma, Katrianne Lehtipalo, Kerrigan Cain,
Lubna Dada, Markku Kulmala, Mathieu Cazaunau, Mike J. Newland,
Mila Ródenas, Peter Wiesen, Spiro Jorga, Spyros Pandis, and Tuukka Petäjä

Abstract In this chapter, we focus on aspects of analysis of typical simulation
chamber experiments and recommend best practices in term of data analysis of simulation chamber results relevant for both gas phase and particulate phase atmospheric
chemistry. The first two sections look at common gas-phase measurements of relative rates and product yields. The simple yield expressions are extended to account
for product removal. In the next two sections, we examine aspects of particulate
phase chemistry looking firstly at secondary organic aerosol (SOA) yields including
correction for wall losses, and secondly at new particle formation using a variety
of methods. Simulations of VOC oxidation processes are important components of
chamber work and one wants to present methods that lead to fundamental chemistry

P. Seakins
University of Leeds, Leeds, United Kingdom

A. Allanic · A. Mellouki · J.-F. Doussin (✉) · M. Cazaunau
Centre National de la Recherche Scientifique, Paris, France
e-mail: jean-francois.doussin@lisa.ipsl.fr

A. Jammoul
Ministry of Agriculture, Bir Hasan, Lebanon

A. Muñoz · M. Ródenas
Fundación Centro de Estudios Ambientales del Mediterráneo, Valencia, Spain

A. R. Rickard · M. J. Newland
University of York, York, UK

J. Kleffmann · P. Wiesen
Bergische Universität Wuppertal, Wuppertal, Germany

J. Kangasluoma · K. Lehtipalo · L. Dada · M. Kulmala · T. Petäjä
University of Helsinki, Helsinki, Finland

K. Cain
National Aeronautics and Space Administration, Washington, DC, USA

S. Jorga
Carnegie Mellon University, Pittsburgh, USA

S. Pandis
Foundation for Research and Technology Hellas, Heraklion, Greece

© The Author(s) 2023
J.-F. Doussin et al. (eds.), *A Practical Guide to Atmospheric Simulation Chambers*,
https://doi.org/10.1007/978-3-031-22277-1_7

and not to specific aspects of the chamber that the experiment was carried out in. We investigate how one can analyse the results of a simulation experiment on a well-characterized chemical system (ethene oxidation) to determine the chamber-specific corrections. Finally, we look at methods of analysing photocatalysis experiments, some with a particular focus on NO_x reduction by TiO_2-doped surfaces. In such systems, overall reactivity is controlled by both chemical processes and transport. Chambers can provide useful practical information, but care needs to be taken in extrapolating results to other conditions. The wider impact of surfaces on photosmog formation is also considered.

7.1 Introduction

Previous chapters have examined various aspects of chamber characterization, preparation, details on how to introduce reagents (stable species, radicals, particulates) and carry out some concentration measurements. In this chapter, we focus on aspects of analysis of typical experiments.

Sections 7.2 and 7.3 focus on the simplest kinds of gas-phase measurements looking at how one conducts a relative rate experiment to determine rate coefficients (Sect. 7.2) and on making gas-phase yield measurements (Sect. 7.3). Yield measurements can be complex if the target product also reacts on a similar timescale as discussed in Sect. 7.3.3.

Sections 7.4 and 7.5 examine aspects of particulate phase chemistry. Section 7.4 focuses on secondary organic aerosol (SOA) yields including correction for wall loss for both SMPS and AMS measurements. Section 7.5 examines new particle formation using a variety of methods.

Simulations of VOC oxidation processes are an important component of chamber work and one wants to present results that reflect the fundamental chemistry and not specific aspects of the chamber that the experiment was carried out in. Section 7.6 addresses how one can analyse the results of a simulation experiment on a well-characterized chemical system (ethene oxidation) to determine the chamber-specific corrections.

Finally, in Sect. 7.7, we examine some studies on photocatalysis. Section 7.7.2 presents protocols for studying photocatalysis, with a particular focus on NO_x reduction by TiO_2-doped surfaces, which could be applied to most chambers and additionally considers some more applied applications that benefit from the accessibility of large chambers such as EUPHORE. The wider impact of surfaces on photosmog formation is considered in Sect. 7.7.3 where a discussion on how to incorporate additional reactions into studies on ethene and propene photo-oxidation is presented.

7.2   Relative Rate Measurements in a Chamber

7.2.1   Introduction

Relative Rate (RR) determinations of rate coefficients are a frequent activity in chambers. The measurements may be made to determine a novel rate coefficient or as a check on the radicals or oxidants present in a chamber. For example, in ozonolysis studies, radical scavengers (Malkin et al. 2010) are often introduced to remove OH; plotting the decay of two alkenes and checking that the ratio of alkene removal is consistent with the literature ratio of rate coefficients is a good test that radical scavenging is effective and that the system is behaving as it should.

The RR method is based on the following analysis of the decays of the test substrate, SH and a reference compound, RH, with a known rate coefficient, where X represents the reactive species, e.g. OH, Cl, NO_3 or O_3.

$$X + RH \rightarrow HX + R$$
$$X + SH \rightarrow HX + S$$
$$ln\left(\frac{[SH]_0}{[SH]_t}\right) = \frac{k_{SH}}{k_{RH}}ln\left(\frac{[RH]_0}{[RH]_t}\right)$$

$$(E7.2.1.1)$$

Therefore, a plot of $ln\left(\frac{[SH]_0}{[SH]_t}\right)$ *versus* $ln\left(\frac{[RH]_0}{[RH]_t}\right)$ should yield a straight line plot with gradient $\frac{k_{SH}}{k_{RH}}$ as shown in Fig. 7.1. Because the analysis involves a ratio of concentrations, we do not actually need the absolute concentrations, but rather something that is proportional to concentration such as GC area or FTIR peak height.

RR measurements are subject to errors, as discussed below, but usefully these errors are different from those involved in a real-time flash photolysis or discharge

Fig. 7.1 Example of a relative rate plot for the reaction of OH with glycolaldehyde using diethyl ether as a reference compound (Hutchinson, M. MSc University of Leeds 2022)

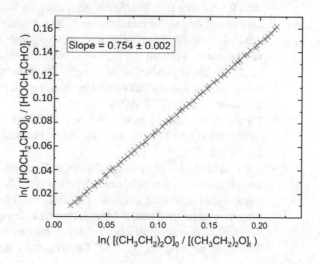

flow experiment (Seakins 2007) and therefore one can have confidence in the accuracy of a rate coefficient if there is good agreement between RR and real-time measurements.

7.2.2 Procedures

(1) *Choice of Reference Compounds*—The reference compound should have a similar rate coefficient to that predicted (e.g. via structure–activity relationships), (Atkinson 1987) otherwise the ratio measurements will be imprecise if one reagent has hardly reacted while the other has almost disappeared. The reference rate coefficient should be well defined and ideally have been reviewed in an evaluation (e.g. IUPAC Kinetics Evaluation). Databases such as EUROCHAMP and NIST are other sources of reference information if evaluated data are not available. Avoid using rate coefficients that are themselves derived from relative rate measurements. A major source of error in RR measurements is if a reagent is removed by another species, so in a study of OH reacting with a saturated species, where OH is generated from CH_3ONO photolysis (Atkinson et al. 1981; Jenkin et al. 1988) there is a potential for O_3 formation, so using an alkene as the reference compound (hence removal by OH and O_3) would not be a good choice. Use a photolysis database (e.g. Mainz Photolysis Database) to ensure that neither the reference nor substrate is predicted to be lost by photolysis (this should always be checked too). It is good practice to use more than one reference compound.

(2) *Experimental method*

- Introduce RH and SH into the chamber to test for wall-loss rates (Chap. 4). Leave for a reasonable period of time, (certainly much longer than the mixing time) sufficient to ensure an accurate estimate of the wall loss. Checks on wall-loss rates should be done on a regular basis as the conditions of the walls may change and wall-loss rates can vary with temperature and pressure. Turn on the lights to check for substrate photolysis, n.b. this could be due to generation of radicals from the walls. This can be checked by having a substrate that cannot be photolysed, but would be lost by radical chemistry. Turn off the lights.
- Introduce the radical precursor to the mix (Chap. 4) to check for any reactions between RH and SH and the substrate. Obviously, this step is not possible for O_3 reactions.
- Turn on the lights to generate the radical species. Make sufficient measurements to ensure a precise relative rate plot. A typical experiment might involve measurements over one half-life, but the exact time will depend on your measurement method. Measurements with significant reagent consumption will be less accurate because one is measuring small concentrations and there may be a higher potential for secondary chemistry. Further additions

of radical precursors may be required. Once the lamps have been turned off, it is often useful to check the wall-loss rate again and use the average value determined from pre- and post-photolysis measurements.

- Correct the reagent concentrations for wall-loss rate (or photolysis loss if applicable).

(3) *Data Analysis*—plotting your data according to equation (E7.2.1.1) should lead to a straight line of the form $y = mx$. Check that the line does indeed pass through the origin. A non-zero intercept could suggest measurement problems and curvature of the plot could be due to the production of additional radical species or that measurements are being compromised (e.g. the reagent FTIR absorption overlaps with a product peak). Measurements using a range of methods (could simply be using several absorptions in an FTIR spectrum to having completely different techniques, e.g. FTIR and PTR-MS) and different reference compounds can identify problems.

When determining the gradient and intercept of the line it is important to weigh the data correctly, i.e. to use a regression analysis that includes errors in both x and y (Brauers and Finlayson-Pitts 1997).

(D) *Reporting Rate Coefficients*—always ensure that you report the ratio $\frac{k_{SH}}{k_{RH}}$, this is your experimental measurements and needs to be available in the literature so that k_{SH} can be re-calculated if there is a revised recommendation or determination of k_{RH}. The reported error in the gradient is primarily going to be statistical from the regression analysis, but the reported error in the absolute measurement must include error in the reference compound.

7.3   Product Yield Measurements

7.3.1   Introduction

Product yields are a common target in simulation chamber measurements. The product yield is defined as

$$\frac{\text{amount of product, P, produced}}{\text{amount of reagent, R, consumed}} = \frac{\Delta P}{|\Delta R|} = Y \tag{E7.3.1.1}$$

Yields can give specific information about one step in a process or the overall yield of a particular product in a process. An example of the first process can be found in the reaction of OH with n-butanol:

$$OH + CH_3CH_2CH_2CH_2OH \rightarrow H_2O + \text{products}$$

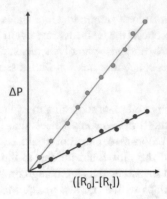

Fig. 7.2 Typical yield plot (formation of products is noted $\Delta[P]$ and concentration of reagent removed ($[R]_0 - [R]_t$)

The abstraction could take place at the α, β, γ, δ or OH sites. Abstraction at the α site leads to the formation of $CH_3CH_2CH_2CHOH$ which reacts with O_2 to give n-butanal. This is the only route to n-butanal formation and therefore n-butanal yield gives the branching ratio (Seakins 2007) for abstraction at the α position. As mentioned, overall yields are also important, for example, while most primary hydrocarbon emissions cannot be detected by satellite measurements, the oxidation products formaldehyde and glyoxal can be detected in the UV. As the yields of formaldehyde and glyoxal are different for different categories of VOC, satellite measurements of the ratio of glyoxal to formaldehyde, RGF, can give information on the primary VOC if the individual RGF is known (Wittrock et al. 2006).

The principles of yield measurements are therefore straightforward, a plot of the concentration (or something proportional to concentration) of product ($\Delta[P]$) versus concentration of reagent removed ($[R]_0 - [R]_t$) should be a straight line with gradient, Y as shown in Fig. 7.2. A good example from Cl-initiated oxidation of n-butanol can be found in Hurley et al. (2009).

However, despite this apparent simplicity, operators should be aware of a number of issues that can affect yield measurements:

- What is the fraction of reagent consumption that occurs via the target channel?
- Accurate measurement of [reagent] and/or [product].
- Consumption of product.

These issues are addressed in the protocol below.

7.3.2 Procedure

(1) Identification of target production pathway—an example system might be looking at yields from a photolysis process. The reagent may be lost via wall

uptake/dilution or via radical loss processes. To deal with wall loss, carry out measurements with just the reagent present and no lights to quantify this non-reactive reagent removal process. The overall concentration of removal will need to be corrected for this process. To avoid complications from removal via radical reactions, then a suitable radical scavenger needs to be present.

If the target process is a radical removal, then carry out tests for photolysis as described in the relative rate protocol. If more than one radical is generated, then it may be difficult to selectively remove radicals, however, it may be that radical concentrations can be measured (or calculated) and if the rate coefficients are known, then the fraction of reagent removed by the target radical can be determined.

(B) Once background checks have been completed the relevant experiment can begin. As with other chamber experiments, selective and specific measurement is required to generate accurate results. As the reaction proceeds, a range of products will be produced and these can interfere (e.g. peak overlap in FTIR or GC measurements, isobaric peaks for MS measurements). Care should be taken to ensure that the calculated yields are independent of the method of measurement. If you are limited to a single method of analysis, make sure that appropriate checks are carried out to test for interference (e.g. for FTIR, measure at several characteristic absorption frequencies, for GC, vary the column conditions to check for underlying peaks).

Complexities in analysis will be minimized at low reagent conversions. Indeed, some yield plots normalize the x- and y-axes measurements, so that the amount of product is determined as a function of the degree of reagent consumption, with the most accurate values being obtained at low conversion, where secondary reactions are minimized, but sufficient reaction needs to occur so that accurate measurements of product production and reagent consumption can be made.

However, depending on the measurement technique used, it is not always possible to make sufficient measurements at low reagent conversion or indeed the target product for the yield measurement is not a primary, but rather a secondary or tertiary product of the reaction and hence may only be produced after significant reagent conversion.

7.3.3 Analysis with Product Consumption

If the target product is consumed by the radical species, then yield plots will tend to curve downwards (red points in Fig. 7.3) as a function of time and can even turnover. Alternatively, if the target product is produced during secondary reactions, then the yield plot (blue points) may curve upwards.

In the study of iso-butanol oxidation, the iso-butanal concentration is determined primarily by the following reactions:

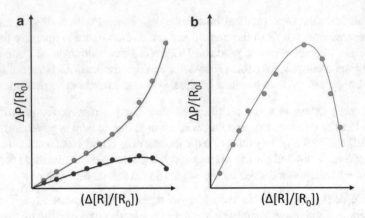

Fig. 7.3 **a** Examples of upward (blue points) and downward (red points) curvature in yield plots at higher reagent consumption. Upward curvature is associated with secondary production, downward curvature occurs with more reactive products. **b** Detailed plot of extensive curvature, so example more reactive iso-butanal produced in the oxidation of iso-butanol, see reaction (R7.3.3.1)

$$Cl + (CH_3)_2CHCH_2OH \rightarrow HCl + (CH_3)_2CHCHOH \qquad (R7.3.3.1a)$$

$$Cl + (CH_3)_2CHCH_2OH \rightarrow HCl + products \qquad (R7.3.3.1)$$

$$Cl + iso\text{-}butanal \rightarrow HCl + products \qquad (R7.3.3.2)$$

Abstraction at the α position (R7.3.3.1a) gives the $(CH_3)_2CHCHOH$ radical which in the presence of sufficient oxygen will react rapidly to give iso-butanal and HO_2, i.e. the rate-determining step in butanal formation is (R7.3.3.1). Under these conditions, it can be shown that

$$\frac{[iso\text{-}butanal]_t}{[iso\text{-}butanol]_0} = \frac{\alpha}{1 - \frac{k_{7.3.3-2}}{k_{7.3.3-1}}}(1-x)[(1-x)^{k_{7.3.3-2}/k_{7.3.3-1}} - 1] \qquad (E7.3.3.1)$$

where

$$x = 1 - \frac{[iso\text{-}butanol]_t}{[iso\text{-}butanol]_0} \qquad (E7.3.3.2)$$

and

$$\alpha = \frac{k_{7.3.3-1a}}{k_{7.3.3-1}} \qquad (E7.3.3.3)$$

In this case, α is essentially the yield of the reaction. Full details on the derivation can be found in the appendix of Meagher et al. (1997) and the yield plot will look similar to that shown in Fig. 7.3.

For more complex situations, it may be necessary to perform a numerical simulation to determine branching ratios or yields in a key reaction. In all circumstances, it is important to measure as many reagents and intermediates as possible as this will reduce the statistical errors in the returned parameters and reduces the chances of systematic errors influencing the results. An example is the study on the branching ratio in the reaction of acetyl peroxy radicals with HO_2 (Winiberg et al. 2016):

$$CH_3C(O)O_2 + HO_2 \rightarrow CH_3C(O)OH + O_3$$

$$CH_3C(O)O_2 + HO_2 \rightarrow CH_3C(O)OOH + O_2$$

$$CH_3C(O)O_2 + HO_2 \rightarrow CH_3 + CO_2 + OH$$

where numerical modelling was used to extract the primary OH yield from the target reaction. A variety of numerical integration packages (e.g. AtChem (see Sect. 7.6.2) or Kintecus) can be used for the numerical fitting.

7.4 Estimating Secondary Organic Aerosol Yields

7.4.1 Introduction

Chemical transport models (CTMs) usually rely on fits to experimentally determine secondary organic aerosol (SOA) yields to model SOA formation in the atmosphere. The SOA mass yield, Y, is defined as the fraction of a volatile organic compound (VOC) that is converted to SOA:

$$Y = \frac{C_{SOA}}{\Delta VOC} \tag{E7.4.1.1}$$

where C_{SOA} is the SOA mass concentration produced and ΔVOC is the amount of VOC reacted, both in $\mu g\ m^{-3}$.

There are two approaches for calculating these yields. The first relies on the concentrations measured in the end of the experiment and the corresponding yield is characterized as "final". This approach results in one measurement per experiment, but it has the advantage that it avoids issues related to the dynamics of the system (e.g. delays in the formation of SOA) given that the system has enough time to equilibrate. The second approach estimates the corresponding yield as a function of time by dividing the corresponding concentrations at a given point. This "dynamic" yield approach provides a range of yield measurement from a single experiment, but may be quite sensitive to the dynamics of the system. For example, the SOA concentration may keep increasing after all the initial VOC has reacted resulting in multiple yield values for the same ΔVOC. Comparison of the results of the two approaches can help ensure that the estimated dynamic yields are not influenced by time delays in the SOA formation processes. If this is the case in the system, then one should rely only on the final yields for the required SOA yield parameterizations.

The measurement of ΔVOC is straightforward in all cases in which the VOC concentration can be accurately measured. As a result, the accuracy of the measured yield mainly depends on the accuracy of the measurement for total formed SOA mass concentrations. However, the SOA mass concentrations in a Teflon chamber are influenced significantly by particle wall losses and corrections are needed. In experiments in which the measured SOA concentrations have not been corrected for wall losses, the corresponding SOA yields have been underestimated. The rest of this section focuses on methods that can be applied to correct SOA chamber experiments for particle wall-losses.

7.4.2 Particle Wall-Loss Correction Procedure

The procedure outlined below corrects for particle wall losses using data collected in a chamber using both a scanning mobility particle sizer (SMPS) and an aerosol mass spectrometer (AMS). This procedure assumes that a coagulation-corrected particle wall-loss constant as a function of particle size has already been calculated according to the method described in Sect. 2.5. However, in order to correct an experiment for particle wall-losses, the particle wall-loss profile must be applicable to the specific experiment (i.e. the profile was measured before/after the experiment, not changing with time during the experiment, etc.). It is recommended that for every SOA yield experiment, a new particle wall-loss profile is generated to account for small perturbations that can occur from daily chamber maintenance (Wang et al. 2018).

Correction of SMPS measurements for particle wall-losses

The correction process includes the following steps:

(1) Acquisition of the coagulation-corrected particle wall-loss profile as a function of particle size.
(2) Correction of the number distribution and of the total number concentration at each time. The corrected particle number concentration at size bin i and time t, $N_i^{tot}(t)$, can be calculated by

$$N_i^{tot}(t) = N_i^{sus}(t) + k_i \int_0^t N_i^{sus}(t)\mathrm{d}t \qquad (E7.4.2.2)$$

where $N_i^{sus}(t)$ is the suspended aerosol number concentration (m^{-3}) of size bin i and time t as measured by the SMPS and k_i is the coagulation-corrected particle wall-loss constant for size bin i. $N_i^{sus}(t)$ includes SOA and seed (if applicable) particles. Once $N_i^{tot}(t)$ is known, the total number concentration at time t, $N^{tot}(t)$, can be calculated by summing the number concentrations at all size bins:

$$N^{tot}(t) = \sum_i N_i^{tot}(t) \qquad \text{(E7.4.2.3)}$$

(3) Calculation of the corrected volume distribution and total volume concentration at each time. The corrected volume concentration at the same size bin and time, $V_i^{tot}(t)$, assuming spherical particles, can be determined by

$$V_i^{tot}(t) = \frac{\pi D_{p,i}^3}{6} N_i^{tot}(t) \qquad \text{(E7.4.2.4)}$$

where $D_{p,i}$ is the particle diameter (m) in size bin i. Similar to the total number concentration, the corrected total volume concentration at time t, $V^{tot}(t)$, can be calculated by

$$V^{tot}(t) = \sum_i V_i^{tot}(t) \qquad \text{(E7.4.2.5)}$$

(4) Calculation of the corrected total mass concentration at each time. If there are no seeds, the mass ($\mu\text{g m}^{-3}$):

$$C_{SOA}(t) = V^{tot}(t)\rho_{SOA} \qquad \text{(E7.4.2.6)}$$

where ρ_{SOA} is the density of the SOA (in $\mu\text{g m}^{-3}$). If there are seeds, the corrected SOA mass concentration can be calculated by

$$C_{SOA}(t) = \left(V^{tot}(t) - V_s\right)\rho_{SOA} \qquad \text{(E7.4.2.7)}$$

where V_s is the corrected seed volume concentration right before SOA formation. In seeded SOA experiments, V_s should be constant after correction for particle wall-losses (Figs. 7.4 and 7.5).

Correction of AMS measurements for particle wall-losses

The correction of the AMS measurements has similarities but also some important differences from the SMPS corrections. More specifically:

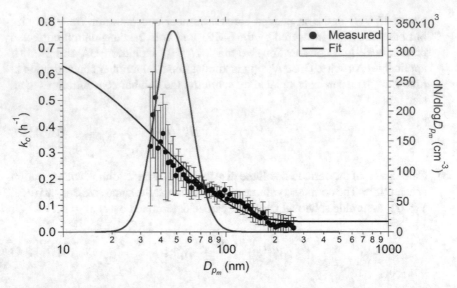

Fig. 7.4 The SMPS derived coagulation-corrected particle wall-loss constants as a function of mobility diameter (black circles, left axis) and an average number distribution measured by the SMPS after 1 h of reaction without being corrected for particle wall losses (red, right axis). The error bars represent the uncertainty of the measured wall-loss constants. The black line is the fit to the measured wall-loss constants extended to encompass the whole diameter range of the SMPS

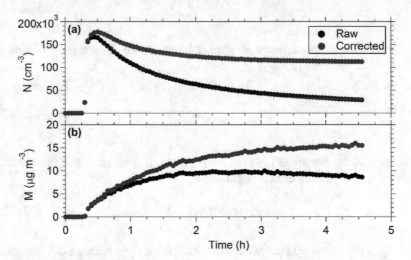

Fig. 7.5 The raw (black circles) and particle wall-loss corrected (red circles) **a** total number and **b** total mass concentrations measured by the SMPS over the course of this experiment. The mass concentrations were determined with a calculated density of 1.23 g cm^{-3}

(1) Conversion of the vacuum aerodynamic diameter measured by the AMS to a mobility diameter. Since the coagulation-corrected particle wall-loss rate constants were measured with an SMPS, the corresponding particle size distribution is based on the electrical mobility diameter, D_{p_m}. However, the AMS measures particle mass distributions based on the vacuum aerodynamic diameter, $D_{p_{va}}$. Therefore, assuming spherical particles, the vacuum aerodynamic diameters from the AMS can be converted to their equivalent mobility diameters using the density of the particles, ρ_p:

$$D_{p_m} = \frac{D_{p_{va}}}{\rho_p} \qquad (E7.4.2.8)$$

The density can be calculated using the AMS size-resolved composition and the corresponding densities.

(B) Calculation of the AMS-specific wall-loss rate constants combining the values measured as a function of the electrical mobility diameter and then converting these values to the corresponding vacuum aerodynamic diameters using Eq. (E7.4.2.8).

(C) Correction of AMS results. The AMS size distributions are split into n size bins. Using the wall-loss constants as a function of the AMS mobility diameters, and the collection efficiency (CE)-corrected AMS mass distributions, the particle wall-loss corrected mass distributions at size bin i and time t, $OA_i^{tot}(t)$, can be calculated by

$$OA_i^{tot}(t) = OA_i^m(t) + k_i \int_0^t OA_i^m(t)dt \qquad (E7.4.2.9)$$

where $OA_i^m(t)$ is the measured mass concentration at each AMS size bin i and time t after correction for the CE and k_i is the coagulation-corrected particle wall-loss constant for size bin i. Once $OA_i^{tot}(t)$ is known, the corrected total mass concentration at time t, $OA^{tot}(t)$, can be calculated by summing the mass concentrations at all size bins:

$$OA^{tot}(t) = \sum_i OA_i^{tot}(t) \qquad (E7.4.2.10)$$

If there are seeds, the corrected SOA mass concentration at time t, $SOA(t)$, can be calculated by

$$SOA(t) = OA^{tot}(t) - M_s \qquad (E7.4.2.11)$$

where M_s is the corrected total seed mass concentration right before SOA formation. Again, after correction, M_s should be a constant. If there are no seeds, Eq. (E7.4.2.11) provides the corrected SOA mass concentration.

The above process can be simplified if the determined wall-loss rate constant is approximately constant in the range covered by the AMS mass size distribution. In this case, an average wall-loss constant, k, can be chosen and the corrected total SOA mass concentration at time t, $SOA(t)$, can be calculated using the expression from (Pathak et al. 2007):

$$SOA(t) = OA^m(t) + k \int_0^t OA^m(t)dt - M_s \qquad (E7.4.2.12)$$

where $OA^m(t)$ is the measured AMS organic aerosol (OA) mass concentration after correction for the CE and M_s is the corrected total seed mass concentration right before SOA formation (Figs. 7.6 and 7.7).

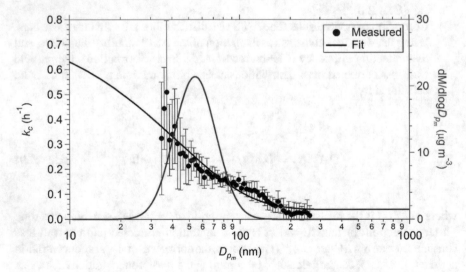

Fig. 7.6 The SMPS derived coagulation-corrected particle wall-loss constants as a function of mobility diameter (black circles, left axis) and an average mass distribution measured by the AMS after 1 h of reaction without being corrected for particle wall losses (red, right axis). The AMS vacuum aerodynamic diameters have been converted to mobility diameters with a calculated density of 1.23 g cm^{-3} and the distribution has been corrected with a calculated CE of 0.55

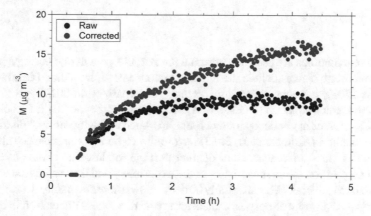

Fig. 7.7 The raw (black circles) and particle wall-loss corrected (red circles) total mass concentration measured by the AMS over the course of this experiment. The measurements have been corrected with a calculated CE of 0.55

7.5   New Particle Formation

New particle formation is a secondary particle formation process by which low volatility vapours cluster and form particles under suitable conditions in the absence of any seed particles. In order to define the intensity of the new particle formation process, the rate at which particles are formed per volume per time can be derived by accounting for the change in particle concentration as a function of time while considering the particle losses, namely, the formation rate. Another measure of the strength of a specific new particle formation event or happening, the growth rate is calculated, which is a measure of how fast particles grow per unit time. In the next section, we provide methods on how to derive the particle formation and growth rates from new particle formation.

In the following sections, we explain in detail how to calculate the variables characterizing the new particle formation (NPF) process, i.e. the particle formation rate (J) at a certain size (dp) and the particle growth rate (GR), further details can be found in Dada et al. (2020). We start by describing different methods how to determine the particle growth (Sect. 7.5.1) and formation rates (Sect. 7.5.2), followed by the relevant processes for chamber experiments, which are needed to determine particle formation rates (Sect. 7.5.2) and finally how to estimate the error in the calculations (Sect. 7.5.5).

7.5.1   Determination of Particle Growth Rates (GR)

The particle growth rate (GR) is defined as the change of the diameter, d_p, as a function of time representing the growing mode:

$$GR = \frac{dd_p}{dt} \tag{E7.5.1.1}$$

Different methods are used to determine the particle growth rate during a particle formation event. These include the maximum concentration method (Lehtinen and Kulmala 2003), the appearance time method (Lehtipalo et al. 2014) and different general dynamics equation (GDE)-based methods (Kuang et al. 2012; Pichelstorfer et al. 2018). Other methods reported in literature, such as the log-normal distribution function method (Kulmala et al. 2012), are found to be incompatible for chamber experiments, due to the absence of distinct particle modes. The choice of the *GR* method depends on the characteristics of the experiment and the available size distribution data. In general, *GR*s can usually be determined more accurately from chamber experiments than from atmospheric measurements due to less fluctuation in the data as well as more accurate particle size distribution measurements. However, several studies compared the different growth rate methods using measurement and simulation data, and found a reasonable agreement within the error bars (Pichelstorfer et al. 2018; Yli-Juuti et al. 2011; Leppa et al. 2011; Li and McMurry 2018). Estimating uncertainties in *GR*s is explained in Sect. 7.5. It is worth mentioning here that *GR* is usually size dependent, and therefore it is useful to calculate the *GR* for several different size ranges rather than one growth rate for the individual particle formation event.

Maximum concentration method

Determine the times, $t_{max,i}$, when the concentration in each size bin, i, of mean diameters of the size bins, $d_{p,mean,i}$, reaches the maximum. See Fig. 7.8 for an example of applying this method to chamber experiment data. To obtain the *GR* using the maximum concentration method:

- Fit a Gaussian function to the time series of size classified particle concentration to obtain $t_{max,i}$ as the time of maximum concentration per size bin of mean diameter ($d_{p,\,mean,\,i}$).
- Plot the mean diameters, $d_{p,mean,i}$, as a function of the maximum times $t_{max,i}$.
- Apply a linear fit to the size range at which the GR is determined.
- Obtain GR as a slope of the linear fit (Fig. 7.8).

Appearance time method

For the 50% appearance time method, determine the times, $t_{app50,\,i}$, when the concentration in each size bin i reaches 50% of the maximum concentration (Leppa et al. 2011; Lehtipalo et al. 2014; Dal Maso et al. 2016). An example of $t_{app50,i}$ determined from the size bin data is shown in Fig. 7.8. To obtain the *GR* using the 50% appearance time method:

- Fit a sigmoidal function to the time series of size classified particle concentration to obtain $t_{app50,i}$ as the time when 50% of maximum concentration per size bin of mean diameter ($d_{p,mean,i}$) is reached.

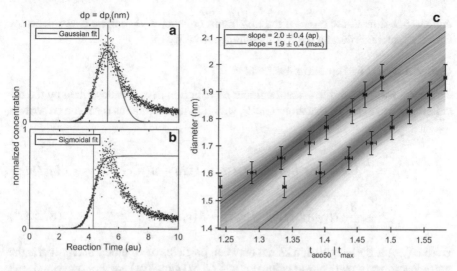

Fig. 7.8 Calculating growth rates from chamber experiments using the maximum concentration method and the appearance time method. In Panel A, the concentration in a size bin is normalized by dividing with the maximum concentration reached during the experiment and then fitted using a Gaussian fit. The same is repeated for all size bins for which a growth rate is calculated. The time corresponding to maximum concentration is then plotted as diameter versus time (tmax) as shown in magenta in Panel C. X-axis uncertainty is the $\pm 1\sigma$ fit uncertainty from the Monte Carlo simulations of 10 000 runs, and Y-axis uncertainty is estimated instrumental sizing uncertainty. GR is obtained as the slope of the linear fit to dp versus tmax data; GR = 1.9 nm/h $\pm$ 0.4. The GR uncertainty is $\pm 1\sigma$ from the Monte Carlo simulations. In Panel B, the concentration in a size bin is normalized by dividing with the maximum concentration reached during the experiment and then fitted using a sigmoidal fit. The same is repeated for all size bins for which a growth rate is calculated. The midpoint of the fits is then plotted as diameter versus time (tapp50) as shown in blue in Panel C. GR is obtained as the slope of the linear fit to dp versus tapp50, GR = 2.0 nm/h $\pm$ 0.3. Note that the maximum concentration method gives the GR at a later time during the experiment, so particle size distribution and gas concentrations in the chamber might have changed. Adapted with permission from Springer Nature: Nature Protocols, Dada et al. copyright 2020. All Rights Reserved

- Plot the mean diameters of the size bins, $d_{p,mean,i}$, as a function of the appearance times $t_{app50,i}$ or $t_{app,i}$.
- Apply a linear fit to the size range at which the GR is determined.
- Obtain GR as a slope of the linear fit (Fig. 7.8).

Note that the GR might change with size, especially during the beginning of the growth process (Tröstl et al. 2016), in this case, using a linear fit is a good assumption only in a narrow size range. It is also possible to determine $t_{app50,i}$ and $t_{app,i}$ from the total concentration measured with a CPC (Riccobono et al. 2012), instead of using the concentration in a certain size bin. Lehtipalo et al. (2014) compared different methods to determine appearance times and concluded that the most robust method is to either determine $t_{app50,i}$ from size bin data or $t_{app,i}$ from total concentration data. Instead of determining the appearance time at 50% of the maximum concentration $t_{app50,i}$, the appearance time at the onset of the maximum concentration can be determined by, for example, determining the 5% appearance time $t_{app5,i}$.

General Dynamic Equation methods

The time-evolution of the aerosol number distribution $n(v, t)$ is described by the so-called general dynamic equation (GDE), which in its continuous form can be written as.

$$\frac{\partial n(v, t)}{\partial t}7 = \frac{1}{2}\int_0^v K(v - q, q)n(v - q, t)n(q, t)dq - n(v, t)\int_0^\infty K(v, q)n(q, t)dq$$

$$- \frac{\partial}{\partial v}(I(v)n(v, t)) + Q(v, t) - S(v, t). \tag{E7.5.1.2}$$

Here $K(v,q)$ is the coagulation kernel between particles of volume v and q, $I(v)$ is the particle volume growth rate at volume v, and $Q(v,t)$ and $S(v,t)$ are the source and sink terms for particle with volume v. In a typical chamber experiment, the only source of particles is nucleation and the sink term arises from wall deposition. The time evolutions of $n(v,t)$ and $K(v,q)$ are known from the measurements.

Find the growth rate $I(v,t)$ and source rate $Q(v,t)$ corresponding to the optimal match between the measured data and the solution to the GDE. This can be done by using different approaches, e.g. (Lehtinen et al. 2004; Verheggen et al. 2006) and (Kuang et al. 2012). In practical applications to measurement data, the parts of the GDE needed are always turned into a discrete form, in addition, particle diameter is used instead of particle volume as a primary variable. Indeed, Pichelstorfer et al. (2018) developed a hybrid method in which $GR(d_p,t)$ was estimated by fitting the evolution of regions of the size distribution to measured data, combined with solving the other microphysical processes from the GDE using process rates from theory. None of these methods, however, are suitable to estimate the error in GR (or Q) rigorously.

7.5.2 Particle Formation Rate

The rate of new particle formation, J_{dp}, is associated with the net flux of particles across the lower detection limit (d_p) of the particle counter. The rate of formation of particles (dN/dt) is obtained by integrating the GDE from the instrument detection limit up to infinity:

$$\frac{dN}{dt} = J_{dp} - S_{dil} - S_{wall} - S_{coag} \tag{E7.5.2.1}$$

Equation (E7.5.2.1) accounts for the loss processes of particles once they have crossed the threshold for detection. dN/dt (preferably measured close to 1.5 nm) can readily be calculated from the total particle number concentration measured with a PSM or other CPC. The detection threshold d_p will be instrument dependent and depends on the cut-off size of the instrument, which is assumed to be a step function. A simple rearrangement leads to Kulmala et al. (2012).

$$J_{dp} = \frac{dN}{dt} + S_{dil} + S_{wall} + S_{coag} [cm^{-3}s^{-1}] \tag{E7.5.2.2}$$

where dN/dt is the time-derivative of the total particle concentration and S_{dil}, S_{wall} and S_{coag} are the loss rate of particles, described in detail in Sect. 7.5.3.

Figure 7.9 shows data from a typical chamber experiment. J_{dp} is variable, particularly at the beginning and end of the experiments as conditions (e.g. lamp fluxes, precursor concentrations) are changing rapidly. Representative values should be taken from the region of constant conditions and experiments should be adjusted so that these conditions, demonstrated by "steady" in Fig. 7.9, are maintained for as long as possible.

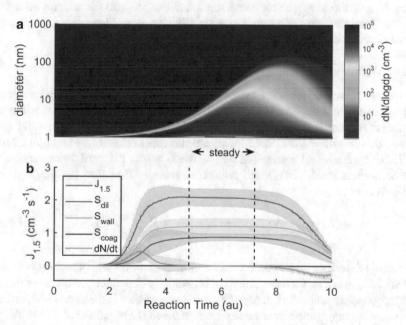

Fig. 7.9 Anticipated results from an NPF experiment performed in a chamber. Panel A shows the simulated time-evolution of particle size distribution during the experiment. Panel B shows the particle formation rate (J1.5) and its different components. Shaded areas correspond to ±1σ uncertainty obtained from the Monte Carlo simulations of 10 000 runs. The time between the dashed lines shows the time with the stable formation rate of particles (steady state), for which the average particle formation rate should be calculated. The magnitude of the components and time scales varies depending on the chamber specifications, experimental plan (gas concentrations, etc.) and particle formation and growth rates (affecting the particle size distribution). Adapted with permission from Springer Nature: Nature Protocols, Dada et al. copyright 2020. All Rights Reserved

7.5.3 Determination of Loss Processes

Determination of dilution losses

Dilution losses are to be accounted for in case the chamber is operated in continuous mode during which synthetic clean air is continuously flowing into the chamber and the instruments are continuously sampling from the chamber. This operation mode causes an artificially lower particle concentration in the chamber due to dilution which needs to be corrected, S_{dil}, in Eqs. (E7.5.3.1) and (E7.5.3.2). The dilution loss rate is determined as follows:

$$S_{dil} = N_{>dp} \cdot k_{dil}[cm^{-3}s^{-1}] \qquad (E7.5.3.1)$$

$$\text{with } k_{dil}[s^{-1}] = \frac{Flow_{\text{synthetic air}}}{V_{\text{chamber}}} \qquad (E7.5.3.2)$$

where $N_{>dp}$ is the total particle concentration above the size for which you want to calculate particle formation rate, k_{dil} is the dilution rate, $Flow_{\text{synthetic air}}$ is the flow rate of clean air and V_{chamber} is the volume of the chamber.

Determination of wall losses

Diffusional losses of particles to the chamber walls (S_{wall}) are chamber specific (e.g. geometry and materials) and have been discussed earlier (Chap. 2). The rate coefficient for loss is inversely proportional to the mobility diameter in a size range below 100 nm where diffusional losses are the most critical (Seinfeld and Pandis 2012). This means that corrections can be made across the particle size range, see also Schwantes et al. (2017) and references therein. Equation (E7.5.3.3) defines wall-loss rates k:

$$S_{\text{wall}}(T) = \sum_i N_{dpi-dpi+1} \cdot k_{\text{wall}}(d_p, T)[cm^{-3}s^{-1}] \qquad (E7.5.3.3)$$

Here $N(d_p)$ describes the number concentration of particles with a mobility diameter (d_p) while k_{wall} is a factor determined experimentally dependent on chamber mixing, chamber conditions and dark decay of the reference species in the absence of particles. The wall-loss rate coefficient can also be calculated (Lehtipalo et al. 2018; Wagner et al. 2017) theoretically, from the temperature dependence of diffusion coefficient, as $D \sim (T/T_{\text{ref}})^{1.75}$ (Poling et al. 2001) and the wall-loss dependence on diffusion coefficient, $k_{\text{wall}} \sim (D)^{0.5}$. For a particle size less than ~ 100 nm on average (McMurry and Rader 1985), k_{wall} is given by

$$k_{\text{wall}}(d_p, T) = F \cdot \left(\frac{T}{T_{\text{ref}}}\right)^{0.875} \cdot \left(\frac{d_{p,\text{ref}}}{d_p}\right)[s^{-1}]. \qquad (E7.5.3.4)$$

where F is a factor determined experimentally based on chamber mixing and other conditions in the chamber as well as dark decay of the reference species in the absence of particles. The mobility diameter of the reference species, the reference temperature at which the experimental loss rate was determined and the studied chamber temperature are given by $d_{p,ref}$, T_{ref} and T, respectively.

Determine the coagulation sink

The loss rate of formed particles to the background particles available in the chamber is known as the coagulation sink (S_{coag}). The pre-existing particles can either be introduced into the chamber for the purpose of studying polluted environments or can result from the growth of particles formed via nucleation processes. In the latter case, the coagulation sink is often negligible early in the experiment but increases gradually as the particles grow to larger sizes while more particles are formed in the chamber (Fig. 7.8). The coagulation sink is calculated as follows:

$$S_{coag}(d_p) = \int k_{coag}\left(d_p, d'_p\right) n\left(d'_p\right) dd'_p \cong \sum_{d'_p=d_p}^{d'_p=max} k_{coag}\left(d_p, d'_p\right) N_{d'_p} [\text{cm}^{-3}\text{s}^{-1}]$$

$$(E7.5.3.5)$$

where $k_{coag}(d_p, d'_p)$ is the Brownian coagulation coefficient for particles sizes d_p and d'_p. It is usually calculated using the Fuchs interpolation between continuum and free-molecule regimes (Seinfeld and Pandis 2016).

7.5.4 Ion Formation Rate

The ion size distributions can be used to calculate the ion formation rates (Kulmala et al. 2012), which allows for studying the importance of charging in the NPF process. When determining the formation rate of charged particles, additional terms need to be added to Eq. (E7.5.3.3) to account for the loss of ions due to their neutralization via ion–ion recombination (S_{rec}) and the production of ions by charging of neutral particles (S_{att}) (Manninen et al. 2009). Since the calculation of recombination and charging between all size bins is rather complicated, it is suggested that the charged formation rates are calculated from a size bin between diameters dp and upper diameter du. The loss of ions out of the studied size bin due to their growth (S_{growth}) needs to be determined. Dada et al. (2020) describe other methods of evaluating ion formation rates and calculate the charged formation rate for positive and negative ions (superscript + and −, respectively) as

$$J_{dp}^{\pm} = \frac{dN_{dp-du}^{\pm}}{dt} + S_{dil} + S_{wall} + S_{growth} + S_{coag} + S_{rec} - S_{att} [\text{cm}^{-3}\text{s}^{-1}]$$

$$(E7.5.4.1)$$

Here $\frac{dN^{\pm}_{dp-du}}{dt}$ is the time-derivative of the ion concentration in a defined size bin. The loss terms of ions due to dilution (S_{dil}), deposition on chamber walls (S_{wall}) and coagulation (S_{coag}) are calculated as given in Eqs. (E7.5.3.1)–(E7.5.3.5) for ions in a size bin between d_p and d_u instead of calculating them for all the particles larger than a certain threshold size.

Determine the growth out-of-the-bin losses

$$S_{growth} = \frac{N}{(d_u - d_p)} \times GR \qquad (E7.5.4.2)$$

where the growth rate of ions out of the size bin is given by GR, and is determined from the ion size distribution.

Determine ion–ion recombination losses

$$S_{rec} = \alpha N^{\pm}_{dp-du} N^{\mp}_{<dp} \qquad (E7.5.4.3)$$

where the ion–ion recombination coefficient (α) is usually assumed to be constant at 1.6×10^{-6} cm^3 s^{-1} (Bates 1985) although the recombination coefficient can depend on the size of the ions and their chemical composition as well as the temperature and relative humidity in the chamber (Franchin et al. 2015).

Determine the production rate of ions

$$S_{att} = \chi N_{dp-du} N^{\pm}_{<dp} \qquad (E7.5.4.4)$$

Here χ is the ion–aerosol attachment coefficient, which, similar to recombination coefficient, may depend on particle size and environmental conditions. χ is usually assumed to be equal 0.01×10^{-6} cm^3 s^{-1} (Hoppel and Frick 1986).

7.5.5 Estimation of Errors

Determination of the error in the growth rate

Uncertainties on the growth rate when using the appearance time and maximum concentration methods are the result of uncertainty in the particle diameter measured by the particle counter and the uncertainty in the fits used for determining the appearance or maximum concentration times.

- In the case that one of either uncertainty is substantially larger than the other, a weighted least square fit on the variable with smaller error as an explanatory variable can be applied. The growth rate and error estimate can then be directly calculated based on the fit.

- In the case that both variables contain a similar magnitude of uncertainty, a fitting method allowing for error on both variables can be used, e.g. total least squares or geometric mean regression. In this case, the error on the *GR* can be determined using a numerical method, e.g. Monte Carlo simulation. Here, the statistical error on the growth rates using the Monte Carlo method can be estimated by reproducing the measurement data 10,000 times with the estimated uncertainties. The GR can be reproduced for all data sets by assuming normally distributed errors including random and systematic errors.
- The *GR* can be reported as the median value and the uncertainty as $\pm$ one standard deviation.

Determine the error in the formation rate

As with the growth rates, the Monte Carlo method for the error estimation can be applied on the formation rate as set out below:

- First, given that the instrumental cut-off diameter affects the detected particle number concentration above a given cut-off diameter, the relation between the cut-off diameter and detected particle concentration can be estimated.
- Assume independent uncertainties for the various parameters: cut-off diameter, N, k_{dil}, k_{wall} and k_{coag}. Assume that these uncertainties are normally distributed and should include random and systematic error.
- The uncertainty on k_{dil} can be estimated from the dilution flow rate on k_{wall} from a decay experiment to which the decay rate can be fitted, and on k_{coag} by assuming 10% error on the size distribution.
- Monte Carlo runs can be constructed so that the first cut off diameter is selected from the cut-off distribution, which determines N, for which the uncertainty is normally distributed and randomly selected.
- Reproduce the formation rate 10,000 times at the plateau value (see Sect. 7.5.2 and Fig. 7.9), from which formation rate is usually determined. The J_{dp} can be reported as the median value with uncertainty as $\pm$ one standard deviation.

7.6 Analysis of Experiments and Application of Chamber-Specific Corrections

7.6.1 Introduction

When running complex experiment in simulation, chamber-specific box modelling can be an important tool for providing detailed chemical insight into chamber experiments.

This type of activities can be modelling exercises to design optimum conditions before specific chamber experiments. It often includes the exploration [oxidant]/[VOC] ratios or the [VOC]/[NO$_x$] ratios sensitivities or simulating précursors reactivity with respect to timescales of experimental systems as well as the formation and loss of target products or intermediates.

Modelling is also extremely valuable to aid interpretation of chamber experiments. It often proceeds by comparisons between temporal profiles of modelled and measured concentrations of not only O$_3$, NO$_x$ and the precursor VOC, but also of a wide range of intermediates and products. These comparison request efficient chamber-specific auxiliary mechanisms and in-turn the use of modelling to interpret data provides meaningful interpretation and evaluation of the auxiliary mechanism.

Finally, chamber evaluation is key to the development and optimization of chemical mechanisms. It is indeed a central process in the knowledge transfer of our chemical understanding with real atmosphere models, linking fundamental laboratory and theoretical chemical understanding through to the chemical mechanisms used in science and policy models.

State-of-science detailed "benchmark" mechanisms are needed for fundamental chemical understanding and the development and optimization of reduced mechanisms, underpinning a range of atmospheric modelling activities. Mechanisms for individual VOCs in benchmark mechanisms are often tested using data from highly instrumented smog chambers. These experiments have not only been used to evaluate the mechanisms, but also to develop them further and to indicate, where necessary, the need for additional experimental measurements. Evaluation studies help to identify gaps and uncertainties in the mechanism where some revision or updating is necessary and to test new experimental data and theory.

A number of mechanisms, used widely in policy models, have been and continue to be developed and optimized on the basis of chamber data (e.g. SAPRC (Carter 2010)), and it is important that the benchmark chemical mechanism is evaluated alongside these, often "reduced" mechanisms, both in relation to the chamber and for atmospheric conditions. An example of such a detailed state-of-science benchmark mechanism is the Master Chemical Mechanism.

The Master Chemical Mechanism (MCM) is a near-explicit chemical mechanism that describes the detailed gas-phase degradation of a series of primary emitted VOCs. It is extensively employed by the atmospheric science community in a wide variety of science and policy applications where chemical detail is required to assess issues related to air quality and climate. The current version, MCMv3.3.1, treats the degradation of 143 emitted VOCs and currently contains about 17,500 elementary reactions of 6,900 closed-shell and radical species, constructed manually based on the mechanism development protocols (Jenkin et al. 1997; Saunders et al. 2003; Jenkin et al. 2015). The MCM is available to all, along with a series of interactive tools to facilitate its usage at the following websites: http://mcm.york.ac.uk, http://mcm.leeds.ac.uk/MCM/ and http://mcm.york.ac.uk.

The MCM has been extensively evaluated, optimized and developed using a wide range of smog chamber experiments. Examples include

- Development of MCMv3.1 aromatic chemistry was evaluated and optimized using an extensive range of photo-oxidation chamber experiments carried out at the highly instrumented EUPHORE chamber (Bloss et al. 2005).
- Chamber-specific box models have been used in the evaluation of the MCMv3.1 1,3,5-trimethylbenzene mechanism, and to investigate potential gas-phase precursors to the secondary organic aerosol (SOA) formed in photo-oxidation experiments carried out at the PSI aerosol chamber (Rickard et al. 2010).
- The performance of the MCMv3.2 β-caryophyllene mechanism, and its ability to form SOA in coupled gas-to-aerosol partitioning model was evaluated using a series of ozonolysis and β-caryophyllene/NO$_x$ chamber experiments carried out at the University of Manchester aerosol chamber (Jenkin et al. 2012).

The following section describes how an MCM chamber-specific box model is constructed and run, how chamber-specific parameters are applied and how they can be used in the analysis of chamber experimental data.

7.6.2 General Approach

At the core of a zero-dimensional box model is the chemical mechanism, which describes the chemical system that is being modelled. At a mathematical level, the chemical mechanism is a system of coupled ordinary differential equations (ODE) which can be solved versus time using an appropriate numerical integrator. A number of open sources, free to use modelling toolkits designed to be used with the MCM are available, include the following:

- **AtChem Online** (Sommariva et al. 2020)—https://atchem.leeds.ac.uk/webapp/.
- **AtChem2** (Sommariva et al. 2020)—https://github.com/AtChem/AtChem2.
- **DSMACC** (Emmerson and Evans 2009)—http://wiki.seas.harvard.edu/geos-chem/index.php/DSMACC_chemical_box_model.
- **Kintecus**—http://www.kintecus.com/.
- **Chemistry with Aerosol Microphysics in Python (PyCHAM) box model** (O'Meara et al. 2021)—https://github.com/simonom/PyCHAM.

The AtChem online website contains tutorial material and a number of examples. Any chemical mechanism can be integrated by these tools, as long as they are in an appropriate format.

In general, the following parameters need to be defined to run a basic chemical box model:

- model variables and constraints and solver parameters;
- environmental variables and constraints;
- photolysis rates;
- initial concentrations of chemical species and lists of output variables.

No two chambers are the same and they exhibit unique and evolving chemical characteristics. As such, chamber-specific "auxiliary mechanisms" are needed in chamber models in order to take into account the background reactivity of the chamber. This allows separation of the chamber-specific chemical processes from the underlying processes that are being studied in experiments. These auxiliary mechanisms are essential to make results from experiments carried out in different chambers comparable and transferable to the atmosphere.

Chamber auxiliary mechanisms mainly take into account chemical processes occurring at the chamber walls, which depend on the specific experimental conditions and recent chemical history (Rickard et al. 2010). Important chemical factors that need to be considered include the following:

- Rapid cycling of reactive NO_{x-y} species (especially with respect to HONO formation) to/from the chamber walls.
- Chamber wall sources of reactive species, which can significantly contribute to the radical budget throughout the experiment.
- Losses of reactive gas/aerosol species to the chamber walls.
- Chamber dilution effects via leaks and/or gas removal by instruments.
- Characterization and ageing of different types of UV lamp systems used to simulate photochemically important areas of the solar actinic spectrum.

Chamber auxiliary mechanisms can be evaluated and optimized in a range of chamber experiments using well-defined and simple photochemical systems (e.g. ethene or propene photo-oxidation (Chap. 2)).

7.6.3   Building a Chamber Box Model

Examples of how to build a chamber box model are given in the MCM/AtChem tutorial available via the MCM website (http://mcm.york.ac.uk/atchem/tutorial_int ro.htt).

The modelling tool chosen to run the chamber model was AtChem Online (Sommariva et al. 2020). The complete MCM v3.3.1 ethene mechanism, along with the appropriate inorganic reaction scheme, was extracted from the MCM website using the "subset mechanism extractor" (http://mcm.york.ac.uk/extract.htt). The model was initiated using the values listed in Examples of how to build a chamber box model are given in the MCM/AtChem tutorial available via the MCM website (http://mcm.york.ac.uk/atchem/tutorial_intro.htt). Below, we will look at an example of building a chamber box model for a simple "high NOx" ethene photo-oxidation experiment carried out at the EUPHORE outdoor environmental chamber in Valencia, Spain on the 01/10/2001 (Zádor et al. 2005). Table 7.1 shows the initial conditions and other important parameters needed to initialisze the model.

Table 7.1 "Clear sky" photolysis rates were calculated according to a set of empirical parameterizations (http://mcm.york.ac.uk/parameters/photolysis_para m.htt), defined for each photolysis reaction as described in Jenkin et al. (1997) and

Table 7.1 Initial concentrations and other parameters needed for initialization of the chamber box model for the 01/10/2001 EUPHORE ethene "high NO$_x$" experiment

	01/10/2001 (high NO$_x$)
Start time (hh:mm)	10:05
End time (hh:mm)	16:00
C$_2$H$_4$ (ppbv)	613
NO (ppbv)	175
NO$_2$ (ppbv)	23
O$_3$ (ppbv)	0.5
HONO (ppbv)	0.5
HCHO (ppbv)	0.5
CO (ppbv)	423.8
H$_2$O (ppbv)	3.8×10^{-5}
T$_{average}$ (°C)	30.6
Dilution rate (s^{-1})	1.64×10^{-5}

Saunders et al. (2003). The model was started at the time the chamber was opened and output every 5 min until the end of the experiment.

Base Model Run

Figure 7.10 shows the model-measurement comparison of the temporal evolution of C$_2$H$_4$ (ethene), NO$_2$, O$_3$ and HCHO. The pink lines show the base model run results, i.e. not constrained to dilution or the chamber auxiliary chemistry. The decay of ethene is substantially under-predicted, while all the product concentration profiles are over-predicted. The ozone peak is over-predicted by about 30% and has probably not yet peaked in the simulation.

Chamber Dilution Effects

The blue lines in Fig. 7.10 show the model run results when dilution of species has been taken into account. Chamber dilution at EUPHORE is characterized by injecting SF$_6$ and measuring its concentration throughout the experiment by FTIR. The measured first-order dilution rate is given in Examples of how to build a chamber box model are given in the MCM/AtChem tutorial available via the MCM website (http://mcm.york.ac.uk/atchem/tutorial_intro.htt). Below, we will look at an example of building a chamber box model for a simple "high NO$_x$" ethene photo-oxidation experiment carried out at the EUPHORE outdoor environmental chamber in Valencia, Spain on the 01/10/2001 (Zádor et al. 2005). Table 7.1 shows the initial conditions and other important parameters needed to initialisze the model.

Table 7.1 as 1.64×10^{-5} s^{-1}. Unsurprisingly, including dilution in the model significantly improves the profiles of all species.

Effects of Chamber Auxiliary Chemistry

A base case auxiliary mechanism was constructed from EUPHORE characterization experiments and literature data adapted to EUPHORE conditions (Zádor et al. 2005;

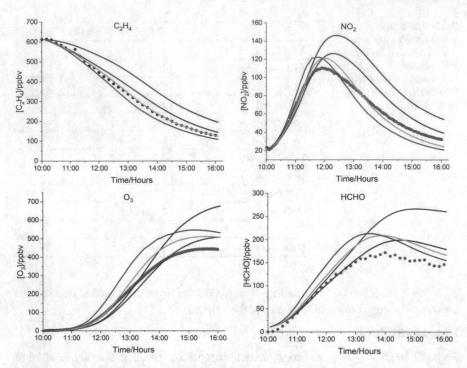

Fig. 7.10 Model-measurement comparisons of the temporal evolution of C_2H_4, NO_2, O_3 and HCHO in the 01/10/2001 EUPHORE ethene "high-NO_x" photo-oxidation experiment. Four model scenarios are shown: red lines = base model run; blue lines = dilution effect included; green lines = dilution + tuned chamber auxiliary chemistry included; magenta lines = dilution + auxiliary chemistry + constrained to measured j(NO_2)–JFAC scaling–included. The black circles are the measured data

Bloss et al. 2005). Discrepancies between the modelled and measured data and a detailed sensitivity analysis were used to derive a tuned auxiliary mechanism which is listed in Table 7.2.

The green lines in Fig. 7.10 show the model run results when dilution and the above chamber-specific auxiliary chemistry are added to the model. The model now gives an excellent prediction of the ethene decay, with the temporal profiles of all

Table 7.2 Parameters from the tuned auxiliary mechanism used to assess the impact of chamber-related processes on the ethene experiments (Bloss et al. 2005)

Process	Tuned rates
NO_2 = HONO	0.7×10^{-5} s^{-1}
NO_2 = wHNO$_3$	1.6×0^{-5} s^{-1}
O_3 = wO$_3$	3.0×10^{-6} s^{-1}
Initial HONO	NO_x dependent

modelled species coming more into line with the measurements. Peak ozone is now only over-predicted by 10%.

Radiation Effects

Radiation effects have been discussed in Chap. 2. All the calculated photolysis processes apply chamber-specific scaling factors (F_x) in order to take into account radiation effects of transmission through the chamber walls, backscatter from the aluminium chamber floor and cloud cover (Bloss et al. 2005; Sommariva et al. 2020). In addition, the photolysis rate of nitrogen dioxide, $j(NO_2)$, is routinely measured in chamber A at EUPHORE and these data are available for the experiment above. Variations in actinic flux from day to day and during the experiment resulting from short temporal-scale variations in cloud cover are accounted for by considering the difference between the measured and clear sky calculated $j(NO_2)$ at any given time during the experiment. This variable scaling factor, *JFAC,* is applied to all calculated photolysis rates along with F_x.

Figure 7.11 shows the temporal profile of the measured $j(NO_2)$ for the 01/10/2001 EUPHORE ethene photo-oxidation experiment, along with the clear sky model calculated parameterized $j(NO_2)$ and the calculated JFAC values (JFAC $= j(NO_2)_{measured}/$ $j(NO_2)_{calculated}$). The magenta lines in Fig. 7.10 show the model run results when dilution, chamber–specific auxiliary chemistry and constraints to the photolysis rate scaling factor JFAC have been added to the model. The timing of most of the profiles has improved further. However, owing to the measured $j(NO_2)$ being generally higher than the calculated $j(NO_2)$, the profiles are all slightly increased (with increased ethene decay) owing to the slight increase in the photo-reactivity of the system.

7.7 Use Simulation Chambers for the Assessment of Photocatalytic Material for Air Treatment

7.7.1 Introduction

Despite considerable progress in the past decades, ambient air pollution and, more specifically, fine particles, nitrogen dioxide and ozone, cause around 400.000 premature deaths each year in the EU (EEA Report 2019).

Photocatalysis has been shown to be a potential process for reducing atmospheric pollutants (Ângelo et al. 2013; Schneider et al. 2014; Boyjoo et al. 2017). Photocatalysis may be used for reducing pollutant levels in outdoors as well as indoors and has been applied mainly for reducing NO_2 concentrations outdoors. More specifically, one of the proposed measures is the photocatalytic degradation of NO_x on titanium oxide (TiO_2) containing surfaces, leading to the formation of adsorbed nitric acid (HNO_3) or nitrate (NO_3^-), which is washed off by rain (Laufs et al. 2010). While photocatalytic nitrate formation has been critically reviewed, photocatalysis could help to improve urban air quality due to a variety beyond the simple reduction in

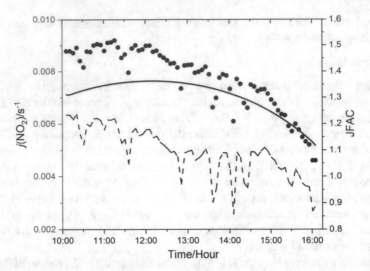

Fig. 7.11 Measured (5-min averages) and calculated j(NO₂) values from the 01/10/2001 EUPHORE ethene "high-NOₓ" photo-oxidation experiment. JFAC scaling factor = j(NO₂) measured/j(NO₂) calculated. Red circles = 5-min average j(NO₂) measured values; blue line = calculated "clear sky" j(NO₂) values using the MCM parameterization (Saunders et al. 2003); Black dotted line = JFAC scaling factor (j(NO₂) measured/j(NO₂) calculated)

NO_x. Firstly, the removal of NO_x reduces direct O_3 production as NO_2 photolysis is reduced and any photocatalytic VOC removal will indirectly reduce O_3 and smog formation. Secondly, while photocatalysis does not reduce the total amount of HNO_3 formation, nitric acid is formed and retained on the surface until washed off and hence will not damage plants or cause respiratory damage. Finally, total nitrate in the rain wash-off can be reduced if treated in wastewater plants.

However, poorly designed photocatalysts can have some negative effects such as the formation of nitrous acid, HONO, photolysis of which can accelerate photochemical smog formation (Laufs et al. 2010; Monge et al. 2010a; Gandolfo et al. 2015) or the production of HCHO or other oxygenated VOCs (Mothes et al. 2016; Toro et al. 2016; Gandolfo et al. 2018). In addition, nitrates need to be regularly removed to maintain efficiency and to prevent photocatalysis of the adsorbed nitrate (Monge et al. 2010a, b). Photocatalytic surfaces at best will only contribute to NO_x reduction; they are not the sole solution and should be considered as part of a wider range of solutions to the issue of poor air quality (Gallus et al. 2015; Kleffmann 2015).

TiO_2 can be found on the market in different formats for environmental purposes, for example, as paints, concrete, pavement stones, granules for asphalt surfaces, roof tiles, window glass, etc. Its effectiveness depends not only on the support (paints, textiles, etc.) but also on the impregnation method (layer, embedded, etc.). Nevertheless, a science-based approach is needed to assess the performance of this process before it is promoted as an effective solution and enters the market.

Atmospheric simulation chambers are well equipped to study the reduction potential of selected photocatalytic surfaces under well-defined atmospheric conditions.

Simulation chamber studies can provide investigators and companies with large-scale assays helping them in developing efficient products and in reducing potentially problematic behaviour as well as providing a basis to encourage local authorities and stakeholders to adopt a more integrated approach to urban air quality management. While atmospheric simulation chambers have many advantages, initial studies can also be carried out in smaller photo-reactors that can characterize the uptake and are useful for screening before considering larger scale measurements (Ifang et al. 2014). These reactors are also the only way to determine the uptake kinetic parameter, i.e. uptake coefficients (γ) for fast photocatalytic reactions (see below). In contrast, in larger smog chambers, fast uptake will be limited by the transport to the active surfaces. However, in smaller flow reactors, secondary chemistry and the impact of photocatalysis on the complex chemistry of the atmosphere, e.g. on summer smog formation, cannot be investigated. Here larger simulation chambers are necessary.

The present section describes experimental approaches using atmospheric simulation chambers for the testing of different photocatalytic materials. Section 7.7.2 presents a protocol for the study of enhanced uptake, exemplified by looking at the removal of NO_x by TiO_2-doped surfaces, along with a number of examples. In Sect. 7.7.3, we look at how surface chemistry can be incorporated into more complex photosmog simulations and finally Sect. 7.7.4 provides recommendations for rigorously using simulation chambers in order to study the photocatalytic activity of material and the effect of their deployment on atmospheric composition.

7.7.2   Photocatalytic Activity Determination Using a Simulation Chamber

Introduction

Simulation chambers can be very useful tools to determine the photocatalytic activity of potential depolluting materials. The principle for the photocatalytic activity measurement can include both NO_x reduction and the production of intermediates such as HONO or oxygenated VOC. One of the assets of this approach, in contrast to more compact testbeds, is the ability to have more realistic conditions and to consider the production of a wider range of compounds. It must nevertheless be kept in mind that atmospheric chambers are not suitable to measure the uptake kinetic parameter for fast photocatalytic processes. Indeed, in a chamber, even if fans are used for efficient mixing, transport to the surface is most of the time the limiting parameter for active samples, at least with $\gamma > 10^{-4}$. This indicative value for γ will depend on mixing efficiency, available reactive surfaces and the volume of the chamber.

In the following sub-sections, we outline an experimental protocol using examples from studies at CNRS-Orléans, considering experimental procedures and data analysis. Finally, we briefly discuss these results highlighting considerations relevant for other studies.

Experimental protocol with an example of glass surfaces

(1) *Sample Preparation*

For the present example study, a photoactive glass was compared to an equivalent area of normal glass. The tested glass consisted of two sets of pieces with different surfaces. Both types of glass are commercially available; the non-treated glass was standard windows glass, while the treated glass was Pilkington™ Activ™ self-cleaning glass. Each test piece consisted of panels of a surface area of 0.39 m^2 (0.88 m × 0.44 m). The preparation of the test samples prior to the experiment consisted of washing with deionized water, and then placing it into the chamber to be flushed with purified air for at least 1 h. In order to ensure the absence of contamination emissions from the materials, air samples were taken prior to the introduction of NO and NO_2. For the present study, both indoor and outdoor atmospheric simulation chambers have been used.

(B) *Chamber descriptions*

(a) *Indoor chamber*

The indoor chamber setup consisted of a 275 L Teflon cube that was used as a static stirred reactor. The experiments were performed at room temperature (25 ± 3 °C) and 760 Torr in dry air (RH < 5%). In the present example, dry conditions were chosen for mechanism investigation purposes. It must be noted that dry conditions are not so relevant for the atmosphere and that photocatalysis is highly dependent on the availability of water molecules adsorbed on the material, which is a function of the relative humidity. It is generally recommended in standard procedures (e.g. ISO 22197-1 2007) to work at 50% RH. The UV exposure unit consisted of an ULTRA-VITALUX 300 W (® OSRAM) lamp used to simulate solar radiation. The test piece was laid flat on the middle of the floor of chamber to be exposed to pollutants. The desired amounts of nitrogen oxide (NO) and nitrogen dioxide (NO_2) were introduced into the chamber via a 20 L min^{-1} air stream. The mixing ratios were measured periodically at regular intervals using a NO_x monitor. During the entire duration of the experiments, a slight airstream 100 mL min^{-1} was added into the chamber in order to compensate the loss from the sampling volume and to maintain a slight overpressure to prevent the outside air from entering the setup.

(b) *Outdoor chamber*

The outdoor chamber was a cube of 1.5 m edge with a volume of 3.4 m^3 made of a 200 μm PTFE film. In addition to the NO_x and O_3 monitors, it was equipped with pressure, temperature and relative humidity sensors. The solar intensity was measured using a $J(NO_2)$ radiometer. A fan positioned inside the chamber gave homogeneous mixing within the chamber in <2 min. The chamber could be covered by a black and opaque cloth that could be rapidly removed. As with the indoor chamber, NO and NO_2, were introduced via a 20 L/min air stream and their mixing ratios were measured continuously and a dilution flow was used to maintain a slight overpressure.

(C) *Experimental procedure*

NO and NO_2 can be introduced into the environmental chambers (indoor and outdoor) in the desired concentration (e.g. to simulate "high" or "low" NO_x conditions) with initial concentrations in the range 43.3–170 and 11.45–50 ppbv, respectively, in this particular example. The system was allowed to stabilize for at least 1 h and the chamber was then exposed to radiation for 4 h. The NO_x concentration–time profiles were measured continuously.

Photocatalytic Efficiency Determination

The photocatalytic activity is studied here exemplarily by measurement of the NO_x loss. However, this loss can be due to combinations of

(i)   wall loss and dilution,
(ii) adsorption on the surface of the sample,
(iii) photolysis by UV light (for NO_2),
(iv) photocatalysis by TiO_2 in the presence of UV light.

Therefore, the measurement of the concentration–time profiles of NO and NO_2 can give information on the TiO_2-material activity providing that the above side effects (i–iii) are considered. Hence, before performing the photocatalytic experiments, blank tests (chamber without material and in the presence of a material without TiO_2) were carried out in order to estimate the loss of NO_x.

The estimation of the catalytic activity of the materials is often represented through various parameters that are all arising from different approaches of various levels of scientific robustness.

(i)   the percentage of NO_x photo-removed ($\%NO_{x(photo-removed)}$),
(ii) the photocatalytic/oxidation rate (PR, $\mu g\ m^{-2}\ s^{-1}$),
(iii) the photocatalytic deposition velocity (v_{photo}),
(iv) the uptake coefficient (γ).

The percentage of NO_x photocatalytically removed is calculated by the following equation:

$$\%NO_{x\ photo-removed} = \left(\frac{[NO_x]_{UV} - [NO_x]_{blank}}{[NO_x]_{UV}} \times 100 \right) \qquad (E7.7.2.1)$$

where $[NO_x]_{UV}$ and $[NO_x]_{blank}$ represent the amount of NO_x (ppb) removed, respectively, during the irradiation of TiO_2 containing sample and that removed during the blank experiment due to side effects.

While sometimes used to compare different material activities under similar conditions and time horizon, using a percentage of reduction is not compatible with kinetic theory. Here zero-order kinetic is applied to a typical first-order photocatalytic reaction at atmospheric relevant pollutant levels. The result is a parameter that can be time dependent in a smog chamber and that is not linearly correlated to the photocatalytic activity (see Ifang et al. 2014).

The photocatalytic/oxidation rate (PR, $\mu g\ m^{-2}\ s^{-1}$) is calculated, taking the sample surface, the chamber volume and the duration of the experiment into consideration. Thus, it provides a more precise estimation of the cleansing capacity of a material than the percentage of photo-removal. However, the PR is directly proportional to the pollutant concentration investigated and can be only applied to the atmosphere, if the PR is normalized to atmospheric conditions. In addition, in this simplified formalism, zero-order conditions are again assumed, for typical first-order photocatalytic reactions. While often used by the industry to advertise the efficiency of depolluting products, this measure is not scientifically robust. Except when the experiments are performed under realistic concentration conditions, it can even be misleading. Indeed, as the experiments are often conducted at much higher NO_x condition than in the real atmosphere (e.g. at 1 ppm NO level recommended by ISO 22197-1 2007), the photolytic oxidation rates are derived often leading to unrealistically high values. It is not recommended to use this formalism unless the NO_x level of the experiment is systematically provided together with the PR values.

The photooxidation rate (PR) is given by the following equation:

$$PR = \left(\frac{V \cdot [NOx]_{TiO_2UV}}{A \cdot t} \right) \tag{E7.7.2.2}$$

where $[NOx]_{TiO_2UV}$ is the concentration of NO_x photocatalytically removed due to the TiO_2 effect ($\mu g\ m^{-3}$), A is the sample surface (m^2), t is the irradiation time (s), and V (m^3) = the volume of the experimental chamber (V = 3.4 or 0.275 m^3).

The deposition velocity was also calculated in order to describe the photocatalytic activity independently, avoiding the influence of the pollution concentration. The photocatalytic velocity (PV) can be approximated by the following equation:

$$PV = \left(\frac{PR}{\frac{[NOx]_{in} + [NOx]_{UV}}{2}} \right) \tag{E7.7.2.3}$$

where PR is the photocatalytic rate ($\mu g\ m^{-2}\ s^{-1}$), $[NO_x]_{in}$ is the initial amount of NO_x ($\mu g\ m^{-3}$) before irradiation and $[NO_x]_{UV}$ is the amount of NO_x ($\mu g\ m^{-3}$) removed during the irradiation of the TiO_2 containing sample. Here again the main issue lies in the kinetic representation of the studied phenomenon. PV expresses itself as first-order kinetic parameter applied to a first-order process but calculated from a zero-order parameter (PR) and this mixed approach cannot be recommended.

The most robust approach is certainly to remain under the first-order kinetic assumption all along the data analysis process as recommended by Ifang et al. (2014). A first-order rate coefficient (k_{rxn}) can be obtained from experimental data only if either (a) there is an absence of secondary chemistry which may be achieved in a fast flow system, or (b) if a rigorous approach is taken to modelling secondary chemistry (e.g. NO_2 photolysis) or processes such as wall loss. In the absence of secondary processes:

$$k_{rxn} = \frac{\ln \frac{[NO_x]_t}{[NO_x]_0}}{t} \qquad (E7.7.2.4)$$

As a first-order rate coefficient, k_{rxn} will be independent of the NO_x concentration and it is recommended to repeat experiments at a range of concentrations to verify this. Of course, k_{rxn} depends on the geometry of the sample and reactor and will scale with the S_{active}/V ratio where S_{active} is the surface area (m^2) of the photocatalytic sample and V is the gas-phase volume (m^3) over the sample. This dependence on reactor configuration means that values of k_{rxn} cannot be directly compared; the dimensionless reactive uptake coefficient (γ) (Finlayson-Pitts and Pitts 2000) however can be compared. (γ) is defined as the ratio of the number of collisions that lead to reaction over all collisions of the gas-phase reactant with a reactive surface and is calculated from Eq. (E7.7.2.5).

$$\gamma = \frac{4 \cdot k_{rxn} \cdot V}{\bar{v} \cdot S_{active}}, \qquad (E7.7.2.5)$$

where $\bar{v}$ is the mean molecular velocity of the reactant (m s^{-1}) defined by kinetic theory:

$$\bar{v} = \sqrt{\frac{8 \cdot R \cdot T}{\pi \cdot M}}, \qquad (E7.7.2.6)$$

in which R is the ideal gas constant (R = 8.314 J mol^{-1} K^{-1}), T is the absolute temperature (K) and M is the molecular mass of the reactant (kg mol^{-1}). When the uptake coefficient is known this can be easily converted into the photocatalytic deposition velocity (v_{surf} in m s^{-1}):

$$v_{surf} = \frac{\gamma \cdot \bar{v}}{4}. \qquad (E7.7.2.7)$$

It has to be highlighted that the photocatalytic deposition velocity is not similar to the deposition velocity, typically used in flux modelling. It represents only the inverse of the surface resistance (r_C) in flux approaches. However, when the resistances for turbulent transport (r_A) and diffusion (r_B) are known, deposition velocities can be easily calculated, from which flux densities (molec. m^{-2} s^{-1}) can be derived in atmospheric models by multiplying with the concentration (molec. m^{-3}).

Examples of Photocatalytic Efficiency Results

(a) *Degradation on self-cleaning window glass*

Typical concentration–time profiles of NO and NO$_2$ during the experiment conducted in the outdoor chamber are presented in Fig. 7.12.

In high NO$_x$ concentration (186–200) ppbV experiments, the loss in 4 h in the presence of a non-treated material under irradiation was (69–75) ppbV and was

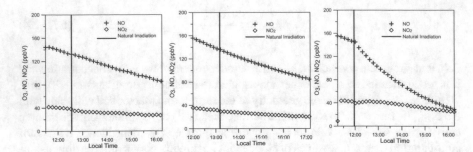

Fig. 7.12 NO and NO$_2$ mixing ratios under natural irradiation at high NO$_x$ concentration in the absence of any surface (left), in the presence of a non-treated glass surface (middle) and finally in the presence of a TiO$_2$-treated glass surface (right)

very similar to that of the loss in the absence of any material (60 ppbV) showing a negligible impact of the non-treated glass surface. The loss with low NO$_x$ in the presence of a non-treated glass surface was found to be 13 ppbV, while that in the presence of TiO$_2$-based material was in the range 41–50 ppbV.

The decay of NO in the absence of any surface was 29% of the initial concentration over 4 h. In the presence of non-treated surface, it was equal to 28–39% showing that the non-treated material had an insignificant effect on the NO$_x$ removal. Therefore, the removal was considered negligible and the experiments in the presence of a non-treated glass material were taken as reference to deduce the TiO$_2$ activity. In all the experiments, the presence of TiO$_2$ showed a significant role in the removal of NO$_x$. In Fig. 7.12, we observe a slight increase in the NO$_2$ which confirms the photocatalytic process of oxidation of NO$_x$ according to the sequence: NO $\rightarrow$ NO$_2$ $\rightarrow$ HNO$_3$ (Laufs et al. 2010).

While being a quite illustrative example in a simulation chamber, such complex experiments can only be evaluated by using model description considering gas-phase photolysis of NO$_2$ (J(NO$_2$)), wall loss, dilution, in addition to the considered photocatalytic chemical mechanism. Through adjustment of the model with the experiment will lead to the first-order rate coefficients (k_{rxn}) for the NO and NO$_2$ reactions on the photocatalytic material, which may be converted into γ (see Eq. E7.7.2.5) by using the S/V ratios of the chambers.

(b) *Test of TiO$_2$ impregnated fabrics in the EUPHORE chamber*

The large volume of the EUPHORE chamber ($\sim$200 m^3) allows for the easy installation of a range of bulky samples as illustrated in Fig. 7.13. For example, 24 m^2 of a TiO$_2$ impregnated fabric was installed in a vertical position and 13 m^2 on the floor, with an S/V ratio of 0.185 m^{-1}. These studies could be carried out over extended time periods (e.g. 36 h), thus allowing a range of solar conditions to be sampled. If necessary, NO$_x$ levels in the chamber could be controlled to simulate a typical diurnal profile with morning and evening rush-hour peaks. As with other chambers, relative humidity can be controlled, but obviously there is less control over temperature and solar radiation.

As an example of studying the effectiveness of pollution reduction by photocatalytic outdoor furniture, a surface of 4.4 m^2 of the photocatalytic material functionalized as furniture was installed, with a surface-to-volume ratio (S_{active}/V) of 0.022 m^{-1}. At such conditions, 50 ppb of NO and 60 ppb of NO$_2$ were introduced into the EUPHORE chamber. Figure 7.14 shows results of the NO$_x$ evolution when both the photocatalytic and the non-photocatalytic materials (blank experiment) were exposed to the solar radiation.

At 120 ppb of NO$_x$ under comparable condition, an initial NO$_x$ reduction of 23.6% in 1 h was found with the photocatalytic materials, while only 7.4% h^{-1} was derived with the non-photocatalytic materials. The quantification of NO$_2$ is more

Fig. 7.13 Photocatalytic materials in the EUPHORE chamber. Left: textiles on structures and on the ground. Right: outdoor furniture

Fig. 7.14 NO$_x$ temporal evolution with non-photocatalytic (pink squares) and photocatalytic material (blue triangles)

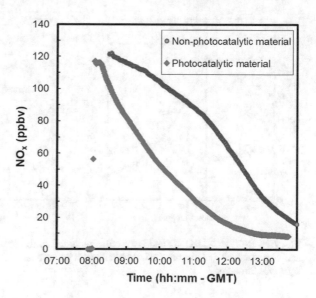

complicated due to secondary reactions by exposing both types of materials to the sunlight. Here NO degradation forms NO_2 and NO_2 photolyse back to NO in the gas phase.

Aware of the limitation of the percentage approach (see above), calculation of the uptake coefficient for NO_x was performed using the region where a first-order kinetic decay could be fitted. This resulted in a γ value of $(5.3 \pm 0.3) \times 10^{-5}$. As a reference, Gandolfo et al. (2015) reported uptake coefficient values of the order of 1.6×10^{-5}, which is lower than the photocatalytic material used here.

(c) *HONO formation on self-cleaning window glass in the CESAM chamber*

The next example considers experiments in the CESAM chamber, where again the NO_x uptake on TiO_2-doped glass surfaces was examined, but with an additional focus on nitrous acid (HONO) detection via FTIR measurements. Samples were prepared and the experiments were carried out using a similar protocol (see *Experimental protocol with an example of glass surfaces*), which included experiments with uncoated glass with the same surface area. The initial NO concentration ranged from 20 to 100 ppb and the relative humidity was varied from 0 to 40% RH. Additional experiments were also carried out in an outdoor Teflon chamber.

Figure 7.15 shows experimental results for the treated and untreated surfaces. After introduction of synthetic air and NO into the CESAM chamber, the concentrations of NO, NO_2, HONO and O_3 were monitored in the dark for 1 h. Then the artificial illumination was turned on and the chemical system was again monitored for 90 min.

In agreement with previous studies, NO uptake on the TiO_2-coated glass was enhanced under irradiation, decreasing with time in both experiments. The NO_2 concentration profile exhibited a maximum under illumination, suggesting that it

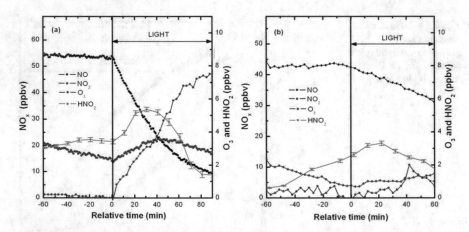

Fig. 7.15 NO, NO_2, HONO and O_3 profiles recorded in the presence of **a**: a TiO_2-coated glass and **b**: a standard glass in the CESAM chamber. The vertical line indicates the moment when the light was turned on

is formed from NO photocatalytic oxidation and then converted into HNO_3 on the surface. At the same time, a significant accumulation of ozone was observed. In agreement with previous studies, HONO production was enhanced under irradiation in the presence of TiO_2. In contrast, when a standard glass was analysed no ozone formation was detected. The NO_2 and NO concentration profiles were similar for the standard glass and the empty outdoor chamber with no evidence of additional photochemical effect. These results indicate that O_3 formation cannot be explained by the gas-phase chemistry (NO_2 photolysis) occurring in the chambers. The differences between the O_3 profiles obtained for the blank experiments and the coated glass suggested that TiO_2 should be involved in the reaction mechanism leading to O_3 formation via heterogeneous reaction. Using a complementary experimental approach (flow tube), a chemical mechanism explaining the formation of ozone has been suggested, see Monge et al. (2010a, b).

7.7.3 Photosmog Studies in the Presence of TiO_2-doped Surfaces

In contrast to the studies described above on the photocatalysis of pure NO_x, TiO_2-doped materials operating in the real world will be exposed also to VOCs in addition to NO_x and therefore photosmog-type experiments are required to examine the real-world performance.

As discussed in depth in Sect. 7.6, development of a chamber-specific auxiliary mechanism is the first step in a photosmog experiment. The experiments described in Sect. 7.7.2 allow for the adjustment of a dedicated auxiliary mechanism (Table 7.3) aimed at describing the effect of photocatalytic materials on the NO_x air chemical system.

Table 7.3 Chemical reactions involved a simplified NO_x chemistry in the presence of TiO_2 containing glass and used for the box modelling described below. Pseudo-first-order rate constants are given for standard glass and TiO_2-doped glass and are only relevant for the CESAM chamber and the available surface of active material used in these experiments

Reactions	Rate constant used for standard glass (s^{-1})	Rate constant used for TiO_2-doped glass (s^{-1})
$NO + h\nu \rightarrow NO_{ads}$	$(1.5–2) \times 10^{-5}$	$(1.5–2) \times 10^{-4}$
$NO_{ads} + h\nu \rightarrow HONO$	$(2–3) \times 10^{-5}$	$(4–6) \times 10^{-5}$
$H_2O \rightarrow H_2O_{ads}$ (fraction adsorb.: 0.1) $H_2O_{ads} + h\nu \rightarrow OH$ (on TiO_2 only)		3×10^{-9}
$NO_{ads} + h\nu \rightarrow NO_{2ads}$ (on TiO_2 only)		$(4–5) \times 10^{-5}$
$NO_{2ads} + h\nu \rightarrow O_3$ (on TiO_2 only)		$(4–5) \times 10^{-4}$

Experiments were performed in the CESAM chamber with the propene–NO$_x$–air system which was then irradiated for 3 h under dry conditions (RH < 1%) following an initial equilibration period of approximately 45 min in the dark. The simulated data were obtained using the standard glass model described in the previous section, combining MCM propene chemistry and the CESAM and standard-glass modules determined for the NO$_x$–air–light system. All kinetic parameters pertaining to NO$_x$ heterogeneous chemistry were kept unchanged.

In spite of a very good ability of the initial model to capture the concentration of NO, NO$_2$, HONO and ozone, the propene loss remained constantly underestimated as shown in Fig. 7.16. This means that propene undergoes some degree of heterogeneous photocatalytic decomposition in the presence of TiO$_2$, probably triggered by the presence of hydroxyl groups formed from adsorbed water vapour molecules following photocatalytic site activation at the interface. As the formaldehyde and acetaldehyde buildup was also significantly underestimated, a simplified surface conversion reaction was added to the TiO$_2$-glass module accordingly:

$$C_3H_6 + h\nu \rightarrow CH_3CHO + HCHO$$

The kinetics rate constant for this photocatalytic process was found to lead to the best fits when set to $(8.7 \pm 0.3) \times 10^{-5}$ s^{-1} under dry condition and to $(1.9 \pm 0.7) \times 10^{-4}$ s^{-1} at 45% relative humidity.

The comparison of Figs. 7.17 and 7.18 shows both the enhancement of propene removal in the presence of TiO$_2$-doped surfaces and the good agreement with the modified models. Such photosmog experiments are useful in assessing real-world

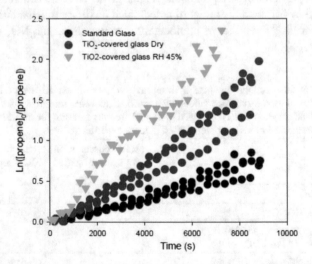

Fig. 7.16 Pseudo-first-order propene loss under similar conditions during photo-oxidation propene/NO$_x$/light experiments in the presence of various surfaces

performance and the studies have shown a significant reduction in propene loss with
relative humidity.

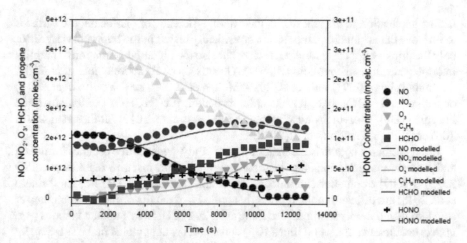

Fig. 7.17 Concentration–time profiles for monitored species during an experiment with standard
glass/propene/NO$_x$/light under dry conditions. Lights were switched on at t = 1800 s and switched
off at t = 10300 s

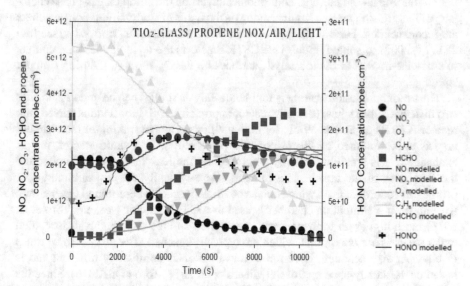

Fig. 7.18 Concentration–time profiles for monitored species during an experiment TiO$_2$-
glass/propene/NO$_x$/light under dry conditions. Lights were switched on at t = 1800 s and switched
off at t=10300 s

7.7.4 Recommendations for the Use of Simulation Chambers in Photocatalysis

Before application in the real world, the overall efficiency of photocatalytic surfaces to improve the urban air quality has to be critically tested in the laboratory. Here not only the primary uptake should be studied, but also the potential formation of harmful intermediates, like nitrous acid (HONO) (Gustafsson et al. 2006; Ndour et al. 2008; Beaumont et al. 2009; Laufs et al. 2010; Monge et al. 2010a; Gandolfo et al. 2015) or oxygenated VOCs, like, for example, formaldehyde (HCHO) (Salthammer and Fuhrmann 2007; Auvinen and Wirtanen 2008; Geiss et al. 2012; Mothes et al. 2016; Toro et al. 2016; Gandolfo et al. 2018).

If only primary uptake is the major focus of study, consecutive product formation and pure material emissions should be studied, small-scale fast flow reactors (e.g. ISO 22197-1 2007) or smaller continuous stirred tank reactors (CSTR, see, e.g. Minero et al. 2013) are efficient tools, if properly applied (for details of the shortcomings of standard flow reactors/methods, see Ifang et al. 2014). Here fast uptake kinetics (e.g. uptake coefficients of several times 10^{-4}) and small product yields in the sub-percent range can be determined for atmospheric conditions when using sensitive analytical instrumentation.

However, when slower heterogenous chemistry, more complex secondary chemistry or the impact of heterogeneous photocatalysis on the complex gas-phase chemistry of the atmosphere (see summer smog) is the focus of study, simulation chambers are recommended. Here small flow reactors with reaction times from <1 s (see ISO 22197-1 2007) to some minutes (see CSTR reactors) are not suitable, for example, to study the impact of photocatalytic surfaces on the O_3-formation during summer smog.

The use of simulation chambers for photocatalysis studies can be undertaken with two different approaches: (a) static reactor approach and (b) continuous stirred tank reactor (CSTR) approach. While for both methods efficient mixing of the chamber air has to be obtained by the use of fans (see below), in a static reactor, the air exchange rate is small and mainly controlled by the leak rate of the chamber and the flow rates of the attached instruments. Here typically the concentration time profiles are recorded, from which uptake coefficients can be determined (see below). In contrast, if a simulation chamber is used as a CSTR reactor (see, e.g. Toro et al. 2016) much higher air exchange rates are applied and the photocatalysis is studied under steady-state conditions, which are typically reached after at least three times of chamber air exchange. Here the data evaluation is completely different and is based on the steady-state approach (Minero et al. 2013; Toro et al. 2016). Since the simulation chambers during the EUROCHAMP projects were used as static reactors, the CSTR concept will not be further considered here. For larger chambers, like, for example, EUPHORE, the CSTR concept is practically not possible, due to either the extremely high air exchange rates necessary or the long duration of experiments, for which stable conditions (e.g. photon flux) are not available.

If a simulation chamber is used in photocatalysis to quantify the initial uptake, there is a kinetic limitation caused by the transport of the reactants to the active surfaces. Here, in heterogeneous chemistry, a high surface-to-volume (S_{active}/V) ratio is recommended, for which unwanted side reactions in the gas phase (e.g. NO_2 photolysis) are minimized. In contrast, for simulation chambers, which are originally aimed to study gas-phase reactions, a small S/V is practical, leading to smaller wall-loss rates compared to the rates of gas-phase reactions. To overcome this mismatch, strong fans have to be used in simulation chambers to ensure an efficient mixing of the air and to minimize the transport limitation to the surfaces under investigation. However, typically, this mixing is not strong enough to study fast uptake kinetics ($\gamma > 10^{-5}$) and the kinetic transport limit has to be determined individually in each chamber. Here instead of the photocatalytic surface, a perfect surface sink for a gas tracer has to be used. As an example, potassium-iodide-coated surfaces and the heterogeneous uptake of ozone can be used, for which close to unity uptake coefficients can be assumed. If the measured first-order uptake rate coefficient for this O_3 uptake (see Eq. E7.7.2.4) in the chamber, corrected for wall losses and the leak rate is converted into an uptake coefficient (see Eq. E7.7.2.5), this represents the upper limit transport coefficient $\gamma_{transport}$. If now the photocatalytic uptake is studied, the measured uptake coefficients ($\gamma_{measured}$) will approach this transport limit for fast true uptake kinetics (γ_{true}). Here the measured uptake can be described by the resistance approach:

$$\frac{1}{\gamma_{measured}} = \frac{1}{\gamma_{transport} + \gamma_{true}}, \qquad (E7.7.4.1)$$

for which the measured uptake is converging to $\gamma_{transport}$ for very active photocatalytic surfaces. From the fixed $\gamma_{transport}$ and the measured uptake, the true photocatalytic uptake can be calculated, if γ_{true} is not much higher than $\gamma_{transport}$. The upper limit for this method to determine high values of γ_{true} depends on the precision of both the measurements of $\gamma_{measured}$ and $\gamma_{transport}$. If, for example, $\gamma_{transport}$ is 10^{-5} and the combined precision error in the smog chamber is 10%, the limit of γ_{true} will be 9 × 10^{-5}, which can be calculated by Eq. (E7.7.4.1) from a measured uptake coefficient of 9 × 10^{-6} (only 10% lower than the transport limit). Since the error of γ_{true} will reach 100% at this limit, it is not recommended to study uptake kinetics more than five times faster than the individual transport limit in the chamber. If γ_{true} is higher, only a lower limit value should be specified and additional measurements in fast flow reactors are recommended.

If a more complex reaction system is studied in a smog chamber, simple analytical evaluation using Eq. (E7.7.2.4) is not possible and the use of a chemical box model is strongly recommended. Here all photocatalytic reactions involved should be implemented. The use of a chemical model for the interpretation of a simulation chamber experiment is especially necessary, if the impact of photocatalysis on the complex chemistry of the atmosphere, e.g. O_3 formation during summer smog, is to be investigated. Here the heterogeneous photochemical reactions have to be implemented as first-order rate coefficients into the existing model tools (e.g. MCM) after parameterization considering the S_{active}/V ratio inside the chamber. Only the use of such modified models can help to understand the observations inside the chamber.

As the simplest example, here the photocatalysis of pure NO_2 mixtures inside a simulation chamber is presented. While in a fast flow reactor, the uptake kinetics of NO_2 can be simply described in Eq. (E7.7.2.4) and only the formation of the side product HONO in the gas phase has to be considered besides the main reaction product of adsorbed HNO_3 (nitrate), the situation is much more complex in a simulation chamber.

Here the following simplified major processes will impact the concentration time profiles (for details regarding the main photocatalytic reactions, see Laufs et al. 2010; minor reactions are still missing below, e.g. wall loss of NO or HNO_3 photolysis):

Photocatalysis of NO_2	$NO_2 + TiO_2 + h\nu \rightarrow$ nitrate
Photocatalytic formation of HONO by NO_2	$NO_2 + TiO_2 + h\nu \rightarrow$ HONO
Wall loss of NO_2	$NO_2 +$ wall $\rightarrow$ products
Heterogeneous formation of HONO on the chamber walls	$NO_2 +$ wall $\rightarrow$ HONO
Gas-phase photolysis of NO_2 (Leighton)	$NO_2 + h\nu \rightarrow NO + O(^3P)$
	$O(^3P) + O_2 \rightarrow O_3$
	$O_3 + NO \rightarrow NO_2 + O_2$
Photocatalysis of NO	$NO + TiO_2 + h\nu \rightarrow NO_2$
Photocatalysis of O_3	$O_3 + TiO_2 + h\nu \rightarrow$ products
Wall loss of O_3	$O_3 +$ wall $\rightarrow$ products
Photocatalytic formation of O_3	nitrate $+ TiO_2 + h\nu \rightarrow O_3$
Photocatalytic formation of HONO by NO	$NO + TiO_2 + h\nu \rightarrow$ HONO
Photocatalysis of HONO	HONO $+ TiO_2 + h\nu \rightarrow$ products
Wall loss of HONO	HONO $+$ wall $\rightarrow$ products
Adsorption of HONO to the catalyst	HONO $+ TiO_2 \rightarrow$ nitrite
Gas-phase photolysis of HONO	HONO $+ h\nu \rightarrow NO + OH$
Gas-phase oxidation of NO	$NO + OH \rightarrow$ HONO
Gas-phase oxidation of NO_2	$NO_2 + OH \rightarrow HNO_3$
Wall loss of HNO_3	$HNO_3 +$ wall $\rightarrow$ products
Adsorption of HNO_3 to the catalyst	$HNO_3 + TiO_2 \rightarrow$ nitrate
And dilution of all gas-phase species given by the air exchange rate	

Although this is the simplest example of a photocatalytic experiment inside a simulation chamber, it is obvious that such a complex system can only be solved by numerical simulation in a box model and additional blank experiments, e.g. by measurements of the chamber wall-loss rates or the dark adsorption rates on the TiO_2. In contrast, any simple analytical evaluation will lead to a misinterpretation of the results, for example, to an overestimation of the photocatalytic uptake of NO_2 by the simultaneous gas-phase photolysis.

In the following, some final recommendations to the general conditions and the experimental requirements are given for photocatalytic simulation chamber experiments.

First, besides the analytical instrumentation necessary to detect all important species in the gas phase (in the above example: NO, NO_2, HONO, HNO_3, O_3), also the measurement of adsorbed species on the photocatalytic surfaces (in the above example: nitrite and nitrate) is helpful for the interpretation of the experimental results. For this purpose, extraction of smaller photocatalytic test surfaces by a suitable solvent and offline analysis is recommended (e.g. extraction by pure water and ion chromatography analysis, see Laufs et al. 2010).

Second, while the measurements of the spectral actinic fluxes is necessary to account for photochemical reactions in the gas phase, the spectral irradiance at the photocatalytic surface of interest is also necessary to evaluate photoactivation of the active material. The irradiance is critically depending on the orientation of the surfaces inside the chamber and on the solar zenith angle (SZA). If no analytical device for measuring the irradiance is available, at least the irradiance should be calculated by available models (see, e.g. TUV) This is only applicable if horizontal photocatalytic surfaces are used. The modelled irradiance should then be scaled inside the chamber by the ratio of measured/modelled actinic fluxes (or $J(NO_2)$).

Third, the photocatalytic surfaces should be washed by ultra-pure water and irradiated in clean synthetic air before the experiments, to remove adsorbed impurities and to obtain more reproducible results. In contrast, if, for example, nitrate has accumulated on the surface, the uptake kinetics of NO_x will slow down.

Fourth, if the pure photocatalytic effect should be studied, reference experiments with inactive similar surfaces should be performed under similar experimental conditions ("blank"). Here, for example, photocatalytic glass can be compared with normal glass (see Sect. 7.7.2) or photocatalytic active paints can be compared with similar normal paints (see Laufs et al. 2010).

Finally, the following experimental conditions are recommended for photocatalytic simulation chamber experiments (see also Ifang et al. 2014):

- Pollutants: When the photocatalysis of nitrogen oxides is studied, typically only NO is investigated. Here we recommend in addition to use the environmentally (and legislatively) more important NO_2. We also recommend investigating the different VOC classes, i.e. aromatics (e.g. toluene), unsaturated VOCs (e.g. propene) and biogenic VOCs (e.g. isoprene).
- Concentration: Since the assumed first-order kinetics of photocatalytic reactions are observed only at low reactant concentrations, atmospherically relevant pollution level should be used. Here experiments from typical urban background conditions (e.g. NO_x: 20 ppb, defined here as *low*) to heavily polluted kerbside conditions (e.g. NO_x: 100–200 ppb, defined here as *high*) should be investigated. However, conditions as typically recommended in available standard procedures (e.g. 1 ppm of NO in ISO 22197-1 2007) should not be used, since often the kinetics changes to zero order at such high pollution level.

- Humidity: Since photocatalysis is strongly dependent on the humidity (Laufs et al. 2010) no dry experiments are recommended, but the use of a medium humidity. Here in most standards on photocatalysis (e.g. ISO 22197-1 2007) a relative humidity of 50% is used, which is also recommended here.
- Irradiance: Since typically TiO_2 is used in photocatalysis, only the irradiance at ca. <400 nm has to be considered. This energy is necessary to activate the photocatalyst TiO_2: $TiO_2 + h\nu \rightarrow e_{cb} + h^+_{vb}$ (see Laufs et al. 2010).
- In a smog chamber, often the light source is the sun, for which the UVA irradiance is varying between ca. 10 and 70 W m^2 depending mainly on the SZA and orientation of the sample inside the chamber. In contrast, in indoor chambers, artificial UV light sources are applied, for which a typical UVA irradiance of 10–20 W m^{-2} is recommended in most standard procedures (average of typical ambient values). Care has to be taken, when the uptake kinetics of indoor experiments is compared to outdoor simulation chamber experiments, with often much higher irradiance used.

References

Ângelo, J., Andrade, L., Madeira, L.M., Mendes, A.: An overview of photocatalysis phenomena applied to NO_x abatement. J. Environ. Manag. **129**, 522–539 (2013). https://doi.org/10.1016/j.jenvman.2013.08.006

Atkinson, R., Carter, W.P.L., Winer, A.M., Pitts, J.N.: An experimental protocol for the determination of oh radical rate constants with organics using methyl nitrite photolysis as an OH radical source. J. Air Pollut. Control Assoc. **31**, 1090–1092 (1981). https://doi.org/10.1080/00022470.1981.10465331

Atkinson, R.: A structure-activity relationship for the estimation of rate constants for the gas-phase reactions of OH radicals with organic compounds. Int. J. Chem. Kinet. **19**, 799–828 (1987). https://doi.org/10.1002/kin.550190903

Auvinen, J., Wirtanen, L.: The influence of photocatalytic interior paints on indoor air quality. Atmos. Environ. **42**, 4101–4112 (2008). https://doi.org/10.1016/j.atmosenv.2008.01.031

Bates, D.R.: Ion-ion recombination in an ambient gas. In: Bates, D., Bederson, B. (eds.) Advances in Atomic and Molecular Physics, pp. 1–40. Academic Press (1985)

Beaumont, S.K., Gustafsson, R.J., Lambert, R.M.: Heterogeneous photochemistry relevant to the troposphere: H_2O_2 production during the photochemical reduction of NO_2 to HONO on UV-illuminated TiO_2 surfaces. ChemPhysChem **10**, 331–333 (2009). https://doi.org/10.1002/cphc.200800613

Bloss, C., Wagner, V., Bonzanini, A., Jenkin, M.E., Wirtz, K., Martin-Reviejo, M., Pilling, M.J.: Evaluation of detailed aromatic mechanisms (MCMv3 and MCMv3.1) against environmental chamber data. Atmos. Chem. Phys. **5**, 623–639 (2005). https://doi.org/10.5194/acp-5-623-2005

Boyjoo, Y., Sun, H., Liu, J., Pareek, V.K., Wang, S.: A review on photocatalysis for air treatment: from catalyst development to reactor design. Chem. Eng. J. **310**, 537–559 (2017). https://doi.org/10.1016/j.cej.2016.06.090

Brauers, T., Finlayson-Pitts, B.J.: Analysis of relative rate measurements. Int. J. Chem. Kinet. **29**, 665–672 (1997). https://doi.org/10.1002/(SICI)1097-4601(1997)29:9%3c665::AID-KIN3%3e3.0.CO;2-S

Carter, W.P.L.: Development of the SAPRC-07 chemical mechanism. Atmos. Environ. **44**, 5324–5335 (2010). https://doi.org/10.1016/j.atmosenv.2010.01.026

Dada, L., Lehtipalo, K., Kontkanen, J., Nieminen, T., Baalbaki, R., Ahonen, L., Duplissy, J., Yan, C., Chu, B., Petäjä, T., Lehtinen, K., Kerminen, V.-M., Kulmala, M., Kangasluoma, J.: Formation and growth of sub-3-nm aerosol particles in experimental chambers. Nat. Protoc. **15**, 1013–1040 (2020). https://doi.org/10.1038/s41596-019-0274-z

Dal Maso, M., Liao, L., Wildt, J., Kiendler-Scharr, A., Kleist, E., Tillmann, R., Sipilä, M., Hakala, J., Lehtipalo, K., Ehn, M., Kerminen, V.M., Kulmala, M., Worsnop, D., Mentel, T.: A chamber study of the influence of boreal BVOC emissions and sulfuric acid on nanoparticle formation rates at ambient concentrations. Atmos. Chem. Phys. **16**, 1955–1970 (2016). https://doi.org/10.5194/acp-16-1955-2016

EEA: Air Quality in EUROPE—2019 report (2019)

Emmerson, K.M., Evans, M.J.: Comparison of tropospheric gas-phase chemistry schemes for use within global models. Atmos. Chem. Phys. **9**, 1831–1845 (2009). https://doi.org/10.5194/acp-9-1831-2009

Finlayson-Pitts, B., Pitts, Jr.: Chemistry of upper and lower atmosphere (2000)

Franchin, A., Ehrhart, S., Leppä, J., Nieminen, T., Gagné, S., Schobesberger, S., Wimmer, D., Duplissy, J., Riccobono, F., Dunne, E.M., Rondo, L., Downard, A., Bianchi, F., Kupc, A., Tsagkogeorgas, G., Lehtipalo, K., Manninen, H.E., Almeida, J., Amorim, A., Wagner, P.E., Hansel, A., Kirkby, J., Kürten, A., Donahue, N.M., Makhmutov, V., Mathot, S., Metzger, A., Petäjä, T., Schnitzhofer, R., Sipilä, M., Stozhkov, Y., Tomé, A., Kerminen, V.M., Carslaw, K., Curtius, J., Baltensperger, U., Kulmala, M.: Experimental investigation of ion–ion recombination under atmospheric conditions. Atmos. Chem. Phys. **15**, 7203–7216 (2015). https://doi.org/10.5194/acp-15-7203-2015

Gallus, M., Ciuraru, R., Mothes, F., Akylas, V., Barmpas, F., Beeldens, A., Bernard, F., Boonen, E., Boréave, A., Cazaunau, M., Charbonnel, N., Chen, H., Daële, V., Dupart, Y., Gaimoz, C., Grosselin, B., Herrmann, H., Ifang, S., Kurtenbach, R., Maille, M., Marjanovic, I., Michoud, V., Mellouki, A., Miet, K., Moussiopoulos, N., Poulain, L., Zapf, P., George, C., Doussin, J.F., Kleffmann, J.: Photocatalytic abatement results from a model street canyon. Environ. Sci. Pollut. Res. **22**, 18185–18196 (2015). https://doi.org/10.1007/s11356-015-4926-4

Gandolfo, A., Bartolomei, V., Gomez Alvarez, E., Tlili, S., Gligorovski, S., Kleffmann, J., Wortham, H.: The effectiveness of indoor photocatalytic paints on NO_x and $HONO$ levels. Appl. Catal. B **166–167**, 84–90 (2015). https://doi.org/10.1016/j.apcatb.2014.11.011

Gandolfo, A., Marque, S., Temime-Roussel, B., Gemayel, R., Wortham, H., Truffier-Boutry, D., Bartolomei, V., Gligorovski, S.: Unexpectedly high levels of organic compounds released by indoor photocatalytic paints. Environ. Sci. Technol. **52**, 11328–11337 (2018). https://doi.org/10.1021/acs.est.8b03865

Geiss, O., Cacho, C., Barrero-Moreno, J., Kotzias, D.D.: Photocatalytic degradation of organic paint constituents-formation of carbonyls. Build. Environ. **48**, 107–112 (2012). https://doi.org/10.1016/j.buildenv.2011.08.021

Gustafsson, R.J., Orlov, A., Griffiths, P.T., Cox, R.A., Lambert, R.M.: Reduction of NO_2 to nitrous acid on illuminated titanium dioxide aerosol surfaces: implications for photocatalysis and atmospheric chemistry. Chem. Commun. 3936–3938 (2006). https://doi.org/10.1039/b609005b

Hoppel, W.A., Frick, G.M.: Ion—aerosol attachment coefficients and the steady-state charge distribution on aerosols in a bipolar ion environment. Aerosol Sci. Technol. **5**, 1–21 (1986). https://doi.org/10.1080/02786828608959073

Hurley, M.D., Wallington, T.J., Laursen, L., Javadi, M.S., Nielsen, O.J., Yamanaka, T., Kawasaki, M.: Atmospheric chemistry of n-butanol: kinetics, mechanisms, and products of Cl Atom and OH radical initiated oxidation in the presence and absence of NO_x. J. Phys. Chem. A **113**, 7011–7020 (2009). https://doi.org/10.1021/jp810585c

Hutchinson, M.: Chemistry. University of Leeds (2022)

Ifang, S., Gallus, M., Liedtke, S., Kurtenbach, R., Wiesen, P., Kleffmann, J.: Standardization methods for testing photo-catalytic air remediation materials: problems and solution. Atmos. Environ. **91**, 154–161 (2014). https://doi.org/10.1016/j.atmosenv.2014.04.001

ISO 22197-1: Fine Ceramics (Advanced Ceramics, advanced Technical Ceramics)—Test method for air-purification performance of semiconducting photocatalytic materials—Part 1: Removal of nitric oxide. Reference number, ISO 22197-1:2016(en), Switzerland (2016)

Jenkin, M.E., Hayman, G.D., Cox, R.A.: The chemistry of CH_3O during the photolysis of methyl nitrite. J. Photochem. Photobiol. A **42**, 187–196 (1988). https://doi.org/10.1016/1010-6030(88)80062-5

Jenkin, M.E., Saunders, S.M., Pilling, M.J.: The tropospheric degradation of volatile organic compounds: a protocol for mechanism development. Atmos. Environ. **31**, 81–104 (1997). https://doi.org/10.1016/S1352-2310(96)00105-7

Jenkin, M.E., Wyche, K.P., Evans, C.J., Carr, T., Monks, P.S., Alfarra, M.R., Barley, M.H., McFiggans, G.B., Young, J.C., Rickard, A.R.: Development and chamber evaluation of the MCM v3.2 degradation scheme for β-caryophyllene, Atmos. Chem. Phys. **12**, 5275–5308 (2012). https://doi.org/10.5194/acp-12-5275-2012

Jenkin, M.E., Young, J.C., Rickard, A.R.: The MCM v3.3.1 degradation scheme for isoprene. Atmos. Chem. Phys. **15**, 11433–11459 (2015). https://doi.org/10.5194/acp-15-11433-2015

Kleffmann, J.: Discussion on "field study of air purification paving elements containing TiO2" by Folli et al. (2015), Atmospheric Environment, **129**, 95–97 (2016). https://doi.org/10.1016/j.atmosenv.2016.01.004

Kuang, C., Chen, M., Zhao, J., Smith, J., McMurry, P.H., Wang, J.: Size and time-resolved growth rate measurements of 1 to 5 nm freshly formed atmospheric nuclei. Atmos. Chem. Phys. **12**, 3573–3589 (2012). https://doi.org/10.5194/acp-12-3573-2012

Kulmala, M., Petäjä, T., Nieminen, T., Sipilä, M., Manninen, H., Lehtipalo, K., Dal Maso, M., Aalto, P., Junninen, H., Paasonen, P., Riipinen, I., Lehtinen, K., Laaksonen, A., Kerminen, V.-M.: Measurement of the nucleation of atmospheric aerosol particles. Nat. Protoc. **7**, 1651–1667 (2012). https://doi.org/10.1038/nprot.2012.091

Laufs, S., Burgeth, G., Duttlinger, W., Kurtenbach, R., Maban, M., Thomas, C., Wiesen, P., Kleffmann, J.: Conversion of nitrogen oxides on commercial photocatalytic dispersion paints. Atmos. Environ. **44**, 2341–2349 (2010). https://doi.org/10.1016/j.atmosenv.2010.03.038

Lehtinen, K.E.J., Kulmala, M.: A model for particle formation and growth in the atmosphere with molecular resolution in size. Atmos. Chem. Phys. **3**, 251–257 (2003). https://doi.org/10.5194/acp-3-251-2003

Lehtinen, K.E.J., Rannik, Ü., Petäjä, T., Kulmala, M., Hari, P.: Nucleation rate and vapor concentration estimations using a least squares aerosol dynamics method. J. Geophys. Res.: Atmos. **109** (2004). https://doi.org/10.1029/2004JD004893

Lehtipalo, K., Leppa, J., Kontkanen, J., Kangasluoma, J., Franchin, A., Wimnner, D., Schobesberger, S., Junninen, H., Petaja, T., Sipila, M., Mikkila, J., Vanhanen, J., Worsnop, D.R., Kulmala, M.: Methods for determining particle size distribution and growth rates between 1 and 3 nm using the particle size magnifier. Boreal Environ. Res. **19**, 215–236 (2014)

Lehtipalo, K., Yan, C., Dada, L., Bianchi, F., Xiao, M., Wagner, R., Stolzenburg, D., Ahonen, L.R., Amorim, A., Baccarini, A., Bauer, P.S., Baumgartner, B., Bergen, A., Bernhammer, A.-K., Breitenlechner, M., Brilke, S., Buchholz, A., Mazon, S. B., Chen, D., Chen, X., Dias, A., Dommen, J., Draper, D. C., Duplissy, J., Ehn, M., Finkenzeller, H., Fischer, L., Frege, C., Fuchs, C., Garmash, O., Gordon, H., Hakala, J., He, X., Heikkinen, L., Heinritzi, M., Helm, J.C., Hofbauer, V., Hoyle, C.R., Jokinen, T., Kangasluoma, J., Kerminen, V.-M., Kim, C., Kirkby, J., Kontkanen, J., Kürten, A., Lawler, M.J., Mai, H., Mathot, S., Mauldin, R. L., Molteni, U., Nichman, L., Nie, W., Nieminen, T., Ojdanic, A., Onnela, A., Passananti, M., Petäjä, T., Piel, F., Pospisilova, V., Quéléver, L.L.J., Rissanen, M.P., Rose, C., Sarnela, N., Schallhart, S., Schuchmann, S., Sengupta, K., Simon, M., Sipilä, M., Tauber, C., Tomé, A., Tröstl, J., Väisänen, O., Vogel, A.L., Volkamer, R., Wagner, A.C., Wang, M., Weitz, L., Wimmer, D., Ye, P., Ylisirniö, A., Zha, Q., Carslaw, K.S., Curtius, J., Donahue, N.M., Flagan, R.C., Hansel, A., Riipinen, I., Virtanen, A., Winkler, P.M., Baltensperger, U., Kulmala, M., Worsnop, D.R.: Multicomponent new particle formation from sulfuric acid, ammonia, and biogenic vapors. Sci. Adv. **4**, eaau5363 (2018). https://doi.org/10.1126/sciadv.aau5363

Leppä, J., Anttila, T., Kerminen, V.M., Kulmala, M., Lehtinen, K.E.J.: Atmospheric new particle formation: real and apparent growth of neutral and charged particles. Atmos. Chem. Phys. **11**, 4939–4955 (2011). https://doi.org/10.5194/acp-11-4939-2011

Li, C., McMurry, P.H.: Errors in nanoparticle growth rates inferred from measurements in chemically reacting aerosol systems. Atmos. Chem. Phys. **18**, 8979–8993 (2018). https://doi.org/10.5194/acp-18-8979-2018

Malkin, T.L., Goddard, A., Heard, D.E., Seakins, P.W.: Measurements of OH and HO_2 yields from the gas phase ozonolysis of isoprene. Atmos. Chem. Phys. **10**, 1441–1459 (2010). https://doi.org/10.5194/acp-10-1441-2010

Manninen, H., Petäjä, T., Asmi, E., Riipinen, N., Nieminen, T., Mikkilä, J., Hõrrak, U., Mirme, A., Mirme, S., Laakso, L., Kerminen, V.-M., Kulmala, M.: Long-term field measurements of charged and neutral clusters using Neutral cluster and Air Ion Spectrometer (NAIS). Boreal Environ. Res. **14**, 591–605 (2009)

McMurry, P.H., Rader, D.J.: Aerosol wall losses in electrically charged chambers. Aerosol Sci. Technol. **4**, 249–268 (1985). https://doi.org/10.1080/02786828508959054

Meagher, R.J., Mcintosh, M.E., Hurley, M.D., Wallington, T.J.: A kinetic study of the reaction of chlorine and fluorine atoms with HC(O)F at 295±2 K. Int. J. Chem. Kinet. **29**, 619–625 (1997). https://doi.org/10.1002/(SICI)1097-4601(1997)29:8%3c619::AID-KIN7%3e3.0.CO;2-X

Minero, C., Bedini, A., Minella, M.: On the standardization of the photocatalytic gas/solid tests. Int. J. Chem. Reactor Eng. **11**, 717–732 (2013). https://doi.org/10.1515/ijcre-2012-0045

Monge, M.E., D'Anna, B., George, C.: Nitrogen dioxide removal and nitrous acid formation on titanium oxide surfaces—an air quality remediation process? Phys. Chem. Chem. Phys. **12**, 8991–8998 (2010a). https://doi.org/10.1039/b925785c

Monge, M.E., George, C., D'Anna, B., Doussin, J.-F., Jammoul, A., Wang, J., Eyglunent, G., Solignac, G., Daële, V., Mellouki, A.: Ozone formation from illuminated titanium dioxide surfaces. J. Am. Chem. Soc. **132**, 8234–8235 (2010b). https://doi.org/10.1021/ja1018755

Mothes, F., Böge, O., Herrmann, H.: A chamber study on the reactions of O3, NO, NO2 and selected VOCs with a photocatalytically active cementitious coating material. Environ. Sci. Pollut. Res. **23**, 15250–15261 (2016). https://doi.org/10.1007/s11356-016-6612-6

Ndour, M., D'Anna, B., George, C., Ka, O., Balkanski, Y, Kleffmann, J., Stemmler, K., Ammann, M.: Photoenhanced uptake of NO2 on mineral dust: Laboratory experiments and model simulations. Geophys. Res. Lett. **35** (2008). https://doi.org/10.1029/2007GL032006

O'Meara, S.P., Xu, S., Topping, D., Alfarra, M.R., Capes, G., Lowe, D., Shao, Y., & McFiggans, G.: PyCHAM (v2.1.1): a Python box model for simulating aerosol chambers. Geosci. Model Dev. **14**, 675–702 (2021). https://doi.org/10.5194/gmd-14-675-2021

Pathak, R.K., Presto, A.A., Lane, T.E., Stanier, C.O., Donahue, N.M., Pandis, S.N.: Ozonolysis of α-pinene: parameterization of secondary organic aerosol mass fraction. Atmos. Chem. Phys. **7**, 3811–3821 (2007)

Pichelstorfer, L., Stolzenburg, D., Ortega, J., Karl, T., Kokkola, H., Laakso, A., Lehtinen, K.E.J., Smith, J.N., McMurry, P.H., Winkler, P.M.: Resolving nanoparticle growth mechanisms from size- and time-dependent growth rate analysis. Atmos. Chem. Phys. **18**, 1307–1323 (2018). https://doi.org/10.5194/acp-18-1307-2018

Poling, B.E., Prausnitz, J.M., O'connell, J.P.: The properties of gases and liquids. Mcgraw-Hill, New York (2001)

Riccobono, F., Rondo, L., Sipilä, M., Barmet, P., Curtius, J., Dommen, J., Ehn, M., Ehrhart, S., Kulmala, M., Kürten, A., Mikkilä, J., Paasonen, P., Petäjä, T., Weingartner, E., Baltensperger, U.: Contribution of sulfuric acid and oxidized organic compounds to particle formation and growth. Atmos. Chem. Phys. **12**, 9427–9439 (2012). https://doi.org/10.5194/acp-12-9427-2012

Rickard, A.R., Wyche, K.P., Metzger, A., Monks, P.S., Ellis, A.M., Dommen, J., Baltensperger, U., Jenkin, M.E., Pilling, M.J.: Gas phase precursors to anthropogenic secondary organic aerosol: using the master chemical mechanism to probe detailed observations of 1,3,5-trimethylbenzene photo-oxidation. Atmos. Environ. **44**, 5423–5433 (2010). https://doi.org/10.1016/j.atmosenv.2009.09.043

Salthammer, T., Fuhrmann, F.: Photocatalytic surface reactions on indoor wall paint. Environ. Sci. Technol. **41**, 6573–6578 (2007). https://doi.org/10.1021/es070057m

Saunders, S.M., Jenkin, M.E., Derwent, R.G., Pilling, M.J.: Protocol for the development of the master chemical mechanism, MCM v3 (Part A): tropospheric degradation of non-aromatic volatile organic compounds. Atmos. Chem. Phys. **3**, 161–180 (2003)

Schneider, J., Matsuoka, M., Takeuchi, M., Zhang, J., Horiuchi, Y., Anpo, M., Bahnemann, D.W.: Understanding TiO_2 photocatalysis: mechanisms and materials. Chem. Rev. **114**, 9919–9986 (2014). https://doi.org/10.1021/cr5001892

Schwantes, R.H., McVay, R.C., Zhang, X., Coggon, M.M., Lignell, H., Flagan, R.C., Wennberg, P.O., Seinfeld, J.H.: Science of the environmental chamber. In: Advances in Atmospheric Chemistry, pp. 1–93. World Scientific, New Jersey (2017)

Seakins, P.W.: Product branching ratios in simple gas phase reactions, Annual Reports Section "C" (Physical Chemistry), 103, 173–222, https://doi.org/10.1039/b605650b, 2007.

Seinfeld, J.H., Pandis, S.N.: Atmospheric chemistry and physics: from air pollution to climate change. Wiley (2012)

Seinfeld, J.H., Pandis, S.N.: Atmospheric chemistry and physics: from air pollution to climate change, 3rd edn. Wiley (2016)

Sommariva, R., Cox, S., Martin, C., Borońska, K., Young, J., Jimack, P.K., Pilling, M.J., Matthaios, V.N., Nelson, B.S., Newland, M.J., Panagi, M., Bloss, W.J., Monks, P.S., Rickard, A.R.: AtChem (version 1), an open-source box model for the master chemical mechanism. Geosci. Model Dev. **13**, 169–183 (2020). https://doi.org/10.5194/gmd-13-169-2020

Toro, C., Jobson, B.T., Haselbach, L., Shen, S., Chung, S.H.: Photoactive roadways: determination of CO, NO and VOC uptake coefficients and photolabile side product yields on TiO_2 treated asphalt and concrete. Atmos. Environ. **139**, 37–45 (2016). https://doi.org/10.1016/j.atmosenv.2016.05.007

Tröstl, J., Chuang, W.K., Gordon, H., Heinritzi, M., Yan, C., Molteni, U., Ahlm, L., Frege, C., Bianchi, F., Wagner, R., Simon, M., Lehtipalo, K., Williamson, C., Craven, J.S., Duplissy, J., Adamov, A., Almeida, J., Bernhammer, A.-K., Breitenlechner, M., Brilke, S., Dias, A., Ehrhart, S., Flagan, R.C., Franchin, A., Fuchs, C., Guida, R., Gysel, M., Hansel, A., Hoyle, C.R., Jokinen, T., Junninen, H., Kangasluoma, J., Keskinen, H., Kim, J., Krapf, M., Kürten, A., Laaksonen, A., Lawler, M., Leiminger, M., Mathot, S., Möhler, O., Nieminen, T., Onnela, A., Petäjä, T., Piel, F.M., Miettinen, P., Rissanen, M.P., Rondo, L., Sarnela, N., Schobesberger, S., Sengupta, K., Sipilä, M., Smith, J.N., Steiner, G., Tomè, A., Virtanen, A., Wagner, A.C., Weingartner, E., Wimmer, D., Winkler, P.M., Ye, P., Carslaw, K.S., Curtius, J., Dommen, J., Kirkby, J., Kulmala, M., Riipinen, I., Worsnop, D.R., Donahue, N.M., Baltensperger, U.: The role of low-volatility organic compounds in initial particle growth in the atmosphere. Nature **533**, 527–531 (2016). https://doi.org/10.1038/nature18271

Verheggen, B., Mozurkewich, M.: An inverse modeling procedure to determine particle growth and nucleation rates from measured aerosol size distributions. Atmos. Chem. Phys. **6**, 2927–2942 (2006). https://doi.org/10.5194/acp-6-2927-2006

Wagner, R., Yan, C., Lehtipalo, K., Duplissy, J., Nieminen, T., Kangasluoma, J., Ahonen, L.R., Dada, L., Kontkanen, J., Manninen, H.E., Dias, A., Amorim, A., Bauer, P.S., Bergen, A., Bernhammer, A.K., Bianchi, F., Brilke, S., Mazon, S.B., Chen, X., Draper, D.C., Fischer, L., Frege, C., Fuchs, C., Garmash, O., Gordon, H., Hakala, J., Heikkinen, L., Heinritzi, M., Hofbauer, V., Hoyle, C.R., Kirkby, J., Kürten, A., Kvashnin, A.N., Laurila, T., Lawler, M.J., Mai, H., Makhmutov, V., Mauldin Iii, R.L., Molteni, U., Nichman, L., Nie, W., Ojdanic, A., Onnela, A., Piel, F., Quéléver, L.L.J., Rissanen, M.P., Sarnela, N., Schallhart, S., Sengupta, K., Simon, M., Stolzenburg, D., Stozhkov, Y., Tröstl, J., Viisanen, Y., Vogel, A.L., Wagner, A.C., Xiao, M., Ye, P., Baltensperger, U., Curtius, J., Donahue, N.M., Flagan, R.C., Gallagher, M., Hansel, A., Smith, J.N., Tomé, A., Winkler, P.M., Worsnop, D., Ehn, M., Sipilä, M., Kerminen, V.M., Petäjä, T., Kulmala, M.: The role of ions in new particle formation in the CLOUD chamber. Atmos. Chem. Phys. **17**, 15181–15197 (2017). https://doi.org/10.5194/acp-17-15181-2017

Wang, N., Jorga, S.D., Pierce, J.R., Donahue, N.M., Pandis, S.N.: Particle wall-loss correction methods in smog chamber experiments. Atmos. Meas. Tech. **11**, 6577–6588 (2018). https://doi.org/10.5194/amt-11-6577-2018

Winiberg, F.A.F., Dillon, T.J., Orr, S.C., Groß, C.B.M., Bejan, I., Brumby, C.A., Evans, M.J., Smith, S.C., Heard, D.E., Seakins, P.W.: Direct measurements of OH and other product yields from the $HO_2 + CH_3C(O)O_2$ reaction. Atmos. Chem. Phys. **16**, 4023–4042 (2016). https://doi.org/10.5194/acp-16-4023-2016

Wittrock, F., Richter, A., Oetjen, H., Burrows, J.P., Kanakidou, M., Myriokefalitakis, S., Volkamer, R., Beirle, S., Platt, U., Wagner, T.: Simultaneous global observations of glyoxal and formaldehyde from space. Geophys. Res. Lett. **33** (2006). https://doi.org/10.1029/2006GL026310

Yli-Juuti, T., Nieminen, T., Hirsikko, A., Aalto, P.P., Asmi, E., Hõrrak, U., Manninen, H.E., Patokoski, J., Dal Maso, M., Petäjä, T., Rinne, J., Kulmala, M., Riipinen, I.: Growth rates of nucleation mode particles in Hyytiälä during 2003–2009: variation with particle size, season, data analysis method and ambient conditions. Atmos. Chem. Phys. **11**, 12865–12886 (2011). https://doi.org/10.5194/acp-11-12865-2011

Zádor, J., Turányi, T., Wirtz, K., Pilling, M.J.: Measurement and investigation of chamber radical sources in the European Photoreactor (EUPHORE). J. Atmos. Chem. **55**, 147–166 (2006). https://doi.org/10.1007/s10874-006-9033-y

Zádor, J., Wagner, V., Wirtz, K., Pilling, M.J.: Quantitative assessment of uncertainties for a model of tropospheric ethene oxidation using the European Photoreactor (EUPHORE). Atmos. Environ. **39**, 2805–2817 (2005). https://doi.org/10.1016/j.atmosenv.2004.06.052

Chapter 8
Application of Simulation Chambers to Investigate Interfacial Processes

Peter A. Alpert, François Bernard, Paul Connolly, Odile Crabeck, Christian George, Jan Kaiser, Ottmar Möhler, Dennis Niedermeier, Jakub Nowak, Sébastien Perrier, Paul Seakins, Frank Stratmann, and Max Thomas

Abstract Earlier chapters of this work have described procedures and protocols that are applicable to most chambers, this chapter has a slightly different focus; we predominantly consider multiphase processes where the applications are on phase transfer of chemical species rather than chemical reactions and the processes are generally occurring in highly specialized chambers. Three areas are described. Firstly, cloud formation processes; here, precise control of physical and thermodynamic properties is required to generate reproducible results. The second area examined is the air/sea interface, looking at the formation of aerosols from nonanoic acid as a surfactant with humic acid as a photosensitizer. The final apparatus described is the Roland von Glasow sea-ice chamber where a detailed protocol for the reproducible formation of sea-ice is given along with an outlook of future work. The systems studied in all three sections are characterized by difficulties in making detailed in situ

P. A. Alpert
Paul Scherrer Institute, Würenlingen, Switzerland

F. Bernard · C. George · S. Perrier
Centre National de la Recherche Scientifique, Paris, France

P. Connolly
University of Manchester, Manchester, UK

O. Crabeck · J. Kaiser
University of East Anglia, Norwich, UK

O. Möhler
Karlsruhe Institute of Technology, Karlsruhe, Germany

D. Niedermeier · F. Stratmann
Leibniz Institut Fuer Troposphaerenforschung, Leipzig, Germany

J. Nowak
University of Warsaw, Warsaw, Poland

P. Seakins (✉)
University of Leeds, Leeds, UK
e-mail: P.W.Seakins@leeds.ac.uk

M. Thomas
University of Otago, Dunedin, New Zealand

© The Author(s) 2023
J.-F. Doussin et al. (eds.), *A Practical Guide to Atmospheric Simulation Chambers*,
https://doi.org/10.1007/978-3-031-22277-1_8

293

observations in the real world, either due to the transitory nature of systems or the practical difficulties in accessing the systems. While these specialized simulation chambers may not perfectly reproduce conditions in the real world, the chambers do provide more facile opportunities for making extended and reproducible measurements to investigate fundamental physical and chemical processes, at significantly lower costs.

Chapters 2–7 have described procedures and protocols that are applicable to most chambers, although there will be variations dependent on the chemical application of the chamber (e.g., whether the primary focus is gas- or aerosol-phase chemistry). This chapter has a slightly different focus; we predominantly consider non-chemical applications in a few more specialized chambers. Section 8.1 describes the study of cloud processes focusing particularly on the AIDA (Karlsruhe) and LACIS (Leipzig) chambers. Sections 8.2 and 8.3 consider air–ocean and air–sea-ice interactions. The systems studied in all three sections are characterized by difficulties in making detailed in situ observations in the real world, either due to the transitory nature of systems (e.g., clouds or the continual breaking and reformation of the sea surface layer) or the practical difficulties in accessing the systems (e.g., use of aircraft for cloud systems or the logistics of polar expeditions). While these specialized simulation chambers may not perfectly reproduce conditions in the real world, the chambers do provide more facile opportunities for making extended and reproducible measurements, at significantly lower costs.

The chapter also gives a wider flavor of the roles that simulation chambers can play in atmospheric science. While the main focus is on physical aspects, chemical measurements in the gas, liquid, aerosol, and solid phases can also be made. Providing some background in non-conventional uses of simulation chambers may give practitioners in more conventional chambers some ideas about how their work can be extended or made more realistic by considering additional interactions or processes. Many chambers run access programs, either through ACTRIS or their institutions, and so it may be possible to access these chambers to further develop your ideas.

8.1   Application of Simulation Chambers to the Study of Cloud Processes

8.1.1   Introduction

Clouds are the source of precipitation, contribute significantly to the Earth's radiation budget, and are therefore an important player for both the weather and the climate. In the last few decades, comprehensive research activities have been conducted to understand cloud processes and the associated interactions (Mason and Ludlam 1951; Hobbs 1991; Kreidenweis et al. 2019) which lead to an increase of the quantitative knowledge of these systems. Nevertheless, not all processes and their influence on weather and climate are yet sufficiently well understood and quantified (Quaas et al. 2009; Seinfeld et al. 2016; Kreidenweis et al. 2019).

A reason for our limited understanding is that atmospheric clouds are highly complex systems. Clouds are transient and usually occur in places that are not easily accessible, making an extensive characterization of clouds very difficult. Furthermore, the observation of a large number of clouds is required as no cloud is like another. The study of atmospheric clouds is therefore very ambitious, expensive, and sometimes even impossible (Stratmann et al. 2009). In consequence, laboratory investigations under well-defined and reproducible conditions are needed in addition to atmospheric observations in order to better understand and quantify cloud processes and related interactions (List et al. 1986; Stratmann et al. 2009; Kreidenweis et al. 2019).

Over the last 40 years, a number of laboratory facilities such as expansion cloud chambers, continuous-flow systems, and wind tunnels have been developed and applied to aerosol and cloud research under controlled, reproducible, and atmospherically relevant conditions (see Chang et al. (2016, 2017) and Cziczo et al. (2017) for details about specific laboratory facilities). The results obtained with these laboratory facilities have already filled in gaps in the puzzle of understanding aerosol–cloud interactions (Chang et al. 2016).

Cloud simulation facilities have been developed and used during the previous few decades to investigate a wide range of processes of relevance for the formation and life cycles of atmospheric clouds. Experiments at these facilities can be conducted over a wide range of simulated and well-controlled thermodynamic, physical, and chemical conditions of relevance for large variety of atmospheric cloud types and climatic regions. Among the addressed research topics are aerosol-cloud condensation nuclei (CCN) processes, ice-nucleation (IN) processes, ice crystal processes, and turbulence effects in cloud microphysics.

The advantage to using a cloud chamber for the study of ice nucleation is that it is a close analogy for how ice nucleation occurs in the atmosphere. The AIDA chamber has been used for experiments on the ice-nucleation activity of a variety of aerosols such as mineral dust (Möhler et al. 2006), ammonium sulfate (Abbatt et al. 2006), bacteria (Möhler et al. 2008), or soil dust (Steinke et al. 2016). A new ice-nucleation active surface site (INAS) density was developed as a result of these cloud chamber experiments (Connolly et al. 2009; Niemand et al. 2012) and was used in models to predict the atmospheric abundance of INPs (Ullrich et al. 2017). The Manchester Ice Cloud Chamber (MICC) has been used to quantify ice nucleation on mineral particles and secondary organic aerosol using the INAS metric to quantify ice-nucleation efficiency (see Emersic et al. 2015; Frey et al. 2018). In general, chamber expansion techniques agree with other techniques to quantify ice nucleation, such as drop-freezing cold stages, at temperatures lower than −25 °C; however, at temperatures higher than −20 °C, we tend to see higher INAS values when using the chamber method (Hiranuma et al. 2015; DeMott et al. 2018; Emersic et al. 2015). The reasons for this are unknown at present. A further advantage to using a cloud chamber is that competition effects between aerosol external mixtures can be measured and understood, therefore improving models (e.g., Simpson et al. 2018).

The Leipzig Aerosol Cloud Interaction Simulator (LACIS) (Stratmann et al. 2004; Hartmann et al. 2011)—which is a laminar flow tube at TROPOS (Leipzig, Germany)—has been used to study aerosol–cloud interaction processes under controllable and reproducible conditions. With LACIS, both the hygroscopic growth and droplet activation of various inorganic (Wex et al. 2005; Niedermeier et al. 2008) and organic materials such as HULIS (humic-like substances; Wex et al. 2007; Ziese et al. 2008), soot (Henning et al. 2010; Stratmann et al. 2010), and secondary organic aerosol particles (Wex et al. 2009; Petters et al. 2009) could be consistently described. Furthermore, LACIS has been comprehensively applied for the investigation and quantification of the immersion freezing behavior of biological (Augustin et al. 2013; Hartmann et al. 2013), mineral dust (Niedermeier et al. 2010; Augustin-Bauditz et al. 2014; Hartmann et al. 2016), and ash particles (Grawe et al. 2016, 2018).

The fall-speed of ice particles within atmospheric clouds has a strong impact on climate feedback (Mitchell et al. 2011). Ice particle fall-speed is governed by the size and shape of the particles (Heymsfield and Westbrook 2010). In the atmosphere, a fundamental process in the generation of large precipitating particles is the coming together and subsequent aggregation of two or more ice crystals. Many in situ observations have confirmed that ice crystal aggregation is important over a large range of ambient temperatures between $0\,°C$ and $-60\,°C$ (e.g., Connolly et al. 2005; Gallagher et al. 2012); however, the rates that aggregation occurs at different ambient conditions are very uncertain. Laboratory experiments including cloud chambers are able to study and quantify the ice crystal aggregation process and other secondary ice crystal processes under a range of simulated conditions (e.g., Connolly et al. 2012). In such experiments, tall chambers, such as the 10-m-high Manchester Ice Cloud Chamber, are desirable to enable sufficient time for the ice crystals to sediment and interact with each other.

Ice clouds make a major contribution to radiative forcing in the atmosphere, both trapping IR radiation and reflecting solar radiation. The balance between these processes determines whether clouds have a net warming or cooling, and depends on the macro-physical and optical properties of the cloud, in particular, the ice water content and the cloud optical depth. Furthermore, accurate retrieval of ice cloud properties using remote sensing platforms requires knowledge of the light scattering properties of ice crystals (such as their backscatter and volume extinction). In contrast to liquid clouds, there is a gap in knowledge about the backscatter and volume extinction properties of atmospheric ice clouds. Moreover, irregularities on the ice crystal surfaces can affect their general light scattering properties (e.g., Liu et al. 2013) and there is increasing evidence that ice particles within cirrus clouds are dominated by ice particles that have substantially roughened surfaces (e.g., Ulanowski et al. 2014). Cloud chambers can play a vital role in narrowing the gap in our knowledge by providing direct measurements of the light scattering properties of ice crystals under a range of conditions. These measurements aid in the development of parameterizations and provide data with which to test advanced scattering models (e.g., Smith et al. 2015, 2016).

Atmospheric clouds are often inhomogeneous, nonstationary, and intermittent. They cover a huge range of spatial and temporal scales with cross-scale interactions between turbulent fluid dynamics and microphysical processes that influence the development and the behavior of clouds (Bodenschatz et al. 2010). Turbulence drives mixing and entrainment in clouds leading to strong fluctuations in temperature, water vapor, and consequently (super-)saturation as well as aerosol particle concentration affecting cloud droplet/ice crystal formation and growth/decay (Siebert et al. 2006; Chandrakar et al. 2016; Siebert and Shaw 2017). On the other hand, the phase transformation processes can introduce bulk-buoyancy effects and influence cloud dynamics (Stevens 2005; Malinowski et al. 2008; Bodenschatz et al. 2010).

To date, there are very few laboratory facilities for the study of aerosol–cloud–turbulence interaction processes because of the high requirements for accuracy and reproducibility of experimental parameters. One example is the turbulent moist-air wind tunnel LACIS-T (turbulent Leipzig Aerosol Cloud Interaction Simulator, Niedermeier et al. 2020) at TROPOS. LACIS-T can specifically be used to study the influence of turbulence on cloud microphysical processes, such as droplet and ice crystal formation. The investigations take place under controlled and reproducible flow and thermodynamic conditions. The temperature range of warm, mixed-phase, and cold clouds (i.e., -40 °C $<$ T $<$ 25 °C) can be covered. The continuous-flow design of LACIS-T allows for the investigation of processes occurring on small spatial (micrometer to decimeter scale) and temporal scales (up to a few seconds), with a Lagrangian perspective. A specific benefit of LACIS-T is the well-defined location of aerosol particle injection directly into the turbulent mixing zone as well as the precise control of the respective initial and boundary flow velocity and thermodynamic conditions.

8.1.2 Design of Expansion-Type Cloud Chambers to Study Cloud Microphysical Processes

Expansion-type cloud chambers are capable of simulating processes occurring in air parcels that undergo steady cooling, e.g., in updrafting air parcels related to convective or lee wave cloud formation. Cooling in such chambers is induced by active pumping to the cloud chamber. The rate of pressure reduction is related to a well-defined adiabatic cooling rate, and thus to an increase of relative humidity. The operation of an expansion-type cloud simulation facility requires a clean and vacuum-tight cloud chamber with precise temperature and pressure control. Two such facilities are operated in Europe, the AIDA cloud chamber (Aerosol Interactions and Dynamics in the Atmosphere) at the Karlsruhe Institute of Technology (KIT) (https://www.imk-aaf.kit.edu/AIDA_facilities.php), and Manchester Ice Cloud Chamber (MICC, http://www.cas.manchester.ac.uk/restools/cloudchamber/) at the University of Manchester. In both facilities, the cloud chamber is located inside a cold room, cooled by air ventilation, and has rigid walls of high heat capacity, but without active wall cooling. This has the advantage of precise and homogeneous temperature control, but the disadvantage is that the wall temperatures remain almost constant, while the gas is cooled adiabatically during the expansion run of a cloud experiment. This results in an increasing temperature difference between walls and the gas volume inside the cloud chamber, thus to an increasing heat flux into the volume causing a steady reduction of the cooling rate at a constant pumping rate. By that, both the super-cooling and the duration of a single cloud run starting at a certain pressure and wall temperature are limited, and it is not possible to operate such cloud chambers for longer time periods at constant cooling rates. Therefore, the new dynamic cloud chamber AIDAd was developed at KIT and came into operation in early 2020. This new cloud simulation chamber has active wall cooling and can therefore be operated with isothermal gas and wall temperature distributions in a wide range of cooling rates and a wide temperature range. The setup, instrumentation, and operation parameters of the three expansion-type chambers AIDA, AIDAd, and MICC will briefly be described in the following sections.

The cloud simulation chamber AIDA

The AIDA chamber was designed and engineered as an atmospheric simulation chamber for long-term aerosol and trace gas chemistry experiments. It came into operation in 1997 and, during the first years of operation, was mainly used for experiments on heterogeneous chemistry (Kamm et al. 1999) and aerosol optical properties (Schnaiter et al. 2005). After an intensive period of polar stratospheric cloud research (Wagner et al. 2005; Zink et al. 2002); Möhler et al. 2006), AIDA was predominantly converted into an expansion-type cloud chamber (Möhler et al. 2003, 2005) and also used for a series of experiments on secondary aerosol formation (Saathoff et al. 2009). More recently, the AIDA chamber was also equipped with an LED light source to simulate the shortwave solar spectrum in the troposphere for experiments on atmospheric photochemistry.

Here, we focus on describing the setup, instrumentation, and operation of AIDA as a cloud simulation chamber. Figure 8.1 shows a schematic view of the facility, with the cloud chamber located in the cold box and surrounded by four platforms with ample space for the operation of instruments that are permanently installed or contributed and operated by partners of specific measurement campaigns. The cloud chamber is made of aluminum, has a height of about 7 m, a diameter of about 4 m, and a volume of 84 m^3. A mixing fan is located about 1.5 m above the chamber floor with a vertical rotational axis co-aligned to the vertical axis of the cylindrical cloud chamber. The fan induces an upward directed air flow and eddy turbulence inside the chamber, and by that provides chamber internal mixing and homogeneity of trace gas, aerosol, and cloud components with a mixing time scale of about 1 min. The mixing time scale and homogeneous distribution of components inside the chamber are critical for the interpretation of experimental results that are obtained with a large number of instruments measuring or sampling at locations. The whole cloud chamber volume can be considered as a large and uniform cloud element or air parcel that experiences, within uncertainty and fluctuation ranges, the same dynamic change of cloud formation variables and processes. Figure 8.2 summarizes the instruments that are coupled to the AIDA chamber.

The new dynamic cloud chamber AIDAd

The main advantage of the new dynamic cloud chamber AIDAd compared to the existing AIDA aerosol and cloud chamber facility is that it will allow one to investigate aerosol-cloud processes at simulated cloud updraft conditions in a wide range of well-controlled cooling rates, moisture content, as well as aerosol and trace gas mixtures and compositions. AIDAd was designed, engineered, and constructed in close collaboration with Bilfinger Noell GmbH, Germany. The vacuum chamber, cooling system, and cloud chamber were installed during 2019. First successful cooling runs were performed in August 2019, and final test runs are conducted during November 2019.

AIDAd has a double-chamber design (Figs. 8.3 and 8.4), similar to the design of the dynamic cloud chambers of the Colorado State University (Demott and Rogers 1990) and the Meteorological Research Institute in Tsukuba, Japan (Tajiri et al. 2013). The cloud chamber is located inside an outer vacuum chamber composed of thin-walled flow channels with rectangular cross section. It is mounted on the top plate of the outer vacuum chamber. The vertical tubes are part of the inner synthetic oil circuits for the temperature control of the five cloud chamber segments. The

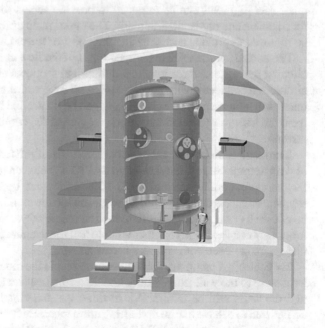

Fig. 8.1 Schematic representation of the AIDA cloud simulation chamber facility. The cloud chamber is located inside a cold box with precise control of the temperature in the range from +60 °C to −90 °C. Spatial and temporal temperature homogeneity within ± 0.3 °C can be achieved inside the cold box and the cloud chamber.
© KIT

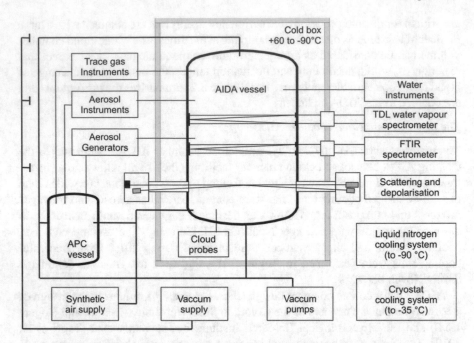

Fig. 8.2 Technical components and instrumentation of the AIDA cloud chamber facility. © KIT

vacuum chamber is also constructed in five segments. The bottom part holds all the coolant supply tubes and sensor feedthroughs, and can therefore not be removed. The other four segments can be removed to provide access to the inner cloud chamber for maintenance or installation work. The upper plate of the cloud chamber can also be removed to provide access to the inner part of the cloud chamber.

Pre-cooled synthetic oil is pumped through the flow channels in five independent circuits: the bottom, three identical cylindrical sections, and the top. The five inner circuits are connected with the pre-cooled oil reservoir through an outer circuit. The inner wall temperature of each segment can be controlled by either adding colder oil from the reservoir (outer circuit) upon request of the cooling system or by electrical heating.

The cloud chamber can either be operated at uniform temperatures in all segments or with a temperature difference of up to ±10 °C between two neighboring segments. Furthermore, cooling rates of up to 10 K min^{-1} can be applied to any of the five segments. In case of wall cooling, the gas inside the cloud chamber can also actively be cooled by controlled pressure reduction inside the vacuum chamber. Connecting tubes between the cloud chamber volume and the vacuum chamber keep the pressure difference between both volumes below a few hPa.

By controlled pumping, the temperature inside the cloud chamber volume can be kept close to the wall temperature, and therefore heat exchange between the volume and the walls will be minimized. In this case, the cloud chamber volume can be considered to behave like an updrafting atmospheric air parcel with adiabatic cooling conditions. Therefore, the AIDAd cloud chamber will be capable of simulation cloud

Fig. 8.3 Design of the cloud chamber with wall cooling located inside a vacuum chamber for cloud simulation experiments by expansion cooling. © KIT

2.5m

4.5 m

processes at simulated and well-controlled adiabatic cooling rates between about 0.1 K min^{-1} and 10 K min^{-1}.

The Manchester Ice Cloud Chamber (MICC)

The Manchester Ice Cloud Chamber (MICC) is a 1 m diameter, 10 m tall chamber situated on three floors of the Centre for Atmospheric Science at the University of Manchester. There are three separate cold rooms enclosing the cloud chamber, capable of being cooled to −55 °C using independently controlled compressors and fans within each enclosure (see Fig. 8.5 for a schematic of the chamber). Thermocouples are used to measure the gas temperature inside the chamber throughout its length.

Clouds are made by two methods. The first method is similar to that described above for AIDA where a quasi-adiabatic expansion of the air inside the chamber is utilized to lead to cloud formation on aerosol particles within the chamber. Aerosol particles can be introduced using a rotating brush generator (see Emersic et al. 2015) or can be generated using the Manchester Aerosol Chamber (MAC) facility (see Frey et al. 2018), which is also in the Centre for Atmospheric Science at the University of Manchester, and pumped into the MICC. Aerosol particles and cloud particle

Fig. 8.4 Side view of the new dynamic cloud chamber AIDAd at KIT. © KIT/Markus Breig

properties are measured during the experiments by sampling the air from the chamber using pumps to draw the air through particle sampling instruments (see Fig. 8.5). These instruments can be placed at portholes on each of the three floors. This method typically creates cloudy conditions for 5–10 min due to a heat flux from the chamber walls, which eventually warms the air to temperatures where the air becomes sub-saturated with respect to water vapor.

The second method of creating ice and mixed-phase clouds is to introduce humid air into the chamber at temperatures below 273.15 K prior to nucleating ice by periodically allowing compressed air to exit a solenoid valve near the top of the chamber. It is possible to create long-lived ice and mixed-phase clouds using method 2. Method 2 has been used to study ice crystal aggregation (Connolly et al. 2012) and the light scattering properties of ice crystals (Smith et al. 2015, 2016).

8.1.3 Design of a Chamber to Study the Influence of Turbulence onto Cloud Microphysical Processes: LACIS-T

LACIS-T (Niedermeier et al. 2020) at TROPOS is a turbulent moist-air wind tunnel. It is a closed-loop system being designed to generate a locally homogeneous and

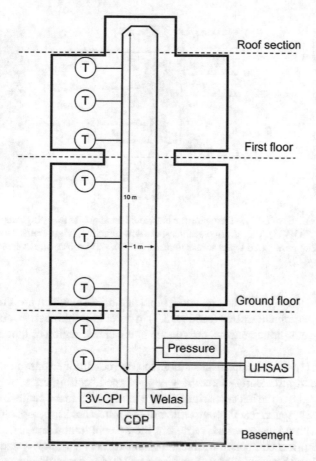

Fig. 8.5 Technical components and some of the instrumentation at the MICC facility. The chamber is housed within three cold rooms that span three floors of the building. Each cold room can be cooled to −55 °C using individual compressors. Instrumentation is variable and can be fitted to ports within each of the three sections

isotropic turbulent airflow. The temperature and water vapor saturation of the airflow can be precisely controlled and aerosol particles—acting as cloud condensation nuclei (CCN) or ice-nucleating particles (INPs)—can be injected into it. Under suitable conditions, cloud droplet formation or heterogeneous ice formation and the subsequent growth can be observed within the turbulent flow.

A schematic of LACIS-T is shown in Fig. 8.6. The main components are radial blowers, particle filters, valves, flow meters, the humidification system, heat exchangers, turbulence grid, the measurement section, and the adsorption dehumidifying system. These components are applied in order to generate two particle-free airflows (approximately 5000 l min^{-1} each) each of which is conditioned to a certain temperature and dew-point temperature (the range is −40 °C < T, Td < 25 °C). These two conditioned particle-free airflows pass passive square-mesh grids (mesh length

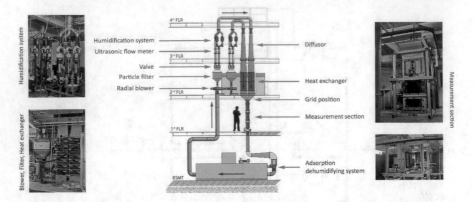

Fig. 8.6 A schematic of LACIS-T showing the individual components (© by Ingenieurbüro Mathias Lippold, VDI; TROPOS). The red arrows indicate the flow direction. (Figure reused with permission from Niedermeier et al. 2020 Open access under a CC BY 4.0 license, https://creativecommons. org/licenses/by/4.0/)

of 1.9 cm, rod diameter of 0.4 cm, and a blockage of 30%) which are situated 20 cm above the measurement section (see Fig. 8.7) in order to create nearly isotropic and, in transverse planes, homogeneous turbulence in the center region of the measurement section.

At the inlet of the measurement section, the two conditioned particle-free airflows are merged and turbulently mixed. A wedge-shaped "cutting edge" separates both airflows right above the inlet (see right picture in Fig. 8.7). In the center of this cutting edge, three rectangular feedthroughs (20 mm × 1 mm each, 1 mm separation between feedthroughs) are located which represent the aerosol inlet. Here, aerosol particles of known chemical composition, size, and number concentration—size selection is conducted via a Differential Mobility Analyzer (DMA, type "Vienna medium")—and particles are counted by means of a condensation particle counter CPC (TSI 3010, TSI Inc., USA)—are injected into the mixing zone, in which cloud droplet formation and/or freezing take place at ambient pressure. A super-saturated environment can be created through the process of isobaric mixing (Bohren and Albrecht 1998). The exact humidity within the turbulent region depends on the temperatures and dew-point temperatures within the two particle-free airflows, as well as the location within the turbulent mixing zone. The mean velocity inside the measurement section can be varied between 0.5 and 2 m s^{-1}.

The measurement section itself is of cuboidal shape. It is 2.0 m long, 0.8 m wide, and 0.2 m deep. The design of aerosol inlet and measurement section reduces wall effects onto the processes of interest occurring in the mixing zone. Furthermore, the measurement section design ensures flexibility in terms of instrument mounting as panels with required access ports can be mounted as well as customized optical windows can be installed. Depending on the experiment, the measurement section can be equipped with different instruments to measure the prevailing turbulence, thermodynamic, and microphysical properties. These include the following:

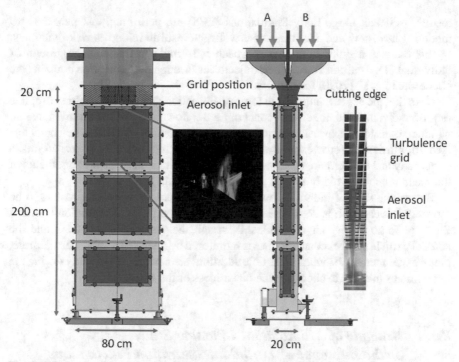

Fig. 8.7 A sketch of the measurement section is shown including its dimensions, the position of the turbulence grids, the cutting edge, and the aerosol inlet (© by Ingenieurbüro Mathias Lippold, VDI; TROPOS). The red box on the right-hand side marks the location where the particles are injected. The picture in the center shows a formed cloud which is illuminated by a green laser light sheet. (Figure reused with permission from Niedermeier et al. 2020 Open access under a CC BY 4.0 license, https://creativecommons.org/licenses/by/4.0/)

- A hot-wire anemometer to measure the mean flow velocity and velocity fluctuations as well as to obtain turbulence characteristics such as turbulence intensity and dissipation rate.
- Several PT100 resistance thermometers and a cold-wire anemometer to obtain mean temperature and temperature fluctuations.
- Two dew-point hygrometers to monitor the mean water vapor concentration in the particle-free airflows as well as in the measurement section.
- Two different optical sensors to determine cloud particle size distributions inside the measurements section: a white-light optical particle spectrometer and a 3D dual-phase Doppler anemometer.

After the measurement section, the whole flow is dried by means of an adsorption dehumidifying system, split up again into two airflows being driven by the radial blowers and cleaned by the particle filters.

Computational fluid dynamics (CFD) simulations accompany and complement the experimental LACIS-T studies. They are used, on the one hand, to determine suitable experimental parameters and, on the other hand, to interpret the experimental results. In detail, Large Eddy Simulations (LES) are performed in OpenFOAM® modeling heat, flow, and mass transfer as well as aerosol and cloud particle dynamics. In this context, a Euler–Lagrange approach is formulated tracking the growth of individual cloud particles along their trajectories through the simulation domain (see Niedermeier et al. (2020) for details).

Note that the experiments on the topic of formation and growth of cloud droplets require individualized conditions concerning the flow field and the thermodynamic parameters inside the measurement section. These conditions need to be characterized prior to the respective experiments. This includes high-resolution measurements of velocity and temperature (on the decimeter to millimeter scale), measurements of the mean relative humidity as well as numerical simulations.

Before the start of individual experiments, the measurement section has to be thoroughly cleaned. It is considered clean when the particle concentration is below 1 cm^{-3}. To do so, dry air is circulated through the system for about 1 h and the aerosol particle number concentration is monitored by means of a CPC. Furthermore, blank experiments (i.e., without particle addition) are performed regularly during the experiments in order to check for the cleanliness of the system.

8.1.4   Example of a Simulation Chamber Study on the Influence of Turbulent Saturation Fluctuations on Droplet Formation and Growth

In the following section, an experimental study on droplet formation and growth using LACIS-T is presented which aimed at investigating how turbulent saturation fluctuations influence the formed droplet size distribution (Niedermeier et al. 2020). The experiment was conducted as follows: a temperature difference of $\Delta T = 16 \text{ K}$ was set between the two particle-free airflows. The temperature and dew-point temperature of the airflows were set to 20 °C in branch A and 4 °C in branch B, respectively, so that RH = 100% in each air flow. Due to the mixing of both saturated air flows in the measurement section, super-saturation occurred. Based on earlier performed characterization experiments and corresponding LES in OpenFOAM® (not shown), the mean relative humidity (RH) was approximately 101.5%. For the investigations, size-selected, monodisperse NaCl particles with dry diameters $D_{p,dry}$ of 100 nm, 200 nm, 300 nm, and 400 nm are applied. For each injected $D_{p,dry}$, the particle concentration is set to 1000 cm^{-3}. The mean flow velocity inside the measurement section was 1.5 m s^{-1}. A Welas 2300 sensor was used for determining droplet size distributions during this type of experiment with the sensor being positioned at center position inside the measurement section at $z = 40$ cm or $z = 80$ cm below the aerosol inlet. For each $D_{p,dry}$, the sizes and numbers of droplets formed were measured for 20 min in order to obtain meaningful counting statistics.

The size distributions determined at the two positions are shown in Fig. 8.8. In both plots, the normalized droplet number versus the particle diameter is displayed.

The following observations can be made: (a) for each $D_{p,dry}$ the formed droplets grow with increasing distance to the aerosol inlet; (b) all size distributions nearly fall together at $z = 80$ cm; (c) the size distributions are negatively skewed; and (d) we also observe a significant number of particles close to $D_p = 300$ nm, i.e., close to the Welas 2300 detection limit.

To start with the interpretation of these observations, we included the critical diameters $D_{p,crit}$ for particle activation which are 1.2 μm, 3.4 μm, 6.3 μm, and 9.7 μm (dotted lines in Fig. 8.8 for $D_{p,dry} = 100$ nm, 200 nm, 300 nm, and 400 nm, respectively). Looking at Fig. 8.8, it can be seen that for $D_{p,dry} = 100$ nm and 200 nm most of the droplets feature sizes above the critical size and are therefore activated. However, the droplets grown on particles with dry sizes $D_{p,dry} = 300$ nm and 400 nm have sizes clearly below the critical size. These droplets are not activated, despite the mean super-saturation being sufficient for their activation. The droplet growth is kinetically limited which is the reason for this observation. In order to reach the respective critical diameter, the particles have to be exposed to a certain level of super-saturation for a given time frame (Chuang et al. 1997; Nenes et al. 2001). For the prevailing super-saturation, the needed time frame is on the order of several tens of seconds for the $D_{p,dry} = 300$ nm and 400 nm particles. However, it takes about 0.5 s to reach $z = 80$ cm inside LACIS-T which is too short for these particles to achieve their respective $D_{p,crit}$. Moreover, the given time frame also limits the growth of the droplets formed on the particles with $D_{p,dry} = 100$ nm and $D_{p,dry} - 200$ nm. Under the prevailing conditions and the sole observation of the grown droplet distributions, it is not possible to distinguish between the activated and non-activated droplet distributions. In other words, the droplet growth is kinetically limited, independent of whether the droplets are in the hygroscopic or dynamic

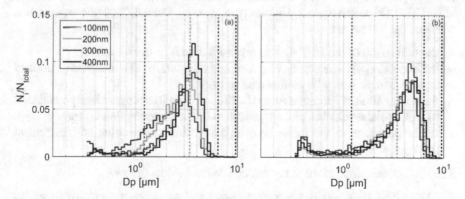

Fig. 8.8 Droplet formation and growth of differently size-selected, monodisperse NaCl particles Dp,dry = 100 nm–400 nm) for $\Delta T = 16$ K measured at two different positions below the aerosol inlet (left figure: $z = 40$ cm below the aerosol inlet and right figure: $z = 80$ cm below the aerosol inlet). The dotted lines represent the critical diameters Dp,crit for particle activation which are 1.2 μm, 3.4 μm, 6.3 μm, and 9.7 μm for Dp,dry = 100 nm, 200 nm, 300 nm, and 400 nm, respectively. (Figure reused with permission from Niedermeier et al. 2020 Open access under a CC BY 4.0 license, https://creativecommons.org/licenses/by/4.0/)

growth mode and the dry particle size is of minor importance for the observed droplet distributions.

For the interpretation of the significant number of particles close to $D_p = 300$ nm and the negative skewness of the size distributions, we consider the results from the respective LES which yield a RH standard deviation in the order of 4% (absolute). It can be concluded from these simulations that the small particles ($D_p < D_{p,crit}$), on the one hand, are hygroscopically grown particles which did not experience super-saturated conditions and, on the other hand, are evaporating droplets because they experienced sub-saturated conditions in the fluctuating saturation field.

In conclusion, the turbulent saturation fluctuations broaden the droplet size distribution toward smaller diameters caused by evaporating droplets or less-grown droplets, i.e., turbulence influences cloud droplet activation and growth/evaporation. The obtained results also imply that droplet activation in a turbulent environment may be inhibited due to kinetic effects/limitations. On the other hand, locally elevated super-saturations may occur due to turbulence which might lead to an increase of the activated droplet number.

8.1.5 Example of Using a Simulation Chamber as Platform for Instrument Test and Intercomparison

One important role for simulation chambers is in providing a well-defined and controllable environment to test instrumentation. A performance evaluation of the Ultrafast Thermometer (UFT) 2.0 under turbulent cloudy conditions was undertaken as part of EUROCHAMP-2020 trans-national access (TNA) activity in 2019 at LACIS-T (PI Jakub Nowak, University of Warsaw, Poland). Specific experiments included the following:

- the calibration of the UFT sensors against a reference thermometer;
- the inspection of the accuracy, response, and orientation dependence in a turbulent flow by comparison with commercial sensors;
- the examination of the character and likelihood of wetting under cloudy conditions as well as the estimation of its dependence on the incidence angle; and
- the investigation of the influence of salt deposition on wetting and instrument performance.

The main conclusions of the study can be summarized as follows:

1. All studied UFT versions provide accurate and consistent temperature readings when being linearly calibrated against the reference sensor. The accuracy and response of the UFTs allow for studying details of mixing between air masses differing in temperature.
2. The effect of the incidence angle, i.e., tilting the sensors by a chosen angle with respect to the mean flow in the LACIS-T, has a negligible effect on the mean temperature but significantly influences the obtained temperature fluctuations.

3. In super-saturated air, water vapor condensation has a major contribution to the sensor wetting in contrast to collisions of cloud droplets. The wetting manifests in a decreasing time response due to the growing total heat capacity.
4. Salt deposited on the sensors does not exert a measurable effect on the temperature measurement and the probability of wetting. However, it might contribute to the mechanical deterioration of the instrument which results in floating calibration and intensifies the chance of entire sensor damage. Excessive salt deposition, although unlikely for atmospheric conditions, can trigger hygroscopic condensation already at the relative humidity of about 76% which is well below saturation level.

This study has provided valuable information concerning the properties and performance of the whole family of the UFT thermometers. It has been the first experiment in which all the versions were systematically compared with a reference and between each other in controlled turbulent flow with well-defined thermodynamic conditions. Further work will involve the improvement of the design, in particular, introducing a mechanism preventing the instrument from condensational wetting, e.g., with hydrophobic coatings or alternate heating in a double-sensor device to periodically evaporate collected water.

8.2 Application of Simulation Chambers to the Study of Processes at the Air–Sea Interface

The coupling between oceans and the atmosphere influences a broad range of processes, from nutrient balance for marine biology, to climate. Logically, as the oceans cover most of the Earth's surface, they also exert a major control on the atmospheric concentration of many trace gases. In fact, air–sea exchanges are key for the atmospheric chemistry, physics, and the biogeochemistry of the oceans (Liss and Johnson 2014). The exchange of trace gases between the oceans and the troposphere is a multifaceted process involving several physical, chemical, and biological processes in each media. In this context, wind speed is a crucial parameter as it influences bubble bursting, waves, rain, and surface films (Garbe et al. 2014). Bubble bursting contributes largely to the marine aerosol budget through the injection of small droplets into the atmosphere (de Leeuw et al. 2011), while surface films influence the air–sea gas exchange via several mechanisms due to their particular characteristics.

The top layer of the oceans operationally defined as the top 1 μm to 1 mm of the ocean is often called the sea surface microlayer (SML). It possesses different chemical and physical properties than the underlying water due to reduced mixing in this region. The SML is enriched in organic and inorganic matter, mainly hydrophobic in nature, but also of associated microorganisms (Cunliffe et al. 2011; Liss and Duce 1997). This surface layer is chemically reactive, as it contains a significant fraction of dissolved organic matter (DOM), containing a high proportion of functional groups such as carbonyls, aromatic moieties, and carboxylic acids (Sempere and Kawamura 2003; Stubbins et al. 2008) and can be conceived as a complex gelatinous film (Cunliffe et al. 2013).

This section will present a chamber-based strategy to investigate the chemical processes, at low wind speed or in other words with a focus on understanding the associated interfacial chemistry without bubble bursting. After designing a dedicated chamber, we will go through some examples showing the impact on VOC emissions and organic aerosol formation.

8.2.1 Design of a Simulation Chamber Dedicated to Air–Sea Processes Study

One of the more significant artifacts in chamber investigations is due to the influence of walls on the observed chemistry. Therefore, investigating the air/water interfacial chemistry in a chamber can be achieved by turning a drawback into an advantage, i.e., by placing the interface of interest on (ideally all) the walls of a given chamber. In this case, we will investigate photochemical processes on a liquid surface mimicking the SML on top of bulk water, based on in situ monitoring of gases and particles, i.e., the experimental samples will be made of bulk water, containing a photosensitizer of interest; the interface, enriched with a given surfactant; and the overlying gas phase.

For this purpose, a $2 \, m^3$ chamber ($1 \, (l) \times 1 \, (L) \times 2 \, (h) \, m$) made of FEP (fluorinated ethylene propylene) film was built for this purpose (Fig. 8.9). To mimic the ocean, a glass container can be placed at the bottom of the chamber giving a reactive surface to be investigated. In the specific example presented here, the glass container had a capacity of 89 L and an exposed surface of $0.64 \, m^2$.

The chemical processes occurring on this surface have to be dominant compared to those occurring on the remaining walls in order to obtain valuable information. Therefore, operating the chamber under clean conditions is essential, but made complicated due to the high intrinsic relative humidity in such experiments (i.e., experimental runs with a significant volume of liquid water).

The experimental chamber and water basin have to be scrubbed using ethanol, then rinsed with water and dried thoroughly before each experiment. After cleaning, the chamber can be flushed with 40 L min^{-1} of N_2 for more than 48 h after which $RH < 5\%$ is maintained. After flushing, background checks have to be performed using a 7.5 L min^{-1} flow of N_2 and turning on and off visible and UV light (see below for the light specifications) and with and without the presence of 0.6–5.0 ppm of O_3 and water. Clean chamber conditions could be considered as met if particle concentration remained below 1 cm^{-3} and the sum of NO and NO_2 concentrations (NO_x) is kept at <0.6 ppb. After cleaning, 30 L of water (resistivity of 18.2 MΩ cm) can be introduced into the basin at a liquid flow rate of ~1 L min^{-1} using a peristaltic pump and Teflon plumbing lines. Background checks can again be performed using UV irradiation and ozone injection before and after filling the basin with pure water.

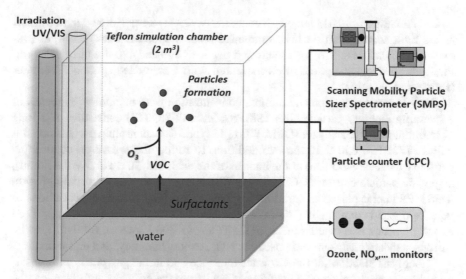

Fig. 8.9 Scheme of the multiphase atmospheric simulation chamber used for the investigation of chemical processes at the air–sea interface

The concentration of NO_x always should remain <1 ppb and particle concentrations <10 cm^{-3} during water injection. After all background levels were established, the basin was emptied, dried, and then refilled with the solution to be investigated.

The actual experiment starts with filling with the basin with ca. 30 L of aqueous samples (i.e., water + photosensitizer) again at a liquid flow rate of · 1 L min^{-1} to which a surfactant can be added, through a septum installed immediately before the basin, at a concentration leading to mono- to multi layer coverage at the air/water interface. These surfactants covered the 0.6 m^2 water surface in excess such that surfactant lenses were in equilibrium with a monolayer at its respective equilibrium spreading pressure. The chamber is then flushed again (40 L min^{-1} N$_2$) for hours or days and returned to experimental conditions (7.5 L min^{-1} N$_2$) for another day before commencing UV irradiation of the surfactant interface. The systems investigated are listed in Table 8.1.

Table 8.1 Seawater surrogates used in simulation chamber experiments involving air–sea interface

Bulk water composition	Added surfactant	References
Water + Humic acid	Nonanol	Alpert et al. (2017)
Water + Humic acid	Nonanoic acid	Alpert et al. (2017), Bernard et al. (2016)
Authentic biofilms	None—surfactant was produced in situ following cell lysis	Bruggemann et al. (2017, 2018)

12 UV lamps (OSRAM lamps, Eversun L80W/79-R) are positioned in two banks as the light source, with six lamps mounted on two opposite sides. UV light irradiated the chamber at 8 W m^{-2} measured between 300 and 420 nm in wavelength. Figure 8.10 shows the photon flux measured with a calibrated spectrophotometer (Barsotti et al. 2015).

Particle size distribution and number concentration were monitored by means of a scanning mobility particle sizer (SMPS Model 3936, TSI) consisting of a long differential mobility analyzer (DMA 3081, TSI) and a condensation particle counter (CPC 3772, TSI, d50 > 10 nm). In addition, to follow the formation of ultrafine particle (diameter > 2.5 μm) at the bottom of the chamber (30 cm above the liquid), a specific particle counter (UCPC 3776, TSI) is used. To observe particle growth, the SMPS inlet is placed an additional 150 cm higher. Gas-phase concentrations of volatile organic compounds (VOCs) are monitored using a high-resolution proton transfer reaction mass spectrometer (PTR-TOF–MS 8000, Ionicon Analytik), while various standard analyzers were used for NO_x, ozone, humidity, and temperature.

Ozonolysis reactions at the air/water interface and in the gas phase with unsaturated compounds, photochemically produced from the air/water interface, can be triggered by injecting ozone into the chamber up to a concentration of 600 ppb. Ozone is either generated using a corona discharge (Biozone Corporation, USA) or a UV light generator (Jelight Model 600). Fast introduction of ozone in the chamber, reaching the desired concentration, is achieved in a few minutes. This concentration is used to initiate SOA particle nucleation and verify the presence of unsaturated VOCs, although it is higher than typically observed at the Earth's surface. The concentration and size distribution of new particle formation over time subsequent to ozone injection can therefore be monitored.

After each experiment, the chamber was cleaned for at least 24 h by flushing purified air at high flow rates in presence of ozone, at several ppm, under maximum

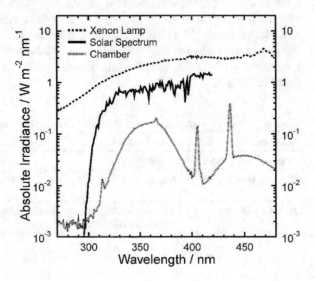

Fig. 8.10 Absolute irradiance and a function of wavelength for the 12 UV fluorescent light tubes used in chamber experiments, the actinic solar spectrum , and a xenon lamp shown as dotted, solid, and dashed lines, respectively

irradiation. The glass container was also evacuated and rinsed with alkaline ([NaOH] = 10 mM) and ultra-pure water several times, in order to promote the dissolution of organic materials.

8.2.2 Example of a Simulation Chamber Study of a Photosensitized Production of Aerosol at the Air–Sea Interface

Using this simulation chamber, we were able to investigate aerosol formation from photosensitized reactions at the air–water interface (Alpert et al. 2017; Bernard et al. 2016; Rosignol et al. 2016). For this purpose, aqueous solutions of humic acid (HA), used as a proxy for dissolved organic matter and hence as photosensitizer, and nonanoic acid (NA) as a surfactant, were introduced into the chamber. Then the lights were switched on to trigger the targeted photochemical processes.

The NA concentrations ranged from 0.1 mM to 10 mM, while humic acid was added in the range from 1 to 10 mg L^{-1}. After introducing 15 L of ultra-pure water (for 20 min), NA was injected. The formation of small "organic islands" of NA was minimized, but not avoided, by using a very slow injection rate. Over time, these islands agglomerated, increasing their size, and simultaneously decreasing their number concentration. At 0.1 mM of NA, they rapidly disappeared after the introduction of the acid. HA was usually injected around 25 min after NA. Allowing enough to the equilibrium time (ca. 90 min) of NA between the gas and liquid phases, lamps were switched on. The actual experiment lasted for at least 14 h. During the irradiation period, temperature and relative humidity were stable at about 300 K and 84%, respectively.

Particle formation

Figure 8.11 shows a typical experiment where photosensitized production of aerosol was observed, while the list of experiments and corresponding initial conditions are summarized in Table 8.2. Particles in the chamber were subject to dilution, wall loss, and coagulation processes, which represented sink processes. Reported particle concentrations were not corrected for these losses. Background concentrations of particles before the irradiation period were found to be in the range of <50–500 cm^{-3}. As shown in Fig. 8.11, a significant production of secondary organic aerosol was observed rapidly after the injection of ozone. It is noteworthy to underline that the initial composition of the gas phase did not carry any chemical functionality that was expected to react through ozonolysis. In other words, compounds reacting with ozone were produced through photosensitized chemistry at the air–water interface.

These observations contrast sharply with our blank experiments. In the absence of any surfactant, HA (20 mg L^{-1}) photochemistry did not produce any particles and their concentration did not exceed background levels. It is important to note

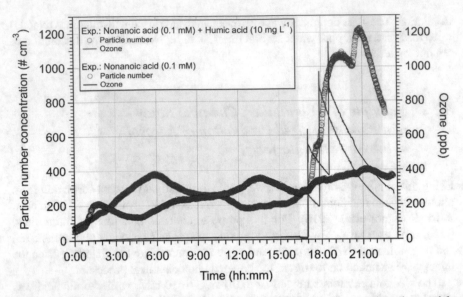

Fig. 8.11 Comparison of particle formation measured with the ultrafine condensation particle counter from smog chamber experiments conducted in the presence of humic acid (HA) and nonanoic acid (NA) (Exp. 2 in Table 8.2) compared to nonanoic acid (NA) only (Exp. 3 in Table 8.2). The yellow sections are the periods when the lights were on

Table 8.2 Experimental initial conditions: dark ozone reaction after UV light processing of the liquid mixture of nonanoic acid (NA) and humic acid (HA)

Experiment	[NA] mM	[HA] mg L^{-1}	[O$_3$] ppb	Particle number^a cm^{-3}
1	–	20	829	112
2	0.1	10	250	285
3	0.1	–	526	64
4	0.1	10	500	396
5	0.1	1	301	68
6	2	–	534	84
7	2	10	461	3057
8	0.5	10	476	568
9	1	10	391	887

^{a}Particle number concentrations were subtracted from the particle background

that the HA concentration used in the blank experiment was higher than for mixed HA and NA experiments. Also, with NA only, no significant dark particle formation (as compared to the results shown below) was observed after introducing ozone. Some residual photochemistry of NA films was observed and avoided by using low concentrations. This highlights the involvement of HA in the photochemical transformation of NA and thus demonstrates that particle formation originated from the photosensitized reaction.

The gaseous temporal profiles of volatile organic compounds were monitored by means of the PTR–ToF–MS instrument. A large number of products were identified with various chemical functionalities, such as saturated aldehydes (C_7–C_9), unsaturated aldehydes (C_6–C_9), alkanes (C_7–C_9), alkenes (C_5–C_9), and dienes (C_6–C_9).

Interestingly, in the absence of ozone, these gas products only lead to a small SOA production, with particle number concentrations ranging from 150 to 700 cm^{-3} and close to the background levels prior to irradiation, showing that direct photochemical processes were not important under our experimental conditions.

Lights were turned off and 30 min after ozone was added in the dark. OH radical formation might be scavenged by gaseous NA ($k_{OH+NA} = 9.76 \times 10^{-12}$ cm^3 molecule^{-1} s^{-1}) (Cui et al. 2019). For all the experiments with the combined presence of NA and HA, new particle formation was observed, confirming the production of SOA precursors among all produced VOCs. The observed maximum background subtracted number concentrations ranged from 68 to 3060 cm^{-3}. The lowest number concentration was observed with the lowest concentration of NA (0.1 mM) and HA (1 mg L^{-1}), while the highest one was logically with the highest concentrations of both NA (2 mM) and HA (10 mg L^{-1}). This highlights the fact that both bulk and surface concentrations are key drivers in the observed SOA formation. The total particle mass concentration (ΔM_0) formed under these experimental conditions did not exceed 1 $\mu g\ m^{-3}$ during the dark ozonolysis reaction. This chemistry led to the formation of condensable organic vapors of volatility low enough to induce the formation of new particles. Such compounds have been referred to as *extremely* low-volatility organic compounds (LVOC) (Ehn et al. 2014). New particle formation is characterized by a significant increase in particle number, with low mass concentrations. The observed SOA production is in agreement with the formation of products, bearing one or several unsaturated sites, which are potential SOA precursors.

SOA formation potential from photosensitized reactions

A surfactant will alter the surface tension of a liquid as described by the Gibbs adsorption isotherm, where the surface excess concentration of nonanoic acid (NA) is as follows (Donaldson and Anderson 1999):

$$\Gamma_{NA} = -\frac{1}{RT} \times \left(\frac{d\gamma}{dC_{NA}} \right) \qquad (E8.2.2.1)$$

where Γ_{NA} is the surface excess concentration of NA (in molecules cm^{-2}), representing the amount of NA at the air–sea interface; C_{NA} the bulk concentration of NA (in mol cm^{-3}); γ is the surface tension (in N m^{-1}); R is the gas constant; and T is the temperature in Kelvin. The surface tensions of the systems used in this study were previously measured (Ciuraru et al. 2015), leading to surface excess concentrations in the range from 1.66×10^{14} to 3.99×10^{14} molecules cm^{-2} in this work.

Assuming that the humic acids are evenly distributed in the solution, at pH ≈ 4, HA are fully soluble in the concentration range used (Klaviņš and Purmalis 2014), and the surface concentration of the surfactant drives the chemical formation, then a correlation between the measured number particles and the chemical formation rate can be expected (Boulon et al. 2013). In fact, assuming that both nonanoic and humic acids were in large access, i.e., constant during the experiment, the chemical production rate of gaseous products P_g (in molecule cm^{-3} s^{-1}), neglecting the influence of mass transport or dilution in the chamber, can be simplified as

$$P_g \propto k \times \Gamma_{NA} \times [HA] \times (A/V) \qquad \text{(E8.2.2.2)}$$

where k (cm^3 molecule^{-1} s^{-1}) corresponds to the overall rate coefficient for the photosensitized reaction of NA in the presence of HA, including reactions kinetics, product yields and phase transfer kinetics, A is the surface area of the liquid (in m^2), and V is the internal volume of the chamber (in m^3).

The amount of condensable vapor (here simply named [LVOC]) is then related to the ozonolysis of unsaturated products (functionalized alkenes). Assuming that the ozonolysis reaction occurred under pseudo-first-order conditions, the concentration of condensable products (in molecule cm^{-3}) can be expressed as

$$[LVOC] \propto k \times \Gamma_{NA} \times [HA] \times t_{irr} \times (A/V) \times \left[1 - \exp(-k_{O3} \times [O_3]) \times t_{ind}\right]$$
$$\text{(E8.2.2.3)}$$

where k_{O3} corresponds to the bimolecular reaction rate coefficient of ozone reacting with unsaturated compounds (in cm^3 molecule^{-1} s^{-1}), t_{irr} is the irradiation time (in s), and t_{ind} is the induction time (in s) corresponding to the time interval between the introduction of ozone and particle measurements.

Hereby, we assumed that the ratio of the SOA precursor concentration to the total amount of products is similar whatever the initial liquid-phase concentrations are. Figure 8.12 shows indeed a correlation between the number of particles and the concentration of condensable vapors, similar to Boulon et al. (2013). Under our experimental conditions, the formation of particles was in fact photochemically controlled by the photochemical interfacial process, and not the ozone concentration, which was always in excess.

This example highlights the peculiar photochemistry chemistry occurring at the SML, which *in fine* affects the emission of oceanic VOC. In the field, decoupling such processes from physical ones (mixing though waves, wind, bubble bursting, etc.) would be quite challenging. In this study, the use of a multiphase atmospheric simulation chamber has proven to be a reliable approach to explore the in situ formation of gases and particles from photo-induced chemical processes at the air–water interface. Therefore, investigating processes occurring at the air–sea interface in a dedicated multiphase chamber opens new routes for characterizing specifically interfacial chemical pathways.

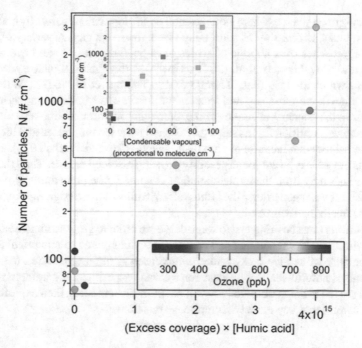

Fig. 8.12 Particle number concentration as a function surface excess coverage of humic acid and ozone concentration, and the number of particles (N) and the estimated levels of condensable vapors (proportional to molecule cm $^{-3}$)

8.3 Application of Simulation Chambers to the Study of Cryosphere–Atmosphere Interface

8.3.1 Introduction

The upper surface of sea ice is one of the important interfaces with the atmosphere (Law et al. 2013) and the role of snow and ice in mediating important aspects of atmospheric chemistry continues to be an important topic (Abbatt et al. 2012). At its lower surface, sea ice forms the boundary with the ocean and hence mediates the transport of a variety of physical (e.g., energy, momentum) and chemical components (gas, particles) between the atmosphere and the polar ocean. Not only is sea ice an interface between the atmosphere and ocean, it is also an important environment in its own right and is the location for many important biological and chemical processes (Fritsen et al. 1994; Garnett et al. 2019; King et al. 2005; Vancoppendle et al. 2013).

Natural sea ice is difficult and expensive to access. Sea ice is also extremely heterogeneous in space and time (Miller et al. 2015) with interesting phenomena occurring during formation and melting. The lack of observational data means that many scientific questions remain to be addressed (Swart et al. 2019). Observing laboratory-grown sea ice, where experimental conditions can be carefully controlled in sea-ice tanks, is one way of addressing this knowledge gap. A variety of experimental approaches over a range of scales have been developed and have recently been reviewed by Thomas et al. (2021). However, the enclosed tank environment poses its own challenges (e.g., Thomas 2018; Thomas et al. 2021). Wall effects can alter the sea-ice freeboard and stresses within the sea ice. Severe super-cooling can occur in the "ocean" of sea-ice tanks, damaging instrumentation and hindering measurements. Additionally, as salt is partitioned between the sea ice and the ocean, the ocean salinity can increase to unrealistic levels (Cox and Weeks 1975).

This protocol is relevant for any facility growing artificial sea ice. Though facility specific issues may limit the implementation of some of the procedures grown here, experimenters will need to keep the issues raised in mind when designing experiments and contextualizing their results.

Tank effects must be mitigated to some degree in order to grow artificial sea ice that is scientifically relevant and that can be reasonably compared to numerical models. The generality of sea-ice tank results is increased if the artificial sea ice closely approximates natural sea ice, at least for relevant experimental parameters. If the artificial sea ice is being observed to evaluate numerical models, then experimenters must be careful that key model assumptions are satisfied.

8.3.2 Preparation of Synthetic Sea-Ice Growth

To grow artificial sea ice, researchers must first make their artificial ocean, with a realistic and quantified salinity and composition.

Secondly, they should have a facility that allows a downward freezing of the ocean surface. The ocean is contained in a tank and cooled to near its freezing point. Further cooling should only affect the ocean surface, which will result in the formation of sea ice at its surface. Additional cooling of this sea-ice layer will result in sea-ice growth and a thickening of the sea-ice layer as the sea-ice/ocean interface advances downward.

There are two important aspects of natural sea-ice growth that are difficult to accomplish in the laboratory. First, the upper part of natural sea ice is exposed to extremely cold atmosphere (≤ -60 °C) while the bottom part is constantly at the seawater freezing (-1.86 (around -2 °C)). In a tank experiment, the challenge is to expose the surface ocean and sea ice to freezing temperature while maintaining the bulk-underlying ocean at or just above the freezing point. Maintaining the ocean above freezing prevents super-cooling effect.

Second, natural sea ice is generally free floating, with a freeboard of around 10% of the sea-ice thickness. In the laboratory, sea ice tends to attach to the tank walls, which increases the hydrostatic pressure into the tank due to volume expansion of the sea-ice ocean system. Non-floating sea ice induces generally an artificial upward movement of seawater moving into the ice, which floods the sea-ice surface (Rysgaard et al. 2014).

To ensure freezing from the seawater surface, eliminate super-cooling and avoid ice formation along the walls, the tank sides need to be heavily insulated from the cold atmosphere and/or slightly heated (Naumann et al. 2012; Wettlaufer et al. 1997; Cox and Weeks 1975). Trial and error is required to find the right level of insulation and heating. Our best methodology is to mount heating pads between the glass and the surrounding instrumentation (Fig. 8.13). If sea ice is observed to creep down the tank sides, or if sea ice forms on instrumentation or the corners/base of the tank, the insulation and heating were not sufficient. If the sea ice forms a bowl shape, with greatly reduced thickness at the sides of the tank, the heating is too strong. Bowl-shaped sea ice and creeping sea ice are both visible in Fig. 8.14. With heating pads placed in direct contact with the water, the heating was too strong and local, where a heating pad broke in this run sea ice can be seen to creep down the side of the tank.

Heating the sides of the tank may be sufficient to maintain free-floating sea ice. Such an approach is particularly effective when the tank sides are smooth (glass, for example) or if they are angled such that the sea ice forms in a wedge shape, wider at the top than the bottom, and so floats up. Free floating sea ice bobs when pushed and when a hole is cut in the surface the water line is shallower than the surface.

We recommend having temperature probes recording the temperature in the atmosphere to ensure that temperature stay below freezing. We advise to also monitor the temperature along the tank walls to have a better control of the heat input and avoid freezing on the wall. Finally, monitoring the seawater salinity and bulk temperature with a CTD is necessary to detect potential super-cooling effect.

Our main limitation is linked to the absence of a dilution reservoir. When sea ice forms, it rejects salt into underlying water, which causes an increase in salinity of

Fig. 8.13 Picture of tank with heating pads placed outside the glass, which is our preferred method, and instrumentation mounted on poles. (Picture from Roland von Glasow Air-Sea-Ice Chamber, Centre for Ocean and Atmospheric Sciences, University of East Anglia)

Fig. 8.14 View of sea ice from below during an early trial run with sub-optimal heating. (Picture from Roland von Glasow Air-Sea-Ice Chamber, Centre for Ocean and Atmospheric Sciences, University of East Anglia)

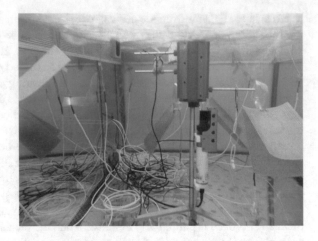

our artificial ocean. To maintain a constant salinity in our artificial ocean during a sea-ice growth experiment, we need to install a dilution system.

8.3.3 Step-by-Step Procedure for Growing Synthetic Sea-Ice

The first step in growing artificial sea ice is to prepare the tank. First, we place heating film (220 W/m^2) on the outside surface of tank walls (Fig. 8.13). Secondly, we insulate the tank sides with quilt insulation and 10 cm of Dow Floormate 500A foam. This setup is sufficient for us to prevent super-cooling in the ocean and to maintain free-floating sea ice up to at least 20 cm thickness.

Ocean instrumentation is mounted on a fixed pole, while sea-ice instrumentation is mounted on a pole that is free to rise in the vertical (Fig. 8.14). As sea ice grows, it is therefore free to rise and maintain a natural freeboard. Cables for all instrumentation are run out of the tank through the ocean and a smaller tank attached to the main tank. These cables therefore do not disturb the sea-ice surface. Pumps are also installed that allow mixing of the ocean. These pumps face each other so at to generate turbulence while minimizing currents (Loose et al. 2011).

Once the insulation and instrumentation are in place, the tank is filled with some artificial ocean. The salinity of this ocean should be realistic (28–35 g kg^{-1}) and the salt composition should be well characterized. Knowing the precise salt composition allows the freezing point of the ocean to be accurately modeled. We often use pure NaCl. When a natural salt composition is required, we use filtered, real seawater, or some aquarium salt mix (Tropic Marin). Salts are mixed with deionised water using the pumps, generally taking around a day to dissolve.

The ocean then needs to be cooled to near its freezing point. We set the coldroom to −20 °C and run the pumps on full during this cooling period to ensure to have a well-mixed seawater before the start of an experiment. When we are ready to start the experiment, we turn the pumps off or set them to their minimum flow rate. Sea-ice formation then begins within an hour or so providing the ocean is within a few tenths of a degree of its freezing point.

During an experiment it is best to enter the coldroom as infrequently as possible. Each time the door is opened there is an influx of warm, moist air into the coldroom, disrupting experimental conditions. Similar to Naumann et al. (2012), when the pumps are on, a layer of grease ice will form and with pumps off nilas will form (Fig. 8.15). A few periodic checks may be necessary, depending on the nature of the experiment. Whether or not the sea ice is free floating can be checked by gently pressing the sea ice at one corner. If it bobs it is free floating. When sampling sea ice, the freeboard can be checked and compared with that expected from the thickness of the sea ice. A shiny wet upper surface is a sign that the sea ice may have fixed to the tank sides and that the surface has flooded. Super-cooling can be inferred by precisely measuring the ocean temperature and salinity, and comparing the in situ temperature to the salinity-dependant freezing point. Severe super-cooling tends to make the ocean salinity and temperature readings increasingly noisy.

In some experiments, sea ice is grown from a cold plate (Wettlaufer et al. 1997; Eide and Martin 1975; Niederauer and Martin 1979; Middleton et al. 2016) in direct contact with the sea-ice surface. The position of the upper interface is defined in this case and the freeboard of the sea ice can only be maintained by adjusting the

Fig. 8.15 Top: Grease ice forming under turbulent growth conditions. Bottom: Nilas forming in quiescent conditions. (Pictures from Roland von Glasow Air-Sea-Ice Chamber, Centre for Ocean and Atmospheric Sciences, University of East Anglia)

ocean volume using a hydrostatic pressure release valve. In small tanks, the heating required to maintain free-floating sea ice may affect the sea-ice growth to such a degree as to be prohibitive.

Future work involves the deployment of atmospheric measurements above the sea ice in order to qualify the impact of sea-ice growth on decay on atmospheric chemistry. The facility is already equipped with several dedicated gas analyzers. A Los Gatos greenhouse gas analyzer (Los Gatos 30R-EP) measures CO_2, CH_4, and H_2O vapor. A T200 UP Teledyne measures NO_x, a T200 U Teledyne measures NO_y, and there is an ozone analyzer (T400 Teledyne) and generator. A lighting rack sits already between 1.5 m above sea-ice tank surface to allow atmospheric photochemical experiments. Solar spectrum LED (FluenceSolar Max), UV-Aa (Cleo performance 100 W), and UV-B (Phillips broadband TL100W) fluorescent bulbs are evenly spaced over the tank in sets of three, with 24 lights in total (Fig. 8.16). Currently, we can create an artificial atmosphere above the main tank by attaching cuboid 50 μm FEP Teflon atmosphere. FEP Teflon is transparent in the visible and UV spectrum, and chemically inert, making it ideal for many photochemical experiments. When the tank is covered with an artificial atmosphere, the temperature and the humidity of the contained headspace increase. The increase of temperature decreases drastically the ice-growing process and the increase of humidity causes ice formation on surfaces in the headspace inducing condensation and refreezing on the Teflon atmosphere. To pursue measurements in the artificial atmosphere, we should need to develop a system extracting the heat and humidity trap in the headspace during ice growth.to extract heat and moisture from the headspace.

Fig. 8.16 Lights on above tank. (Picture from Roland von Glasow Air-Sea-Ice Chamber, Centre for Ocean and Atmospheric Sciences, University of East Anglia)

References

Abbatt, J.P.D., Benz, S., Cziczo, D.J., Kanji, Z., Lohmann, U., Möhler, O.: Solid ammonium sulfate aerosols as ice nuclei: a pathway for cirrus cloud formation. Science 313, 1770–1773 (2006). https://doi.org/10.1126/science.1129726

Abbatt, J.P.D., Thomas, J.L., Abrahamsson, K., Boxe, C., Granfors, A., Jones, A.E., King, M.D., Saiz-Lopez, A., Shepson, P.B., Sodeau, J., Toohey, D.W., Toubin, C., von Glasow, R., Wren, S.N., Yang, X.: Halogen activation via interactions with environmental ice and snow in the polar lower troposphere and other regions. Atmos. Chem. Phys. 12, 6237–6271 (2012). https://doi.org/10.5194/acp-12-6237-2012

Alpert, P.A., Ciuraru, R., Rossignol, S., Passananti, M., Tinel, L., Perrier, S., Dupart, Y., Steimer, S.S., Ammann, M., Donaldson, D.J., George, C.: Fatty acid surfactant photochemistry results in new particle formation. Sci. Rep. 7, 12693 (2017). https://doi.org/10.1038/s41598-017-12601-2

Augustin, S., Wex, H., Niedermeier, D., Pummer, B., Grothe, H., Hartmann, S., Tomsche, L., Clauss, T., Voigtländer, J., Ignatius, K., Stratmann, F.: Immersion freezing of birch pollen washing water. Atmos. Chem. Phys. 13, 10989–11003 (2013). https://doi.org/10.5194/acp-13-10989-2013

Augustin-Bauditz, S., Wex, H., Kanter, S., Ebert, M., Niedermeier, D., Stolz, F., Prager, A., Stratmann, F.: The immersion mode ice nucleation behavior of mineral dusts: a comparison of different pure and surface modified dusts. Geophys. Res. Lett. 41, 7375–7382 (2014). https://doi.org/10.1002/2014GL061317

Barsotti, F., Brigante, M., Sarakha, M., Maurino, V., Minero, C., Vione, D.: Photochemical processes induced by the irradiation of 4-hydroxybenzophenone in different solvents. Photochem. Photobiol. Sci. 14, 2087–2096 (2015). https://doi.org/10.1039/c5pp00214a

Bernard, F., Ciuraru, R., Boréave, A., George, C.: Photosensitized formation of secondary organic aerosols above the air/water interface. Environ. Sci. Technol. 50, 8678–8686 (2016). https://doi.org/10.1021/acs.est.6b03520

Bodenschatz, E., Malinowski, S.P., Shaw, R.A., Stratmann, F.: Can we understand clouds without turbulence? Science 327, 970–971 (2010). https://doi.org/10.1126/science.1185138

Bohren, C.F., Albrecht, B.A.: Atmospheric Thermodynamics. Oxford University Press, New York (1998)

Boulon, J., Sellegri, K., Katrib, Y., Wang, J., Miet, K., Langmann, B., Laj, P., Doussin, J.F.: Sub-3 nm particles detection in a large photoreactor background: possible implications for new particles formation studies in a smog chamber. Aerosol Sci. Technol. 47, 153–157 (2013). https://doi.org/10.1080/02786826.2012.733040

Bruggemann, M., Hayeck, N., Bonnineau, C., Pesce, S., Alpert, P.A., Perrier, S., Zuth, C., Hoffmann, T., Chen, J.M., George, C.: Interfacial photochemistry of biogenic surfactants: a major source of abiotic volatile organic compounds. Faraday Discuss. 200, 59–74 (2017). https://doi.org/10.1039/c7fd00022g

Brüggemann, M., Hayeck, N., George, C.: Interfacial photochemistry at the ocean surface is a global source of organic vapors and aerosols. Nat. Commun. 9, 2101 (2018). https://doi.org/10.1038/s41467-018-04528-7

Chai-Mei, J.: Ambient Air Treatment by Titanium Dioxide (TiO2) Based Photocatalyst in Hong Kong The Chinese University of Hong Kong, Hong KongTender Ref. AS 00–467 (2002)

Chandrakar, K.K., Cantrell, W., Chang, K., Ciochetto, D., Niedermeier, D., Ovchinnikov, M., Shaw, R.A., Yang, F.: Aerosol indirect effect from turbulence-induced broadening of cloud-droplet size distributions. Proc. Natl. Acad. Sci. 113, 14243–14248 (2016). https://doi.org/10.1073/pnas.1612686113

Chang, K., Bench, J., Brege, M., Cantrell, W., Chandrakar, K., Ciochetto, D., Mazzoleni, C., Mazzoleni, L.R., Niedermeier, D., Shaw, R.A.: A laboratory facility to study gas–aerosol–cloud interactions in a turbulent environment: the Π chamber. Bull. Am. Meteor. Soc. 97, 2343–2358 (2016). https://doi.org/10.1175/bams-d-15-00203.1

Chang, K., Bench, J., Brege, M., Cantrell, W., Chandrakar, K., Ciochetto, D., Mazzoleni, C., Mazzoleni, L.R., Niedermeier, D., Shaw, R.A.: A Laboratory facility to study gas–aerosol–cloud interactions in a turbulent environment: the Π chamber. Bull. Am. Meteor. Soc. **97**, 2343–2358 (2017). https://doi.org/10.1175/bams-d-15-00203.1

Chuang, P.Y., Charlson, R.J., Seinfeld, J.H.: Kinetic limitations on droplet formation in clouds. Nature **390**, 594–596 (1997). https://doi.org/10.1038/37576

Ciuraru, R., Fine, L., van Pinxteren, M., D'Anna, B., Herrmann, H., George, C.: Photosensitized production of functionalized and unsaturated organic compounds at the air-sea interface. Sci. Rep. **5**, 12741 (2015). https://doi.org/10.1038/srep12741

Connolly, P.J., Saunders, C.P.R., Gallagher, M.W., Bower, K.N., Flynn, M.J., Choularton, T.W., Whiteway, J., Lawson, R.P.: Aircraft observations of the influence of electric fields on the aggregation of ice crystals. Q. J. r. Meteorol. Soc. **131**, 1695–1712 (2005). https://doi.org/10.1256/qj.03.217

Connolly, P.J., Möhler, O., Field, P.R., Saathoff, H., Burgess, R., Choularton, T., Gallagher, M.: Studies of heterogeneous freezing by three different desert dust samples. Atmos. Chem. Phys. **9**, 2805–2824 (2009). https://doi.org/10.5194/acp-9-2805-2009

Connolly, P.J., Emersic, C., Field, P.R.: A laboratory investigation into the aggregation efficiency of small ice crystals. Atmos. Chem. Phys. **12**, 2055–2076 (2012). https://doi.org/10.5194/acp-12-2055-2012

Cox, G.F., Weeks, W.: Brine Drainage and Initial Salt Entrapment in Sodium Chloride Ice (1975)

Cui, L., Li, R., Fu, H., Li, Q., Zhang, L., George, C., Chen, J.: Formation features of nitrous acid in the offshore area of the East China Sea. Sci. Total Environ. **682**, 138–150 (2019). https://doi.org/10.1016/j.scitotenv.2019.05.004

Cunliffe, M., Upstill-Goddard, R.C., Murrell, J.C.: Microbiology of aquatic surface microlayers. FEMS Microbiol. Rev. **35**, 233–246 (2011). https://doi.org/10.1111/j.1574-6976.2010.00246.x

Cunliffe, M., Engel, A., Frka, S., Gašparović, B., Guitart, C., Murrell, J.C., Salter, M., Stolle, C., Upstill-Goddard, R., Wurl, O.: Sea surface microlayers: a unified physicochemical and biological perspective of the air–ocean interface. Prog. Oceanogr. **109**, 104–116 (2013). https://doi.org/10.1016/j.pocean.2012.08.004

Cziczo, D.J., Ladino, L., Boose, Y., Kanji, Z.A., Kupiszewski, P., Lance, S., Mertes, S., Wex, H.: Measurements of ice nucleating particles and ice residuals. Meteorol. Monogr. **58**, 8.1–8.13, https://doi.org/10.1175/amsmonographs-d-16-0008.1 (2017)

de Leeuw, G., Andreas, E.L., Anguelova, M.D., Fairall, C.W., Lewis, E.R., O'Dowd, C., Schulz, M., Schwartz, S. E.: Production flux of sea spray aerosol. Rev. Geophys. **49**, https://doi.org/10.1029/2010RG000349 (2011)

DeMott, P.J., Rogers, D.C.: Freezing nucleation rates of dilute solution droplets measured between $-30°$ and $-40\,°C$ in laboratory simulations of natural clouds. J. Atmosph. Sci. **47**, 1056–1064 (1990). https://doi.org/10.1175/1520-0469(1990)047%3c1056:fnrods%3e2.0.co;2

DeMott, P.J., Möhler, O., Cziczo, D.J., Hiranuma, N., Petters, M.D., Petters, S.S., Belosi, F., Bingemer, H.G., Brooks, S.D., Budke, C., Burkert-Kohn, M., Collier, K.N., Danielczok, A., Eppers, O., Felgitsch, L., Garimella, S., Grothe, H., Herenz, P., Hill, T.C.J., Höhler, K., Kanji, Z.A., Kiselev, A., Koop, T., Kristensen, T.B., Krüger, K., Kulkarni, G., Levin, E.J.T., Murray, B.J., Nicosia, A., O'Sullivan, D., Peckhaus, A., Polen, M.J., Price, H.C., Reicher, N., Rothenberg, D.A., Rudich, Y., Santachiara, G., Schiebel, T., Schrod, J., Seifried, T.M., Stratmann, F., Sullivan, R.C., Suski, K.J., Szakáll, M., Taylor, H.P., Ullrich, R., Vergara-Temprado, J., Wagner, R., Whale, T.F., Weber, D., Welti, A., Wilson, T.W., Wolf, M.J., Zenker, J.: The Fifth International Workshop on Ice Nucleation phase 2 (FIN-02): laboratory intercomparison of ice nucleation measurements. Atmos. Meas. Tech. **11**, 6231–6257 (2018). https://doi.org/10.5194/amt-11-6231-2018

Donaldson, D.J., and Anderson, D.: Adsorption of atmospheric gases at the air−water interface. 2. C1−C4 alcohols, acids, and acetone. J. Phys. Chem. A **103**, 871–876, https://doi.org/10.1021/jp983963h (1999)

Ehn, M., Thornton, J.A., Kleist, E., Sipilä, M., Junninen, H., Pullinen, I., Springer, M., Rubach, F., Tillmann, R., Lee, B., Lopez-Hilfiker, F., Andres, S., Acir, I.-H., Rissanen, M., Jokinen, T., Schobesberger, S., Kangasluoma, J., Kontkanen, J., Nieminen, T., Kurtén, T., Nielsen, L.B., Jørgensen, S., Kjaergaard, H.G., Canagaratna, M., Maso, M.D., Berndt, T., Petäjä, T., Wahner,

A., Kerminen, V.-M., Kulmala, M., Worsnop, D.R., Wildt, J., Mentel, T.F.: A large source of low-volatility secondary organic aerosol. Nature **506**, 476–479 (2014). https://doi.org/10.1038/nature13032

Emersic, C., Connolly, P.J., Boult, S., Campana, M., Li, Z.: Investigating the discrepancy between wet-suspension- and dry-dispersion-derived ice nucleation efficiency of mineral particles. Atmos. Chem. Phys. **15**, 11311–11326 (2015). https://doi.org/10.5194/acp-15-11311-2015

Frey, W., Hu, D., Dorsey, J., Alfarra, M.R., Pajunoja, A., Virtanen, A., Connolly, P., McFiggans, G.: The efficiency of secondary organic aerosol particles acting as ice-nucleating particles under mixed-phase cloud conditions. Atmos. Chem. Phys. **18**, 9393–9409 (2018). https://doi.org/10.5194/acp-18-9393-2018

Fritsen, C.H., Lytle, V.I., Ackley, S.F., Sullivan, C.W.: Autumn bloom of antarctic pack-ice algae. Science **266**, 782–784 (1994). https://doi.org/10.1126/science.266.5186.782

Gallagher, M.W., Connolly, P.J., Crawford, I., Heymsfield, A., Bower, K.N., Choularton, T.W., Allen, G., Flynn, M.J., Vaughan, G., Hacker, J.: Observations and modelling of microphysical variability, aggregation and sedimentation in tropical anvil cirrus outflow regions. Atmos. Chem. Phys. **12**, 6609–6628 (2012). https://doi.org/10.5194/acp-12-6609-2012

Garbe, C.S., Rutgersson, A., Boutin, J., de Leeuw, G., Delille, B., Fairall, C. W., Gruber, N., Hare, J., Ho, D.T., Johnson, M.T., Nightingale, P.D., Pettersson, H., Piskozub, J., Sahlée, E., Tsai, W.-T., Ward, B., Woolf, D.K., Zappa, C.J.: Transfer across the air-sea interface. In: Liss, P.S., Johnson, M.T. (eds.), Ocean-Atmosphere Interactions of Gases and Particles. Springer, Berlin, Heidelberg, 55–112 (2014)

Garnett, J., Halsall, C., Thomas, M., France, J., Kaiser, J., Graf, C., Leeson, A., Wynn, P.: Mechanistic insight into the uptake and fate of persistent organic pollutants in sea ice. Environ. Sci. Technol. **53**, 6757–6764 (2019). https://doi.org/10.1021/acs.est.9b00967

Grawe, S., Augustin-Bauditz, S., Hartmann, S., Hellner, L., Pettersson, J.B.C., Prager, A., Stratmann, F., Wex, H.: The immersion freezing behavior of ash particles from wood and brown coal burning. Atmos. Chem. Phys. **16**, 13911–13928 (2016). https://doi.org/10.5194/acp-16-13911-2016

Grawe, S., Augustin-Bauditz, S., Clemen, H.C., Ebert, M., Eriksen Hammer, S., Lubitz, J., Reicher, N., Rudich, Y., Schneider, J., Staacke, R., Stratmann, F., Welti, A., Wex, H.: Coal fly ash: linking immersion freezing behavior and physicochemical particle properties. Atmos. Chem. Phys. **18**, 13903–13923 (2018). https://doi.org/10.5194/acp-18-13903-2018

Hartmann, S., Niedermeier, D., Voigtländer, J., Clauss, T., Shaw, R.A., Wex, H., Kiselev, A., Stratmann, F.: Homogeneous and heterogeneous ice nucleation at LACIS: operating principle and theoretical studies. Atmos. Chem. Phys. **11**, 1753–1767 (2011). https://doi.org/10.5194/acp-11-1753-2011

Hartmann, S., Augustin, S., Clauss, T., Wex, H., Šantl-Temkiv, T., Voigtländer, J., Niedermeier, D., Stratmann, F.: Immersion freezing of ice nucleation active protein complexes. Atmos. Chem. Phys. **13**, 5751–5766 (2013). https://doi.org/10.5194/acp-13-5751-2013

Hartmann, S., Wex, H., Clauss, T., Augustin-Bauditz, S., Niedermeier, D., Rösch, M., Stratmann, F.: Immersion freezing of kaolinite: scaling with particle surface area. J. Atmos. Sci. **73**, 263–278 (2016). https://doi.org/10.1175/jas-d-15-0057.1

Henning, S., Wex, H., Hennig, T., Kiselev, A., Snider, J.R., Rose, D., Dusek, U., Frank, G.P., Pöschl, U., Kristensson, A., Bilde, M., Tillmann, R., Kiendler-Scharr, A., Mentel, T.F., Walter, S., Schneider, J., Wennrich, C., and Stratmann, F.: Soluble mass, hygroscopic growth, and droplet activation of coated soot particles during LACIS Experiment in November (LExNo). J. Geophys. Res. Atmosph., 115, https://doi.org/10.1029/2009JD012626 (2010)

Heymsfield, A.J., Westbrook, C.D.: Advances in the estimation of ice particle fall speeds using laboratory and field measurements. J. Atmosph. Sci. **67**, 2469–2482 (2010)

Hiranuma, N., Augustin-Bauditz, S., Bingemer, H., Budke, C., Curtius, J., Danielczok, A., Diehl, K., Dreischmeier, K., Ebert, M., Frank, F., Hoffmann, N., Kandler, K., Kiselev, A., Koop, T., Leisner, T., Möhler, O., Nillius, B., Peckhaus, A., Rose, D., Weinbruch, S., Wex, H., Boose, Y., DeMott, P.J., Hader, J.D., Hill, T.C.J., Kanji, Z.A., Kulkarni, G., Levin, E.J.T., McCluskey, C.S.,

Murakami, M., Murray, B.J., Niedermeier, D., Petters, M.D., O'Sullivan, D., Saito, A., Schill, G.P., Tajiri, T., Tolbert, M.A., Welti, A., Whale, T.F., Wright, T.P., Yamashita, K.: A comprehensive laboratory study on the immersion freezing behavior of illite NX particles: a comparison of 17 ice nucleation measurement techniques. Atmos. Chem. Phys. **15**, 2489–2518 (2015). https://doi.org/10.5194/acp-15-2489-2015

Hobbs, P.V.: research on clouds and precipitation past, present and future. Part II, Bull. Amer. Meteorol. Soc. **72**, 184–191 (1991)

Eide, L.I., Martin, S.: The formation of brine drainage features in young sea ice. J. Glaciol. **14**, 137–154 (1975). https://doi.org/10.3189/s0022143000013460

Kamm, S., Möhler, O., Naumann, K.H., Saathoff, H., Schurath, U.: The heterogeneous reaction of ozone with soot aerosol. Atmos. Environ. **33**, 4651–4661 (1999). https://doi.org/10.1016/S1352-2310(99)00235-6

King, M.D., France, J.L., Fisher, F.N., Beine, H.J.: Measurement and modelling of UV radiation penetration and photolysis rates of nitrate and hydrogen peroxide in Antarctic sea ice: An estimate of the production rate of hydroxyl radicals in first-year sea ice. J. Photochem. Photobiol. A **176**, 39–49 (2005). https://doi.org/10.1016/j.jphotochem.2005.08.032

Klaviņš, M., Purmalis, O.: Surface activity of humic acids depending on their origin and humification degree. Proceedings of the Latvian Academy of Sciences. Section B. Natural, Exact, and Applied Sciences., 67, 493–499, https://doi.org/10.2478/prolas-2013-0083 (2014)

Kreidenweis, S.M., Petters, M., Lohmann, U.: 100 years of progress in cloud physics, aerosols, and aerosol chemistry research. Meteorol. Monogr. **59**, 11.11–11.72, https://doi.org/10.1175/amsmonographs-d-18-0024.1 (2019)

Law, C.S., Brévière, E., de Leeuw, G., Garçon, V., Guieu, C., Kieber, D.J., Kontradowitz, S., Paulmier, A., Quinn, P.K., Saltzman, E.S., Stefels, J., von Glasow, R.: Evolving research directions in surface ocean-lower atmosphere (SOLAS) science. Environ. Chem. **10**, 1–16 (2013). https://doi.org/10.1071/EN12159

Liss, P.S., Duce, R.A.: The Sea Surface and Global Change. Cambridge University Press (1997)

Liss, P.S., Johnson, M.T.: Ocean-Atmosphere Interactions of Gases and Particles. Springer (2014)

List, R., Hallett, J., Warner, J., Reinking, R.: The future of Laboratory Research and Facilities for Cloud Physics and Cloud Chemistry 0003–0007, 1389–1403 (1986)

Liu, C., Lee Panetta, R., Yang, P.: The effects of surface roughness on the scattering properties of hexagonal columns with sizes from the Rayleigh to the geometric optics regimes. J. Quant. Spectrosc. Radiat. Transfer **129**, 169–185 (2013). https://doi.org/10.1016/j.jqsrt.2013.06.011

Loose, B., Schlosser, P., Perovich, D., Ringelberg, D., Ho, D.T., Takahashi, T., Richter-Menge, J., Reynolds, C.M., McGillis, W.R., Tison, J.L.: Gas diffusion through columnar laboratory sea ice: implications for mixed-layer ventilation of CO_2 in the seasonal ice zone. Tellus B: Chem. Phys. Meteorol. **63**, 23–39 (2011). https://doi.org/10.1111/j.1600-0889.2010.00506.x

Malinowski, S.P., Andrejczuk, M., Grabowski, W.W., Korczyk, P., Kowalewski, T.A., Smolarkiewicz, P.K.: Laboratory and modeling studies of cloud–clear air interfacial mixing: anisotropy of small-scale turbulence due to evaporative cooling. New J. Phys. **10**, 075020 (2008). https://doi.org/10.1088/1367-2630/10/7/075020

Mason, B.J., Ludlam, F.H.: The microphysics of clouds. Rep. Prog. Phys. **14**, 147–195 (1951). https://doi.org/10.1088/0034-4885/14/1/306

Middleton, C.A., Thomas, C., De Wit, A., Tison, J.L.: Visualizing brine channel development and convective processes during artificial sea-ice growth using Schlieren optical methods. J. Glaciol. **62**, 1–17 (2016). https://doi.org/10.1017/jog.2015.1

Miller, L.A., Fripiat, F., Else, B.G.T., Bowman, J.S., Brown, K.A., Collins, R.E., Ewert, M., Fransson, A., Gosselin, M., Lannuzel, D., Meiners, K. M., Michel, C., Nishioka, J., Nomura, D., Papadimitriou, S., Russell, L. M., Sørensen, L. L., Thomas, D.N., Tison, J.-L., van Leeuwe, M.A., Vancoppenolle, M., Wolff, E.W., Zhou, J.: Methods for biogeochemical studies of sea ice: the state of the art, caveats, and recommendations. Elementa Sci. Anthropocene **3**, https://doi.org/10.12952/journal.elementa.000038 (2015)

Mitchell, D.L., Mishra, S., Lawson, R.P.: Representing the ice fall speed in climate models: results from tropical composition, cloud and climate coupling (TC4) and the indirect and semi-direct aerosol campaign (ISDAC). J. Geophys. Res. Atmosph. **116**, https://doi.org/10.1029/2010JD 015433 (2011)

Möhler, O., Stetzer, O., Schaefers, S., Linke, C., Schnaiter, M., Tiede, R., Saathoff, H., Krämer, M., Mangold, A., Budz, P., Zink, P., Schreiner, J., Mauersberger, K., Haag, W., Kärcher, B., Schurath, U.: Experimental investigation of homogeneous freezing of sulphuric acid particles in the aerosol chamber AIDA. Atmos. Chem. Phys. **3**, 211–223 (2003). https://doi.org/10.5194/acp-3-211-2003

Möhler, O., Büttner, S., Linke, C., Schnaiter, M., Saathoff, H., Stetzer, O., Wagner, R., Krämer, M., Mangold, A., Ebert, V., Schurath, U.: Effect of sulfuric acid coating on heterogeneous ice nucleation by soot aerosol particles. J. Geophys. Res. Atmosph. 110, https://doi.org/10.1029/200 4JD005169 (2005)

Möhler, O., Field, P.R., Connolly, P., Benz, S., Saathoff, H., Schnaiter, M., Wagner, R., Cotton, R., Krämer, M., Mangold, A., Heymsfield, A.J.: Efficiency of the deposition mode ice nucleation on mineral dust particles. Atmos. Chem. Phys. **6**, 3007–3021 (2006). https://doi.org/10.5194/acp-6-3007-2006

Möhler, O., Georgakopoulos, D.G., Morris, C.E., Benz, S., Ebert, V., Hunsmann, S., Saathoff, H., Schnaiter, M., Wagner, R.: Heterogeneous ice nucleation activity of bacteria: new laboratory experiments at simulated cloud conditions. Biogeosciences **5**, 1425–1435 (2008). https://doi.org/ 10.5194/bg-5-1425-2008

Naumann, A.K., Notz, D., Håvik, L., Sirevaag, A.: Laboratory study of initial sea-ice growth: properties of grease ice and nilas. Cryosphere **6**, 729–741 (2012). https://doi.org/10.5194/tc-6-729-2012

Nenes, A., Ghan, S., Abdul-Razzak, H., Chuang, P.Y., Seinfeld, J.H.: Kinetic limitations on cloud droplet formation and impact on cloud albedo. Tellus B Chem. Phys. Meteorol. **53**, 133–149 (2001). https://doi.org/10.3402/tellusb.v53i2.16569

Niedermeier, D., Wex, H., Voigtländer, J., Stratmann, F., Brüggemann, E., Kiselev, A., Henk, H., Heintzenberg, J.: LACIS-measurements and parameterization of sea-salt particle hygroscopic growth and activation. Atmos. Chem. Phys. **8**, 579–590 (2008). https://doi.org/10.5194/acp-8-579-2008

Niedermeier, D., Hartmann, S., Shaw, R.A., Covert, D., Mentel, T.F., Schneider, J., Poulain, L., Reitz, P., Spindler, C., Clauss, T., Kiselev, A., Hallbauer, E., Wex, H., Mildenberger, K., Stratmann, F.: Heterogeneous freezing of droplets with immersed mineral dust particles—measurements and parameterization. Atmos. Chem. Phys. **10**, 3601–3614 (2010). https://doi.org/10.5194/acp-10-3601-2010

Niedermeier, D., Voigtländer, J., Schmalfuß, S., Busch, D., Schumacher, J., Shaw, R.A., Stratmann, F.: Characterization and first results from LACIS-T: a moist-air wind tunnel to study aerosol–cloud–turbulence interactions. Atmos. Meas. Tech. **13**, 2015–2033 (2020). https://doi.org/10. 5194/amt-13-2015-2020

Niedrauer, T.M., Martin, S.: An experimental study of brine drainage and convection in Young Sea ice. J. Geophys. Res. Oceans **84**, 1176–1186 (1979). https://doi.org/10.1029/JC084iC03p01176

Niemand, M., Möhler, O., Vogel, B., Vogel, H., Hoose, C., Connolly, P., Klein, H., Bingemer, H., DeMott, P., Skrotzki, J., Leisner, T.: A particle-surface-area-based parameterization of immersion freezing on desert dust particles. J. Atmos. Sci. **69**, 3077–3092 (2012). https://doi.org/10.1175/ jas-d-11-0249.1

Petters, M.D., Wex, H., Carrico, C.M., Hallbauer, E., Massling, A., McMeeking, G.R., Poulain, L., Wu, Z., Kreidenweis, S.M., Stratmann, F.: Towards closing the gap between hygroscopic growth and activation for secondary organic aerosol—Part 2: theoretical approaches. Atmos. Chem. Phys. **9**, 3999–4009 (2009). https://doi.org/10.5194/acp-9-3999-2009

Quaas, J., Bony, S., Collins, W.D., Donner, L., Illingworth, A., Jones, A., Lohmann, U., Satoh, M., Schwartz, S.E., Tao, W.-K., Wood, R.: Current understanding and quantification of clouds in the changing climate system and strategies for reducing critical uncertainties. In: Clouds in the Perturbed Climate System, pp. 557–573. MIT Press, Cambridge, MA, USA (2009)

Rossignol, S., Tinel, L., Bianco, A., Passananti, M., Brigante, M., Donaldson, D.J., George, C.: Atmospheric photochemistry at a fatty acid–coated air-water interface. Science **353**, 699–702 (2016). https://doi.org/10.1126/science.aaf3617

Rysgaard, S., Wang, F., Galley, R.J., Grimm, R., Notz, D., Lemes, M., Geilfus, N.X., Chaulk, A., Hare, A.A., Crabeck, O., Else, B.G.T., Campbell, K., Sørensen, L.L., Sievers, J., Papakyriakou, T.: Temporal dynamics of ikaite in experimental sea ice. Cryosphere **8**, 1469–1478 (2014). https://doi.org/10.5194/tc-8-1469-2014

Saathoff, H., Naumann, K.H., Möhler, O., Jonsson, Å.M., Hallquist, M., Kiendler-Scharr, A., Mentel, T.F., Tillmann, R., Schurath, U.: Temperature dependence of yields of secondary organic aerosols from the ozonolysis of α-pinene and limonene. Atmos. Chem. Phys. **9**, 1551–1577 (2009). https://doi.org/10.5194/acp-9-1551-2009

Schnaiter, M., Linke, C., Möhler, O., Naumann, K.-H., Saathoff, H., Wagner, R., Schurath, U., Wehner, B.: Absorption amplification of black carbon internally mixed with secondary organic aerosol. J. Geophys. Res. Atmosph., 110, https://doi.org/10.1029/2005JD006046 (2005)

Seinfeld, J.H., Bretherton, C., Carslaw, K.S., Coe, H., DeMott, P.J., Dunlea, E.J., Feingold, G., Ghan, S., Guenther, A.B., Kahn, R., Kraucunas, I., Kreidenweis, S.M., Molina, M.J., Nenes, A., Penner, J.E., Prather, K.A., Ramanathan, V., Ramaswamy, V., Rasch, P.J., Ravishankara, A.R., Rosenfeld, D., Stephens, G., Wood, R.: Improving our fundamental understanding of the role of aerosol—cloud interactions in the climate system. Proc. Natl. Acad. Sci. **113**, 5781–5790 (2016). https://doi.org/10.1073/pnas.1514043113

Sempéré, R., Kawamura, K.: Trans-hemispheric contribution of C2C10 , -dicarboxylic acids, and related polar compounds to water-soluble organic carbon in the western Pacific aerosols in relation to photochemical oxidation reactions. Global Biogeoch. Cycles Global Biogeochem Cycle, 17, https://doi.org/10.1029/2002gb001980 (2003)

Siebert, H., Franke, H., Lehmann, K., Maser, R., Saw, E.W., Schell, D., Shaw, R.A., Wendisch, M.: Probing finescale dynamics and microphysics of clouds with helicopter-borne measurements. Bull. Am. Meteor. Soc. **87**, 1727–1738 (2006). https://doi.org/10.1175/bams-87-12-1727

Siebert, H., Shaw, R.A.: Supersaturation fluctuations during the early stage of cumulus formation. J. Atmos. Sci. **74**, 975–988 (2017). https://doi.org/10.1175/jas-d-16-0115.1

Simpson, E.L., Connolly, P.J., McFiggans, G.: Competition for water vapour results in suppression of ice formation in mixed-phase clouds. Atmos. Chem. Phys. **18**, 7237–7250 (2018). https://doi.org/10.5194/acp-18-7237-2018

Smith, H.R., Connolly, P.J., Baran, A.J., Hesse, E., Smedley, A.R.D., Webb, A.R.: Cloud chamber laboratory investigations into scattering properties of hollow ice particles. J. Quant. Spectrosc. Radiat. Transfer **157**, 106–118 (2015). https://doi.org/10.1016/j.jqsrt.2015.02.015

Smith, H.R., Connolly, P.J., Webb, A.R., Baran, A.J.: Exact and near backscattering measurements of the linear depolarisation ratio of various ice crystal habits generated in a laboratory cloud chamber. J. Quant. Spectrosc. Radiat. Transfer **178**, 361–378 (2016). https://doi.org/10.1016/j.jqsrt.2016.01.030

Steinke, I., Funk, R., Busse, J., Iturri, A., Kirchen, S., Leue, M., Möhler, O., Schwartz, T., Schnaiter, M., Sierau, B., Toprak, E., Ullrich, R., Ulrich, A., Hoose, C., Leisner, T.: Ice nucleation activity of agricultural soil dust aerosols from Mongolia, Argentina, and Germany. J. Geophys. Res. Atmosph. **121**, 13,559–513,576, https://doi.org/10.1002/2016JD025160 (2016).

Stevens, B.: Atmospheric moist convection. Annu. Rev. Earth Planet. Sci. **33**, 605–643 (2005). https://doi.org/10.1146/annurev.earth.33.092203.122658

Stratmann, F., Kiselev, A., Wurzler, S., Wendisch, M., Heintzenberg, J., Charlson, R.J., Diehl, K., Wex, H., Schmidt, S.: Laboratory studies and numerical simulations of cloud droplet formation under realistic supersaturation conditions. J. Atmos. Oceanic Tech. **21**, 876–887 (2004). https://doi.org/10.1175/1520-0426(2004)021%3c0876:lsanso%3e2.0.co;2

Stratmann, F., Möhler, O., Shaw, R.A., and Wex, H.: Laboratory cloud simulation: capabilities and future directions, in: clouds in the perturbed climate system: their relationship to energy balance, atmospheric dynamics, and precipitation. In: Heintzenberg, J., Charlson, R.J. (eds.) Clouds in the Perturbed Climate System, pp. 149–172. MIT Press, Cambridge, MA, USA (2009)

Stratmann, F., Bilde, M., Dusek, U., Frank, G.P., Hennig, T., Henning, S., Kiendler-Scharr, A., Kiselev, A., Kristensson, A., Lieberwirth, I., Mentel, T.F., Pöschl, U., Rose, D., Schneider, J., Snider, J.R., Tillmann, R., Walter, S., Wex, H.: Examination of laboratory-generated coated soot particles: an overview of the LACIS Experiment in November (LExNo) campaign. J. Geophys. Res. Atmosph., 115, https://doi.org/10.1029/2009JD012628 (2010)

Stubbins, A., Hubbard, V., Uher, G., Law, C.S., Upstill-Goddard, R.C., Aiken, G.R., Mopper, K.: Relating carbon monoxide photoproduction to dissolved organic matter functionality. Environ. Sci. Technol. **42**, 3271–3276 (2008). https://doi.org/10.1021/es703014q

Swart, S., Gille, S.T., Delille, B., Josey, S., Mazloff, M., Newman, L., Thompson, A.F., Thomson, J., Ward, B., du Plessis, M.D., Kent, E.C., Girton, J., Gregor, L., Heil, P., Hyder, P., Pezzi, L.P., de Souza, R.B., Tamsitt, V., Weller, R.A., Zappa, C.J.: Constraining southern ocean air-sea-ice fluxes through enhanced observations. Front. Marine Sci. 6, https://doi.org/10.3389/fmars.2019.00421 (2019)

Tajiri, T., Yamashita, K., Murakami, M., Saito, A., Kusunoki, K., Orikasa, N., Lilie, L.: A novel adiabatic-expansion-type cloud simulation chamber. J. Meteorol. Soc. Japan. Ser. II(91), 687–704 (2013). https://doi.org/10.2151/jmsj.2013-509

Thomas, M.: Brine and pressure dynamics in growing sea ice: measurements and modelling in the Roland von Glasow air-sea-ice chamber. School of Environmental Sciences, University of East Anglia, Doctoral (2018)

Thomas, M., France, J., Crabeck, O., Hall, B., Hof, V., Notz, D., Rampai, T., Riemenschneider, L., Tooth, O.J., Tranter, M., Kaiser, J.: The roland von glasow air-sea-ice chamber (RvG-ASIC): an experimental facility for studying ocean–sea-ice–atmosphere interactions. Atmos. Meas. Tech. **14**, 1833–1849 (2021). https://doi.org/10.5194/amt-14-1833-2021

Ulanowski, Z., Kaye, P.H., Hirst, E., Greenaway, R.S., Cotton, R.J., Hesse, E., Collier, C.T.: Incidence of rough and irregular atmospheric ice particles from Small Ice Detector 3 measurements. Atmos. Chem. Phys. **14**, 1649–1662 (2014). https://doi.org/10.5194/acp-14-1649-2014

Ullrich, R., Hoose, C., Möhler, O., Niemand, M., Wagner, R., Höhler, K., Hiranuma, N., Saathoff, H., Leisner, T.: A new ice nucleation active site parameterization for desert dust and soot. J. Atmos. Sci. **74**, 699–717 (2017). https://doi.org/10.1175/jas-d-16-0074.1

Vancoppenolle, M., Meiners, K.M., Michel, C., Bopp, L., Brabant, F., Carnat, G., Delille, B., Lannuzel, D., Madec, G., Moreau, S., Tison, J.-L., van der Merwe, P.: Role of sea ice in global biogeochemical cycles: emerging views and challenges. Quatern. Sci. Rev. **79**, 207–230 (2013). https://doi.org/10.1016/j.quascirev.2013.04.011

Wagner, R., Benz, S., Möhler, O., Saathoff, H., Schnaiter, M., Schurath, U.: Mid-infrared extinction spectra and optical constants of supercooled water droplets. J. Phys. Chem. A **109**, 7099–7112 (2005). https://doi.org/10.1021/jp051942z

Wettlaufer, J.S., Worster, M.G., Huppert, H.E.: the phase evolution of Young Sea Ice. Geophys. Res. Lett. **24**, 1251–1254 (1997). https://doi.org/10.1029/97GL00877

Wex, H., Kiselev, A., Stratmann, F., Zoboki, J., Brechtel, F.: Measured and modeled equilibrium sizes of NaCl and (NH4)2SO4 particles at relative humidities up to 99.1%. J. Geophys. Res. Atmosph., 110, https://doi.org/10.1029/2004JD005507 (2005)

Wex, H., Hennig, T., Salma, I., Ocskay, R., Kiselev, A., Henning, S., Massling, A., Wiedensohler, A., Stratmann, F.: Hygroscopic growth and measured and modeled critical super-saturations of an atmospheric HULIS sample. Geophys. Res. Lett., 34, https://doi.org/10.1029/2006GL028260 (2007)

Wex, H., Petters, M.D., Carrico, C.M., Hallbauer, E., Massling, A., McMeeking, G.R., Poulain, L., Wu, Z., Kreidenweis, S.M., Stratmann, F.: Towards closing the gap between hygroscopic growth and activation for secondary organic aerosol: Part 1—Evidence from measurements. Atmos. Chem. Phys. **9**, 3987–3997 (2009). https://doi.org/10.5194/acp-9-3987-2009

Ziese, M., Wex, H., Nilsson, E., Salma, I., Ocskay, R., Hennig, T., Massling, A., Stratmann, F.: Hygroscopic growth and activation of HULIS particles: experimental data and a new iterative parameterization scheme for complex aerosol particles. Atmos. Chem. Phys. **8**, 1855–1866 (2008). https://doi.org/10.5194/acp-8-1855-2008

Zink, P., Knopf, D.A., Schreiner, J., Mauersberger, K., Möhler, O., Saathoff, H., Seifert, M., Tiede, R., Schurath, U.: Cryo-chamber simulation of stratospheric H2SO4/H2O particles: Composition analysis and model comparison. Geophys. Res. Lett. **29**, 46–41–46–44, https://doi.org/10.1029/2001GL013296 (2002)

Chapter 9
Conclusions

Jean-François Doussin

In the 1986 edition of their famous monograph, Finlayson-Pitts and Pitts defined experimental atmospheric simulation as "*Perhaps the most direct experimental means of examining the relationship between emissions and air quality*". This statement was, at the time, strongly supported by the enormous amount of research that had been conducted in the laboratory on chemical transformations of pollutants since the early work of Haagen-Smit et al. in the 1950s (Haagen-Smit et al. 1953; Haagen-Smit and Fox 1953).

Some of the early chamber studies, focusing on the interconversion of NO_x in the presence of VOCs and light, made a major contribution to the discovery of the role of the OH radical in the atmospheric photo-oxidation cycle (Heicklen et al. 1969, 1971). They also revealed the mechanism of tropospheric ozone build-up (Weinstock 1971; Niki et al. 1972; Westberg et al. 1971). The indefatigable work of Pitts, Winer, Atkinson, Niki, Becker, Moortgat, Schurath, and others led to the development of a huge database of kinetic reaction parameters that laid the foundations for the first chemical transport models. Subsequently, the use of simulation chambers has led to many other significant breakthroughs in atmospheric chemistry research. Simulation chamber experiments on the atmospheric oxidation of unsaturated hydrocarbons led to the identification of the essential role of biogenic VOCs in rural ozone formation (Abelson 1988). During the 1990s, chamber experiments contributed to the elucidation of the relation between gasoline composition and secondary organic aerosol formation (Odum et al. 1997), and more recently they were key in characterizing the major oxidation routes for organic aerosol in the atmosphere (Jimenez et al. 2009). By revealing that oligomerization processes were occurring in organic aerosol (Kalberer et al. 2006), chamber experiments provided a basis for questioning one of the most established schemes—the assumption that the molecular carbon chains of organic

J.-F. Doussin (✉)
Centre National de la Recherche Scientifique, Paris, France
e-mail: jean-francois.doussin@lisa.ipsl.fr

© The Author(s) 2023
J.-F. Doussin et al. (eds.), *A Practical Guide to Atmospheric Simulation Chambers*,
https://doi.org/10.1007/978-3-031-22277-1_9

pollutants tend to fragment until ultimately CO_2 formation predominantly occurs during atmospheric oxidation.

Over the past few decades, progress in our understanding of the atmospheric oxidation of isoprene—one of the most important biogenic VOCs—illustrates the need for chamber experiments and also epitomizes the synergies between laboratory and field studies. Following the identification of molecular tracers in field samples (Claeys et al. 2004), simulation chamber studies demonstrated that isoprene, C_5H_8, despite only having five carbon atoms, could be oxidized in the atmosphere to form SOA (Kroll et al. 2005).

Soon after, innovative work conducted in the SAPHIR chamber, using emissions from real plants as precursors (Kiendler-Scharr et al. 2009), demonstrated the ability of isoprene to scavenge OH from the forest atmosphere, redirecting the chemistry toward the formation of products less prone to participate in nucleation and almost suppressing new particle formation events. Later on, focusing on gas-phase processes and their consequences for the tropospheric radical budget, Fuchs et al. (2013) detected significantly higher concentrations of hydroxyl radicals than expected based on model calculations, providing direct evidence for a strong hydroxyl radical enhancement due to additional recycling of radicals in the presence of isoprene.

More recently, McFiggans et al. (2019) showed that isoprene can decrease the overall mass yield derived from monoterpenes in mixtures through the scavenging of highly oxygenated monoterpene products by isoprene-derived peroxy radicals. With this discovery, these authors did not only bring important pieces of observation, but they also questioned the additivity of aerosol yields implemented in models. They illustrated that modest aerosol yield compounds are not necessarily net producers and that their oxidation can suppress both particle number and mass for stronger contributors present in the same air mass.

Around the same time as the comments from Finlayson-Pitts and Pitts in the late (1980), the International Union of Pure and Applied Chemistry (IUPAC) set about defining "Smog Chamber in atmospheric chemistry" as follows (Calvert 1990): "*A large confined volume in which sunlight or simulated sunlight is allowed to irradiate air mixtures of atmospheric trace gases (hydrocarbons, nitrogen oxides, sulfur dioxide, etc.) which undergo oxidation. In theory these chambers allow the controlled study of complex reactions which occur in the atmosphere. However, ill-defined wall reactions which generate some molecular and radical species (e.g. HONO, CH$_2$O, OH-radicals, etc.) and remove certain products (H$_2$O$_2$, HNO$_3$, etc.), the use of reactant concentrations well above those in the atmosphere, ill-defined light intensities and wavelength distribution within the chamber, and other factors peculiar to chamber experiments require that caution be exercised in the extrapolation of results obtained from them to atmospheric system*".

This statement may appear to be rather pessimistic when considering the significant progress in atmospheric chemistry that chambers have facilitated. Nevertheless, each of the reservations expressed by IUPAC is not without relevance and the international chamber community has worked hard to address these challenges. In Europe, a coordinated approach has been adopted through the various EUROCHAMP initiatives, which have further investigated way to improve the robustness of experimental simulation results. Even if the 1990 IUPAC definition for "Smog Chamber" was still in use in the last edition of the IUPAC "Gold Book" (Chalk 2019), the readers of the present guide to atmospheric simulation chambers would be able to recognize the considerable advances that have been made in the field over the last 30 years.

9.1 Improving the Robustness of the Simulation Chamber Experiments

For many years, simulation chamber studies were conducted using reactant concentrations several orders of magnitude larger (ppm or hundreds of ppb levels) than those found in the ambient atmosphere (ppb or sub-ppb). However, thanks to the development of more and more sensitive monitoring techniques, working in the ppb range is now relatively standard and the ppt range is also accessible, although still challenging due to sensitivity limitations of measurement techniques. For decades, working with reactant concentrations well above those found in the atmosphere was not considered as a major problem, as long as non-atmospherically relevant radical–radical reactions were kept negligible. However, the highly non-linear nature of secondary aerosol formation and related condensation processes made it more critical to work at realistic concentrations. The conceptual advance brought by Pankow (1994) and Odum et al. (1997) has partially allowed us to take this common drawback of simulation chamber experiments into account when deriving SOA yields. Nevertheless, it was quickly shown (Duplissy et al. 2008) that the chemical composition of the organic aerosol formed, and therefore its physical properties, such as hygroscopicity and CCN activity, depended on the initial concentration of the precursors. Fortunately, demonstration of the critical need for reducing reactant concentrations in chamber experiments arrived around the same time as the introduction of a new generation of very sensitive mass spectrometry techniques such PTR-MS, API-TOF, TOF-CIMS,... which are providing further opportunities for simulation chamber studies to be performed at realistic atmospheric concentrations.

Significant progress has also been made in the characterization of light intensity inside the chamber (including homogeneity) and the provision of a wider range of wavelengths for simulating sunlight-induced atmospheric processes (Chap. 2). Even though UV fluorescent tubes, or so-called "black lights", are still—for cost reasons—the most common light source among the simulation chamber community, and even if the atmospheric relevance of their emission spectrum can be questioned, the recording of related actinic flux information together with chamber data has become a well-accepted practice, thanks to projects such as EUROCHAMP. This good practice is further supported by the information provided in Chap. 2 of this book, which provides a solid basis for homogeneous robust lighting characterization. The availability of this information is indeed critical for any attempt of re-using previous datasets, especially if a modeling approach is planned (which is generally the case).

One of the key steps in the long journey of the scientific community toward a precise understanding of the chemistry at work in atmospheric simulation chambers is the rise of the concept of "chamber chemistry". According to this concept, when the initial precursor concentration is low enough and when the light energy is atmospherically relevant, the observed behavior of chemicals during chamber experiments would be the result of the interplay between the atmospherically relevant chemistry and the chamber effects.

Refusing to consider the chambers as black boxes and hence starting to study the chamber-dependent processes themselves with scientific rigor has led to two complementary efforts. The first of these is the study of wall reactions using the tools of microphysics and surface chemistry. While some aspects of wall reactions are relevant for the understanding of atmospheric processes (see, e.g., Pitts et al. 1984; Rohrer et al. 2005), the true motivation for characterizing them was a different one: the authors were already building what is now known as an "auxiliary mechanism". Thirty years ago, Jeffries et al. (1992) were already recommending that those chamber-dependent reaction sets should be available for each chamber dataset to be simulated. They were pointing out that when evaluating a reaction mechanism in a given chamber, the auxiliary mechanism is combined with a core mechanism which is asserted to be chamber independent, and that misrepresentations in the auxiliary mechanism could induce compensating errors in the core mechanism and so in the derived atmospherically relevant knowledge. This goal has never been that close to being attained, as not only this information has been made available for most of the atmospheric simulation chambers installed in Europe, but also because the present guide, for the first time, provides clear guidance for the building of such auxiliary mechanism (see Chaps. 2 and 3).

The second aspect of chamber-dependent processes is related to the simultaneous exploitation of a large number of datasets arising from various chambers. Based on a quasi-statistic interpretation of the previous concept, it assumes that deconvolution of chamber-dependent processes from atmospherically relevant processes can be significantly enhanced by the parallel analysis of comparable experiments carried out in different simulation chambers. This approach has been proposed as early as in 1999 (Jeffries 1999), but until now it has never been applicable due to the lack of diversity among datasets. In most of the centers developing an experimental atmospheric simulation activity, datasets are indeed carefully stored, forming several databases comprising generally hundreds to thousands of experiments. Nevertheless, the community was missing coherent datasets investigating the same chemical systems by means of very different installations. Thanks to the "multi-chamber experiments" initiative developed in the framework of EUROCHAMP-2020, this approach for three important "standard" chemical systems (propene oxidation, toluene/xylene oxidation, and a-pinene oxidation) may receive its first full-scale validation. Further, the release of the EUROCHAMP database (https://data.eurochamp.org/) has made freely available around 3000 datasets of chamber experiments, generated in more than 20 fully characterized chambers (including an auxiliary mechanism for 16 of them), which exhibit a high diversity of size, type, material, and irradiation.

Together with the present guide, the EUROCHAMP database provides an unparalleled opportunity to ascertain a deeper understanding of these experiments and their implications for atmospheric processes, air quality, and climate. The combined use of data mining and emerging artificial intelligence techniques may further help to stimulate movement toward a thorough reanalysis of experiments in the database.

9.2   Simulating the Complexity of the Real Atmosphere, Working at the Interfaces and Considering Longer Timescale Exposure

For decades, the usual way of operating chambers has been to study a well defined but simplistic starting mixture with the goal of understanding all the mechanistic details of the transformation at work. This approach is still very valuable for characterizing the atmospheric processing and impact of a single compound or to study a well-defined chemical reaction. Leaving aside successive improvements made to the chamber experiments which adopt this "classical" approach to atmospheric simulation, ongoing advances continue to include a resolute movement toward more complex mixtures and more realistic systems, some of which may include including several phases of matter or interactions of chemicals with various surfaces.

From studies of the chemical evolution of emissions from real plants (Joutsensaari et al. 2005; Mentel et al. 2009; Faiola et al. 2018), motor vehicles (Geiger et al. 2002; Platt et al. 2013; Gordon et al. 2014), and wood burning devices (Nordin et al. 2015; Pratap et al. 2019), most of the emitting systems or practices utilize simulations in the attempt to characterize their impact on secondary atmospheric pollution, to evidence interplay between intermediate species arising from various precursors, or to identify tracers for their specific contamination.

Even more challenging is the application of simulation chambers to the interfaces between Earth system compartments. As shown in Chaps. 7 and 8, work on interfacial processes that was historically focused on the gas–aerosol or gas–liquid interfaces is now being extended to air–urban surfaces (Monge et al. 2010), air–sea exchanges (Bernard et al. 2016), as well as interfaces in the cryosphere (Thomas et al. 2021). In this case, not only the complexity of chemical mixture has increased, but the involvement of new surfaces for accommodation and for reaction implies to consider transport to/toward new media and complex mass transfer between phases.

Finally, a third dimension of complexity has recently emerged extending the timescale of simulation experiments. Most of the effects of modern air pollution on health, plants, or cultural heritage are related to long-term exposure: after decades of studies using exposition chambers in which primary pollutants are injected, simulation chambers are now considered as tools of choice to investigate the effect of complex mixtures containing secondary pollutants. For health impact studies, even if very sensitive models (such as pregnant mice or bare epithelial cells) are some-time used, it remains necessary to expose them to simulated polluted air for several days or weeks when the longest experiments in chambers generally last up to 2 to 3 days. To overcome these limits, new protocols are emerging, where chambers are generally operated as slow flow reactors with a constant input of primary pollutants that are allowed to react for an average time equal to the residence time (generally a few hours). When a steady state is attained, such a design can be operated for days providing a steady system (which limits the study of photochemical processes to indoor chambers) and feeds exposition devices where living models are receiving chamber effluents.

This area of research is ongoing and the protocols are still under development but they already benefit from those described in the present guide. Undoubtedly, the present effort in disseminating good practices and harmonizing protocols and the related metadata will have to be continued in the near future. It is probably one of the most critical networking activities that will have to be organized within the ACTRIS European Research Infrastructure.

9.3   Conclusion

The original use of smog chambers for the understanding of chemical transformations in the atmosphere; for the quantification of reaction rates, the extent, and the relevance of various possible pathways; and for the identification of secondary pollutants is still strongly necessary. The models—both operational and research oriented—are still far from an explicit thorough inclusion of all the processes that are required to represent and forecast the actual air quality and climate issues as well as future challenges. At the same time, the field of atmospheric experimental simulation has been extremely active during the past 15 years—and considering the number of new facilities around the world—there is little doubt about its vitality over the next 15 years, and beyond. A number of new methodologies and applications have risen and they will bring the operational capacity of simulation chambers to a new level. This community effort will enable a much broader range of scientific and societal challenges to be addressed, including not only the direct and indirect climate effects of atmospheric pollutants, but also the impact of air composition on health, cultural heritage, and the various compartments of Earth system. Although these applications are still in their early stages, they are quickly growing and are already producing data that will open new ways to consider the interplays between atmospheric transformations and impacts.

References

Abelson, P.H.: Rural and urban ozone. Science **241**, 1569–1569 (1988). https://doi.org/10.1126/science.241.4873.1569

Bernard, F., Ciuraru, R., Boréave, A., George, C.: Photosensitized formation of secondary organic aerosols above the air/water interface. Environ. Sci. Technol. **50**, 8678–8686 (2016). https://doi.org/10.1021/acs.est.6b03520

Calvert, J.G.: Glossary of atmospheric chemistry terms (Recommendations 1990). Pure Appl. Chem. **62**, 2167–2219 (1990). https://doi.org/10.1351/pac199062112167

Chalk, S.J.: Compendium of Chemical Terminology, 2nd ed. (the "Gold Book"). Compiled by A. D. McNaught and A. Wilkinson. Blackwell Scientific Publications, Oxford (1997). Online version (2019) created by S. J. Chalk. IUPAC. ISBN 0-9678550-9-8. https://doi.org/10.1351/goldbook

Claeys, M., Graham, B., Vas, G., Wang, W., Vermeylen, R., Pashynska, V., Cafmeyer, J., Guyon, P., Andreae, M.O., Artaxo, P., Maenhaut, W.: Formation of secondary organic aerosols through photooxidation of isoprene. Science **303**, 1173–1176 (2004). https://doi.org/10.1126/science.1092805

Duplissy, J., Gysel, M., Alfarra, M.R., Dommen, J., Metzger, A., Prevot, A.S.H., Weingartner, E., Laaksonen, A., Raatikainen, T., Good, N., Turner, S.F., McFiggans, G., Baltensperger, U.: Cloud forming potential of secondary organic aerosol under near atmospheric conditions. Geophys. Res. Lett., 35, https://doi.org/10.1029/2007GL031075 (2008)

Faiola, C.L., Buchholz, A., Kari, E., Yli-Pirilä, P., Holopainen, J.K., Kivimäenpää, M., Miettinen, P., Worsnop, D.R., Lehtinen, K.E.J., Guenther, A.B., Virtanen, A.: Terpene composition complexity controls secondary organic aerosol yields from scots pine volatile emissions. Sci. Rep. **8**, 3053 (2018). https://doi.org/10.1038/s41598-018-21045-1

Finlayson-Pitts, B.J., Pitts Jr., J.N.: Atmospheric chemistry: fundamentals and experimental techniques, edited by: Publication, W. I., Wiley, New-York (1986)

Fuchs, H., Hofzumahaus, A., Rohrer, F., Bohn, B., Brauers, T., Dorn, H.P., Häseler, R., Holland, F., Kaminski, M., Li, X., Lu, K., Nehr, S., Tillmann, R., Wegener, R., Wahner, A.: Experimental evidence for efficient hydroxyl radical regeneration in isoprene oxidation. Nat. Geosci. **6**, 1023–1026 (2013). https://doi.org/10.1038/ngeo1964

Geiger, H., Kleffmann, J., Wiesen, P.: Smog chamber studies on the influence of diesel exhaust on photosmog formation. Atmos. Environ. **36**, 1737–1747 (2002). https://doi.org/10.1016/S1352-2310(02)00175-9

Gordon, T.D., Presto, A.A., Nguyen, N.T., Robertson, W.H., Na, K., Sahay, K.N., Zhang, M., Maddox, C., Rieger, P., Chattopadhyay, S., Maldonado, H., Maricq, M.M., Robinson, A.L.: Secondary organic aerosol production from diesel vehicle exhaust: impact of aftertreatment, fuel chemistry and driving cycle. Atmos. Chem. Phys. **14**, 4643–4659 (2014). https://doi.org/10.5194/acp-14-4643-2014

Haagen-Smit, A.J., Bradley, C.E., Fox, M.M.: Ozone formation in photochemical oxidation of organic substances. Ind. Eng. Chem. **45**, 2086–2089 (1953)

Haagen-Smit, A.J., Fox, M.M.: Photochemical ozone formation with hydrocarbons and automobile exhaust. J. Air Pollut. Control Assoc. **4**, 105–108 (1953)

Heicklen, J., Westberg, K., Cohen, N.: Conversion of NO to NO_2 in Polluted Atmospheres, Center for Environmental Studies, Pennsylvania State University, University ParkReport No. 115–69 (1969)

Heicklen, J., Tuesday, C.S. (ed.) Chemical Reactions in Urban Atmospheres, in, American Elsevier, New York, 55–59 (1971)

Jeffries, H.E., Gery, M.W., Carter, W.P.L.: Protocols for evaluating Oxidant Mechanisms for Urban and regional ModelsEPA/600/SR-92/112 (1992)

Jeffries, H.E.: The UNC Chamber Auxiliary Model (Wall Model), US/German—Environmental Chamber Workshop Riverside, CA (1999)

Jimenez, J.L., Canagaratna, M.R., Donahue, N.M., Prevot, A.S.H., Zhang, Q., Kroll, J.H., DeCarlo, P.F., Allan, J.D., Coe, H., Ng, N.L., Aiken, A.C., Docherty, K.S., Ulbrich, I.M., Grieshop, A.P.,

Robinson, A.L., Duplissy, J., Smith, J.D., Wilson, K.R., Lanz, V.A., Hueglin, C., Sun, Y.L., Tian, J., Laaksonen, A., Raatikainen, T., Rautiainen, J., Vaattovaara, P., Ehn, M., Kulmala, M., Tomlinson, J.M., Collins, D.R., Cubison, M.J., Dunlea, J., Huffman, J.A., Onasch, T.B., Alfarra, M.R., Williams, P.I., Bower, K., Kondo, Y., Schneider, J., Drewnick, F., Borrmann, S., Weimer, S., Demerjian, K., Salcedo, D., Cottrell, L., Griffin, R., Takami, A., Miyoshi, T., Hatakeyama, S., Shimono, A., Sun, J.Y., Zhang, Y.M., Dzepina, K., Kimmel, J.R., Sueper, D., Jayne, J.T., Herndon, S.C., Trimborn, A.M., Williams, L.R., Wood, E.C., Middlebrook, A.M., Kolb, C.E., Baltensperger, U., Worsnop, D.R.: Evolution of organic aerosols in the atmosphere. Science 326, 1525–1529 (2009). https://doi.org/10.1126/science.1180353

Joutsensaari, J., Loivamäki, M., Vuorinen, T., Miettinen, P., Nerg, A.M., Holopainen, J.K., Laaksonen, A.: Nanoparticle formation by ozonolysis of inducible plant volatiles. Atmos. Chem. Phys. 5, 1489–1495 (2005). https://doi.org/10.5194/acp-5-1489-2005

Kalberer, M., Sax, M., Samburova, V.: Molecular size evolution of oligomers in organic aerosols collected in urban atmospheres and generated in a smog chamber. Environ. Sci. Technol. 40, 5917–5922 (2006). https://doi.org/10.1021/es0525760

Kiendler-Scharr, A., Wildt, J., Maso, M.D., Hohaus, T., Kleist, E., Mentel, T.F., Tillmann, R., Uerlings, R., Schurr, U., Wahner, A.: New particle formation in forests inhibited by isoprene emissions. Nature 461, 381–384 (2009). https://doi.org/10.1038/nature08292

Kroll, J.H., Ng, N.L., Murphy, S.M., Flagan, R.C., Seinfeld, J.H.: Secondary organic aerosol formation from isoprene photooxidation under high-NO_x conditions. Geophys. Res. Lett., 32, https://doi.org/10.1029/2005GL023637 (2005)

McFiggans, G., Mentel, T.F., Wildt, J. et al.: Secondary organic aerosol reduced by mixture of atmospheric vapours. Nature 565, 587–593 (2019). https://doi.org/10.1038/s41586-018-0871-y

Mentel, T.F., Wildt, J., Kiendler-Scharr, A., Kleist, E., Tillmann, R., Dal Maso, M., Fisseha, R., Hohaus, T., Spahn, H., Uerlings, R., Wegener, R., Griffiths, P.T., Dinar, E., Rudich, Y., Wahner, A.: Photochemical production of aerosols from real plant emissions. Atmos. Chem. Phys. 9, 4387–4406 (2009). https://doi.org/10.5194/acp-9-4387-2009

Monge, M.E., George, C., D'Anna, B., Doussin, J.-F., Jammoul, A., Wang, J., Eyglunent, G., Solignac, G., Daële, V., Mellouki, A.: Ozone formation from illuminated titanium dioxide surfaces. J. Am. Chem. Soc. 132, 8234–8235 (2010). https://doi.org/10.1021/ja1018755

Niki, H., Daby, E.E., Weinstock, B.: Mechanisms of smog reactions. In: Photochemical Smog and Ozone Reactions, Advances in Chemistry, 113, AMERICAN CHEMICAL SOCIETY, 16–57 (1972)

Nordin, E.Z., Uski, O., Nyström, R., Jalava, P., Eriksson, A.C., Genberg, J., Roldin, P., Bergvall, C., Westerholm, R., Jokiniemi, J., Pagels, J.H., Boman, C., Hirvonen, M.-R.: Influence of ozone initiated processing on the toxicity of aerosol particles from small scale wood combustion. Atmos. Environ. 102, 282–289 (2015). https://doi.org/10.1016/j.atmosenv.2014.11.068

Odum, J.R., Jungkamp, T.P.W., Griffin, R.J., Flagan, R.C., Seinfeld, J.H.: The atmospheric aerosol-forming potential of whole gasoline vapor. Science 276, 96–99 (1997). https://doi.org/10.1126/science.276.5309.96

Pankow, J.F.: An absorption model of the gas/aerosol partitioning involved in the formation of secondary organic aerosol. Atmos. Environ. 28, 189–193 (1994). https://doi.org/10.1016/1352-2310(94)90094-9

Pitts JR, J.N., Sanhueza, E., Atkinson, R., Carter, W.P.L., Winer, A.M., Harris, G.W., Plum, C.N.: An investigation of the dark formation of nitrous acid in environmental chambers. Inte. J. Chem. Kinetics 16, 919–939, https://doi.org/10.1002/kin.550160712 (1984)

Platt, S.M., El Haddad, I., Zardini, A.A., Clairotte, M., Astorga, C., Wolf, R., Slowik, J.G., Temime-Roussel, B., Marchand, N., Ježek, I., Drinovec, L., Močnik, G., Möhler, O., Richter, R., Barmet, P., Bianchi, F., Baltensperger, U., Prévôt, A.S.H.: Secondary organic aerosol formation from gasoline vehicle emissions in a new mobile environmental reaction chamber. Atmos. Chem. Phys. 13, 9141–9158 (2013). https://doi.org/10.5194/acp-13-9141-2013

Pratap, V., Bian, Q., Kiran, S.A., Hopke, P.K., Pierce, J.R., Nakao, S.: Investigation of levoglucosan decay in wood smoke smog-chamber experiments: the importance of aerosol loading, temperature, and vapor wall losses in interpreting results. Atmos. Environ. **199**, 224–232 (2019). https://doi.org/10.1016/j.atmosenv.2018.11.020

Rohrer, F., Bohn, B., Brauers, T., Brüning, D., Johnen, F.J., Wahner, A., Kleffmann, J.: Characterisation of the photolytic HONO-source in the atmosphere simulation chamber SAPHIR. Atmos. Chem. Phys. **5**, 2189–2201 (2005). https://doi.org/10.5194/acp-5-2189-2005

Thomas, M., France, J., Crabeck, O., Hall, B., Hof, V., Notz, D., Rampai, T., Riemenschneider, L., Tooth, O.J., Tranter, M., Kaiser, J.: The Roland von Glasow Air-Sea-Ice Chamber (RvG-ASIC): an experimental facility for studying ocean–sea-ice–atmosphere interactions. Atmos. Meas. Tech. **14**, 1833–1849 (2021). https://doi.org/10.5194/amt-14-1833-2021

Weinstock, B.: Tuesday, C.S. (ed.) Chemical Reactions in Urban Atmospheres, in, American Elsevier, New York, 54–55 (1971)

Westberg, K., Cohen, N., Wilson, K.W.: Carbon monoxide: its role in photochemical smog formation. Science **171**, 1013–1015 (1971). https://doi.org/10.1126/science.171.3975.1013